# The Dialogues

# The Dialogues

## Conversations about the Nature of the Universe

Clifford V. Johnson

The MIT Press

Cambridge, Massachusetts

London, England

This book was set in the Blambot fonts inkslinger and crimefighter by Clifford V. Johnson.

Printed and bound in the United States of America.

Library of Congress Cataloging-in-Publication Data

Names: Johnson, Clifford V. (Clifford Victor), 1968- author. | Wilczek, Frank, writer of foreword.
Title: The dialogues : conversations about the nature of the universe / Clifford V. Johnson ; foreword by Frank Wilczek.
Description: Cambridge, Massachusetts ; London, England : The MIT Press, [2017]
Identifiers: LCCN 2017015399 | ISBN 9780262037235 (hardcover ; alk. paper) | ISBN 0262037238 (hardcover ; alk. paper)
Subjects: LCSH: Science. | Cosmology.
Classification: LCC Q172 .J64 2017 | DDC 523.1--dc23 LC record available at https://lccn.loc.gov/2017015399

10 9 8 7 6 5

*To my mother, who always gave me space to explore my full potential.*

*To my son, an explorer.*

# Contents

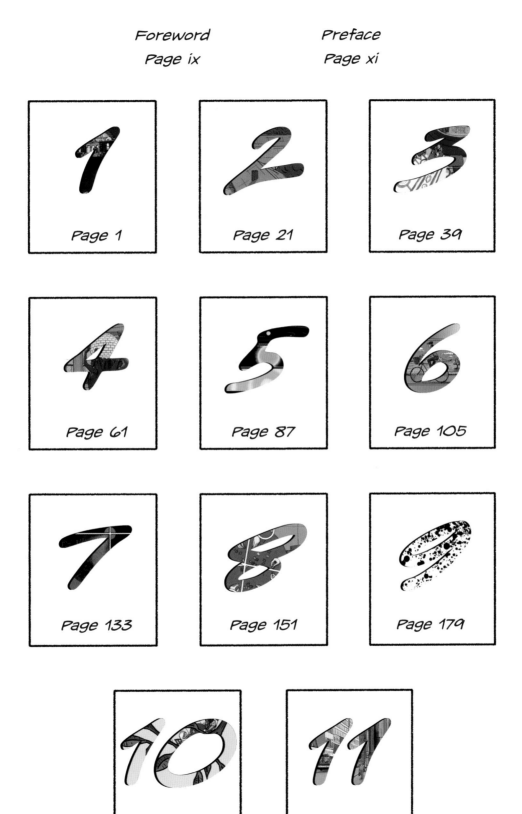

# Foreword

Frank Wilczek is the Herman Feshbach Professor at the Massachusetts Institute of Technology, Chief Scientist at the Wilczek Quantum Center, Shanghai Jiao Tong University, Distinguished Origins Professor, Arizona State University, and Professor of Physics at Stockholm University. He shared the 2004 Nobel Prize in Physics with David J. Gross and H. David Politzer for the discovery of asymptotic freedom in the theory of the strong nuclear interaction. He has written many books and essays, including most recently *A Beautiful Question: Finding Nature's Deep Design* (New York: Penguin Press, 2015).

* * *

I.

Dialogues were Plato's chosen vehicle, through which he framed philosophical debates that continue to this day. David Hume took up the form in his brilliant *Dialogues Concerning Natural Religion*. Hume's work is a classic in the philosophy of science, as well as the philosophy of religion, and its discussion of cosmology, in particular, remains fresh and relevant today.

Galileo, a great admirer of Plato, used dialogue form in two of his masterworks. In *Dialogue Concerning the Two Chief World Systems*, two natural philosophers—Simplicio, a traditional Aristotilean, and Salviati, a Copernican—vie for the allegiance of Sagredo, an intelligent layman. The character of Sagredo is based on a real person, Galileo's friend Giovanni Francesco Sagredo. The scientific debate between Aristotilean physics (with its Earth-centered, Ptolemaic astronomy) and Galilean physics, now called "classical" (with its Sun-centered, Copernican astronomy) is now long settled. The *Dialogue*'s discussion of particular points is dated and, in its extended discussion of the tides, badly misguided. Yet Galileo's *Dialogue* remains alive, and a joy to read, because it brings us in touch with sympathetic human characters, and with a wonderfully attractive interaction among them, wherein they exchange deeply held ideas—and expose them to questioning.

Galileo's *Discourse on Two New Sciences* is a more purely scientific work, in which he presented pioneering ideas in what we'd now call strength of materials and elementary dynamics. The science is brilliant, but of course long superseded. Yet the *Discourse* remains as a beautiful, touching work of literature, again because of the attractive culture it shows us, here featuring the interplay of fundamental science and engineering practice or, more broadly, theory and experiment. In this book Galileo's protagonists, although still named Simplicio, Sagredo, and Salviati, are actually all Galileo himself, at different stages in his intellectual development. Through their debate, he shows us how his ideas evolved.

II.

"Show, don't tell" is advice often given to aspiring playwrights and novelists. When done well, it leads to work that engages us like life itself. Plato, Hume, and Galileo followed that advice, to great effect. For modern scientists who aim to bring their work to the public using printed media it is challenging advice, since their material is, generally speaking, far removed from everyday life.

Clifford Johnson—a practicing scientist, working at the frontiers of physics and cosmology—has, in the book before you, risen to the challenge. He's brought in two very creative innovations.

First, and most strikingly, he's taken "dialogue" to a new level, augmenting text with graphics that depict the protagonists and their surroundings. Graphic novels have become very popular in recent years. Here we have graphic dialogues.

Graphic dialogues add important new dimensions to the dialogue form. They restore some of the richness of face-to-face human communication. Reactions are "shown, not told." The social and cultural context of the conversations is, necessarily, explicit. Facial expressions and body language are in play. Johnson places his interlocutors in interesting, realistic, physical settings. His urban "landscapes" are drawn with a skillful, loving hand.

It is all too easy to forget, in the later dry retelling, that new science is usually the product of attractive young people, who enjoy one another's company. That hidden aspect of scientific life is front and center in Johnson's graphic dialogues.

Second, he's off-loaded heavy scientific details into the brief notes that follow each dialogue. The notes provide, for those who'd like to enter the subjects more deeply, informed pointers to readily accessible sources of additional information.

The dialogues themselves are conversational. They contain, as dialogues worthy of the name should, real give-and-take about issues where the last word has not yet been spoken, and different opinions are possible.

Can there be a theory of everything? Does the concept of a multiverse make sense—and is it useful, or true? What is the value of wildly exploratory research, driven by curiosity and aesthetics rather than tangible goals?

Let's talk about it ...

Frank Wilczek
Cambridge, Massachusetts
March 2017

# Preface

## An Invitation

All around you, people engage with each other in conversations about a variety of topics, reflecting the diversity of interests and concerns that they have about the world around them. You've probably overheard snippets of such conversations, and although you don't necessarily appreciate every aspect of what was said (perhaps because of lack of context or knowledge of the subject), you'll probably agree that it can be fascinating while it lasts! Sometimes, those conversations are about science. They should be, since science is part of our world and our culture, and it has a huge impact on our lives. So it ought to be on the menu right alongside art, music, politics, sports, shopping, celebrities, and all those other things we talk about. Science is also a great source of beauty and wonder. There need be no more reasons beyond that to talk about a topic!

Despite all that, everyday conversation about science seems to be hidden from most of our art, our literature, and other forms of entertainment. It is even largely absent from most presentations of science. This is odd, since such conversations are essential to everyone involved: To unpack the meaning or relevance of some aspect of science, or to simply become more familiar or comfortable with a topic, it helps to talk about it with someone. When scientists are trying to understand each other's work, or uncover some new truth about how nature works, they have conversations. When they try to communicate science ideas or knowledge to the public, it's often best done in the form of a conversation.

This book is therefore an invitation. On the one hand, it's an invitation to eavesdrop on some conversations about science. Like all dialogues, they're rambling, incomplete, and sometimes not fully informed. However, they may be interesting, and they may encourage you to delve a little deeper and find out more about something you got from an exchange. On the other hand, this book is an invitation to join in! Conversation about science shouldn't be left to the experts or to science enthusiasts—it's for everyone. There's no grade given for getting things wrong, for asking questions, for hazarding a guess, or for having an opinion. So while out there in the world, remember these conversations, know that they are happening all around you, and initiate and participate in some of your own. Maybe someone will eavesdrop, start his or her own conversation, and keep the chain of engagement alive.

## About This Book

Let me point out a few things about this book, since it is somewhat unusual. During or after eavesdropping on a conversation, consider consulting the notes that are at the end of each one. They're entirely optional, but they sometimes contain comments about things that were said, and they often supply sources for further reading. Remember also that vast amounts of information are available online to which you can quickly connect by putting a snippet of what was said into a search engine. In doing so, you should be cautious of your sources. I've given a few online sources, but I've focused on books. You should not consider the books listed to be what I regard as the definitive best (or most current) treatments. They are just a mixture of some of my favorites with others that caught my eye, or that seemed suitable to include to help form a collection of entry points into a topic. Read lots of sources, and get several points of view: Remember that science is a human endeavor, and so a book by one scientist on a certain topic might not resonate with you as much as one by another scientist on the very same topic. This is no different from preferring how one poet evokes a summer's day to how another does.

You'll notice that this is a graphic book, or graphic novel, or comic, or whatever the right term is. (There's no universal agreement on the terminology.) The more accurate (but less dexterous) usage is to say that this is a collection of "sequential art," meaning that it is a bunch of images that, taken together (and read with adherence to some agreed upon ordering convention) form a narrative, just like a sentence, or collection of sentences, does. But sequential art—the graphic form—can do so much more than just sentences made of words! In fact, it is particularly well suited to discussing science, especially physics, which is what most of these dialogues are about. This goes beyond the obvious fact that I can simply show shapes and objects that I'd typically have to describe in words in a traditional science book. It also goes beyond the fact that you might be drawn more readily into the conversation by being able to see the interlocutors and their surroundings. Remember that physics has space and time embedded into nearly every aspect of it. In fact, current research into the nature of space and time focuses a lot on how they may arise as a result of the relationships between things. Space and time and the relationships between things are at the heart of how comics work: Images (sometimes contained in panels, but not necessarily) arranged in sequence encourage the reader to infer a narrative that involves the sense of time passing, of movement, and so forth. In this sense, fundamentally, comics are physics! Put this way, upon reflection it is stunning that this graphic form has not been used more to talk about physics, and to communicate what's going on in the fascinating world of physics research. I hope that this book will help change that. I don't do it nearly as much as I could have, but do keep an eye out for places in the book where I tie the physics being discussed to the layout of the images on the page. In the notes I could not resist pointing out some (perhaps) particularly notable cases.

Finally, you'll also notice that there are equations in this book. In popular-level presentations of science ideas, there's a tradition of hiding from the reader the two most powerful tools we use in research: the cartoons and diagrams that we scribble, and the equations that we write. So much of the reasoning we do in physics is visual, and most ideas start that way. Those tools encode so very much in a visual language, and yet authors are encouraged by their editors to replace them with words, to effectively infantalize readers and protect them from the dreaded mathematics. It is no wonder that some topics remain mysterious, obscure, or confusing! It also ends up perpetuating the very fear and unfamiliarity with equations that some people have. To leave out (or minimize the quantity of) equations and diagrams in a book about physics is like writing a book about music and being afraid to talk about or show any musical instruments. You can do it, sure, but you're leaving out so very much of what physics is really like, hiding what makes it attractive to most of its practitioners. So yes, cast your eye over the equations, and don't worry if you don't fully understand them. Appreciate what you can. Having them on the page will get you more comfortable with them, and that familiarity may mean you'll look a second time, and a third, and so on ... and maybe you'll appreciate their meaning a bit more every time.

## *Acknowledgments*

I'm not the first to note how overly specialized we all are. Sometimes it is out of necessity, but all too often it is because we find comfort in apparently getting the measure of someone by being able to classify them, so we all join in the classification game and end up building walls. The apparent wide gulf between my "day job" as a professor of physics and the other things I like to dabble in (aspects of the visual arts, and the humanities) that resulted in this book has meant that most friends and colleagues don't know or haven't really understood what I've been working on, or why. In some cases it has been because it was hard for me to know where to begin explaining, and so I busied myself with getting on with it, hastening the day when I could just point to the end product. So I'd like to express considerable gratitude to everyone who (even if they did not quite understand) was either supportive, or encouraging, or who simply trusted me enough to allow me the space (and time) to work on this (more than seventeen-year-old!) idea and bring it to fruition.

I'm privileged to work at an institution (the University of Southern California) that has many staff and faculty who help keep alive the spirit of genuine cross-disciplinary exploration, and I thank them for that. Notable individuals with words of support and interest include Aimee Bender, Leo Braudy, K. C. Cole, Allison Engel, Karin Huebner, and M. G. Lord. I thank the Fellows and Friends of the Los Angeles Institute for the Humanities at USC for keeping alive the wonderful environment that makes it easy to keep sight of projects like this, and USC's Sidney Harman Academy for Polymathic Study (and especially the late Kevin Starr who helped build and guide it) for providing me with opportunities to inspire a new generation to be explorers. Both institutions let me present the book as a work in progress. I thank Amy Rowat for helpful advice on extra sources to include about food science, and Tameem Albash for patiently listening to me babble about comics. Particular thanks to Nancy Keystone for being an enthusiastic supporter from the moment she heard what I was doing. It is a delight to thank Jessie and Robin French for occasionally allowing me to set up camp in a corner of their home while working on the book. Special thanks go to Tim Morris, my old PhD advisor from almost thirty years ago. I remembered a story that he told about going to the fair with his father, and his interpretation of what happened. He gave me permission to borrow it, and the story told within the last conversation in the book is loosely based on his tale. Thanks to the Aspen Center for Physics (especially Jane Kelly) for being an excellent retreat at which to think about unusual projects. I've completed this book during a sabbatical year, and the Simons Foundation made a full year possible by awarding me a fellowship, for which I am grateful.

It was a challenge to explore the publishing world to try to find a home for this oddball project. I must thank Stephon Alexander and Cecil Castellucci for trying to help by making introductions, even though those avenues did not yield results. I've since learned that such selfless introductions to agents and editors are more rare than I'd have guessed, and so they should be acknowledged. Alice Oven was the first commissioning editor who really listened and "got it" when I found her and explained what I was trying to do, and this book may not have made it to print without her enthusiastically championing it while at IC Press, so a big thank you goes to her. I thank Jermey Matthews for the initiative and enthusiasm that resulted in this book finding a home at the MIT Press, and all the staff at the MIT Press for their warmth and professionalism as we worked on the last stages of this book.

Although I don't see them as often as I'd like, my mother, Delia, and my siblings, Robert and Carol, are with me in every project I undertake and I'm thankful. They, along with my (now departed) father Reginald, have all helped shape who I am. This includes, I believe, having been understanding and supportive of my exploratory nature all my life. I must also thank Robert for (probably) being responsible for me being exposed to comics, sharing them, and helping me track them down back when that was so hard to do, almost four decades ago on that tiny island that we all miss so much.

Most importantly, there's my wife and best friend, Amelia French. My gratitude to her cannot be overstated. It is a cliché to say that without her this project would not have come to fruition, but it is true. I thank her for being the most wonderful partner one could wish for, bringing love, humor, ideas, and the unquestioning and generous support and flexibility that allowed us to negotiate a mutually workable schedule that (I hope) ultimately served each of our respective endeavors well, as well as strengthening the wonderful family we've made. The book is infinitely richer as a result of us having met.

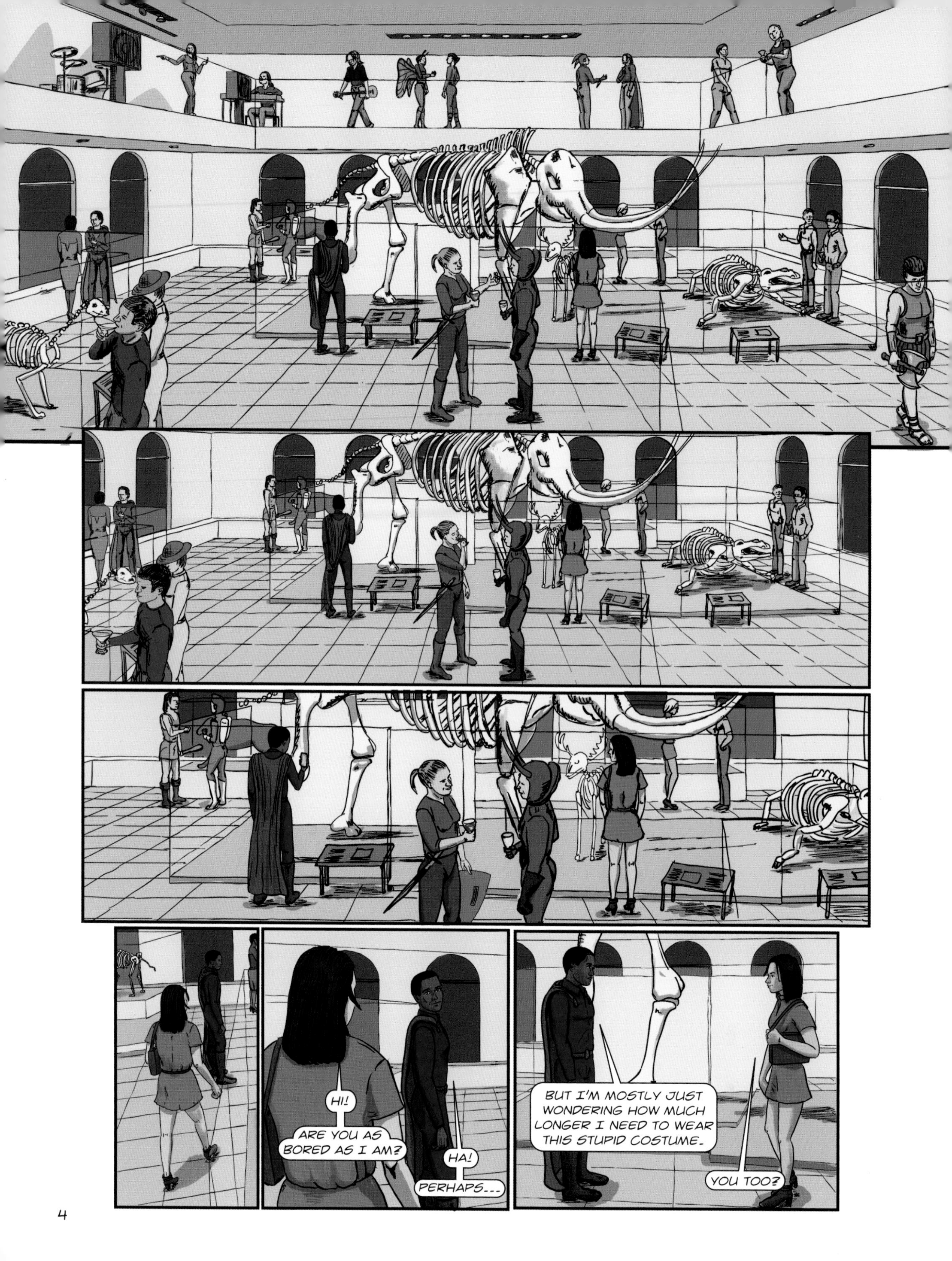
HI!
ARE YOU AS BORED AS I AM?
HA!
PERHAPS...
BUT I'M MOSTLY JUST WONDERING HOW MUCH LONGER I NEED TO WEAR THIS STUPID COSTUME.
YOU TOO?

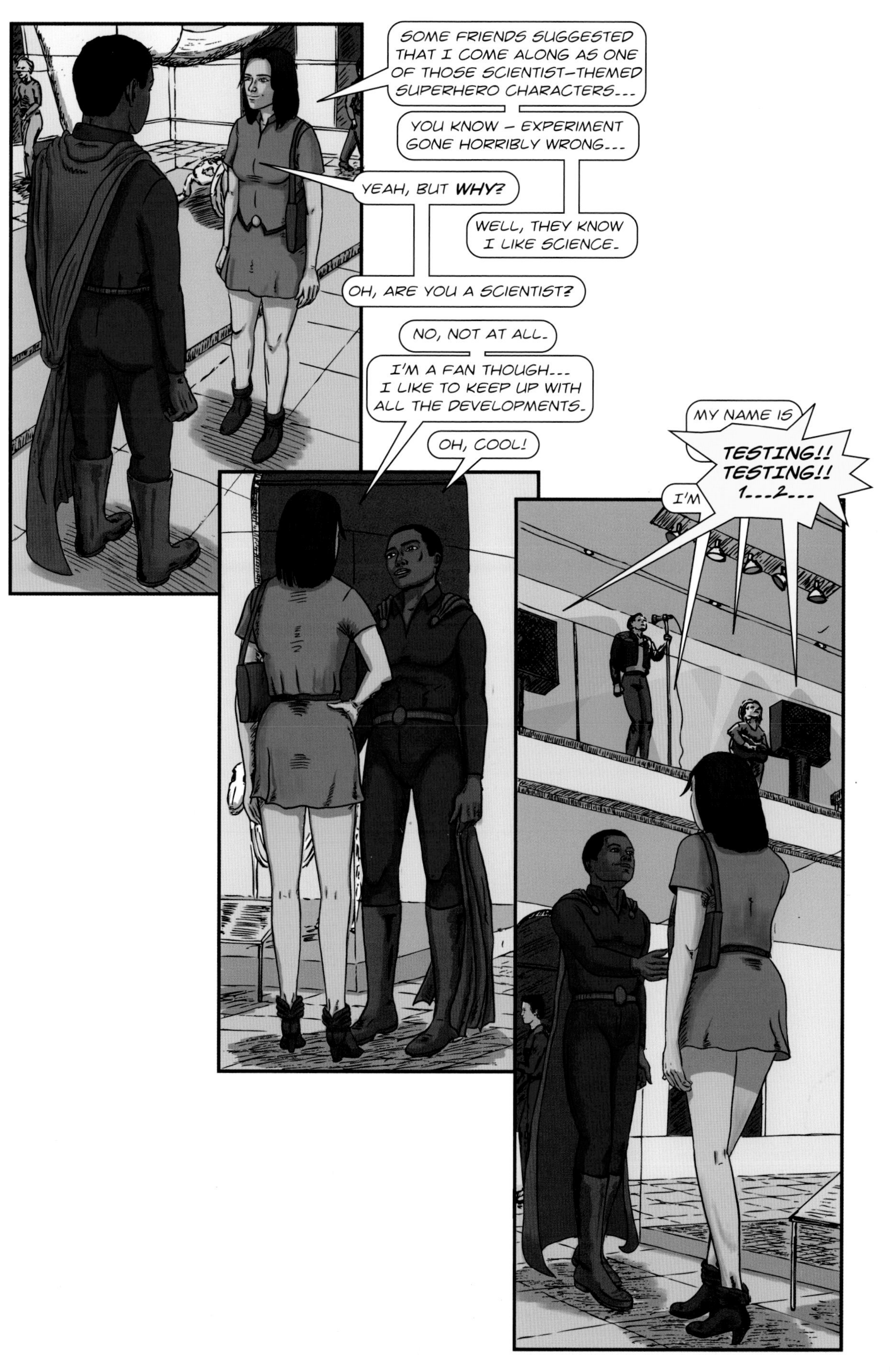
SOME FRIENDS SUGGESTED THAT I COME ALONG AS ONE OF THOSE SCIENTIST-THEMED SUPERHERO CHARACTERS...
YOU KNOW – EXPERIMENT GONE HORRIBLY WRONG...
YEAH, BUT **WHY?**
WELL, THEY KNOW I LIKE SCIENCE.
OH, ARE YOU A SCIENTIST?
NO, NOT AT ALL.
I'M A FAN THOUGH... I LIKE TO KEEP UP WITH ALL THE DEVELOPMENTS.
OH, COOL!
MY NAME IS
I'M
TESTING!! TESTING!! 1...2...

YOU KNOW, I SHOULD HAVE USED MY SUPERHERO NAME!
I'M **CATALYST**. WHO DID YOU COME AS?
CATALYST? GREAT COSTUME!
I'M **ULTRAVIOLET**.
I DON'T HAVE ANY FANCY CHEMISTRY POWERS...
I THINK ALL MY CHARACTER DOES IS FLY, AND IS SOMEWHAT STRONG AND FAST.
YOU KNOW, STANDARD ATTRIBUTES.
FAST HEALING FACTOR?
ER, NO.
ISN'T THAT MORE FOR THE ONES WITH ANIMAL POWERS?
MAYBE.
WELL, NICE CAPE.
THANKS! THEY'RE MAKING A COMEBACK, YOU KNOW.
HA! RIGHT!
SO, WHY DID YOU COME AS ULTRAVIOLET?
YOU DON'T SEEM VERY ENTHUSIASTIC.
IT WAS THE ONLY COSTUME THEY HAD LEFT IN THE STORE.
I DID NOT MIND WHICH HERO IT WAS...
BUT THE SUIT TURNED OUT TO BE SO ITCHY AND **HOT**!

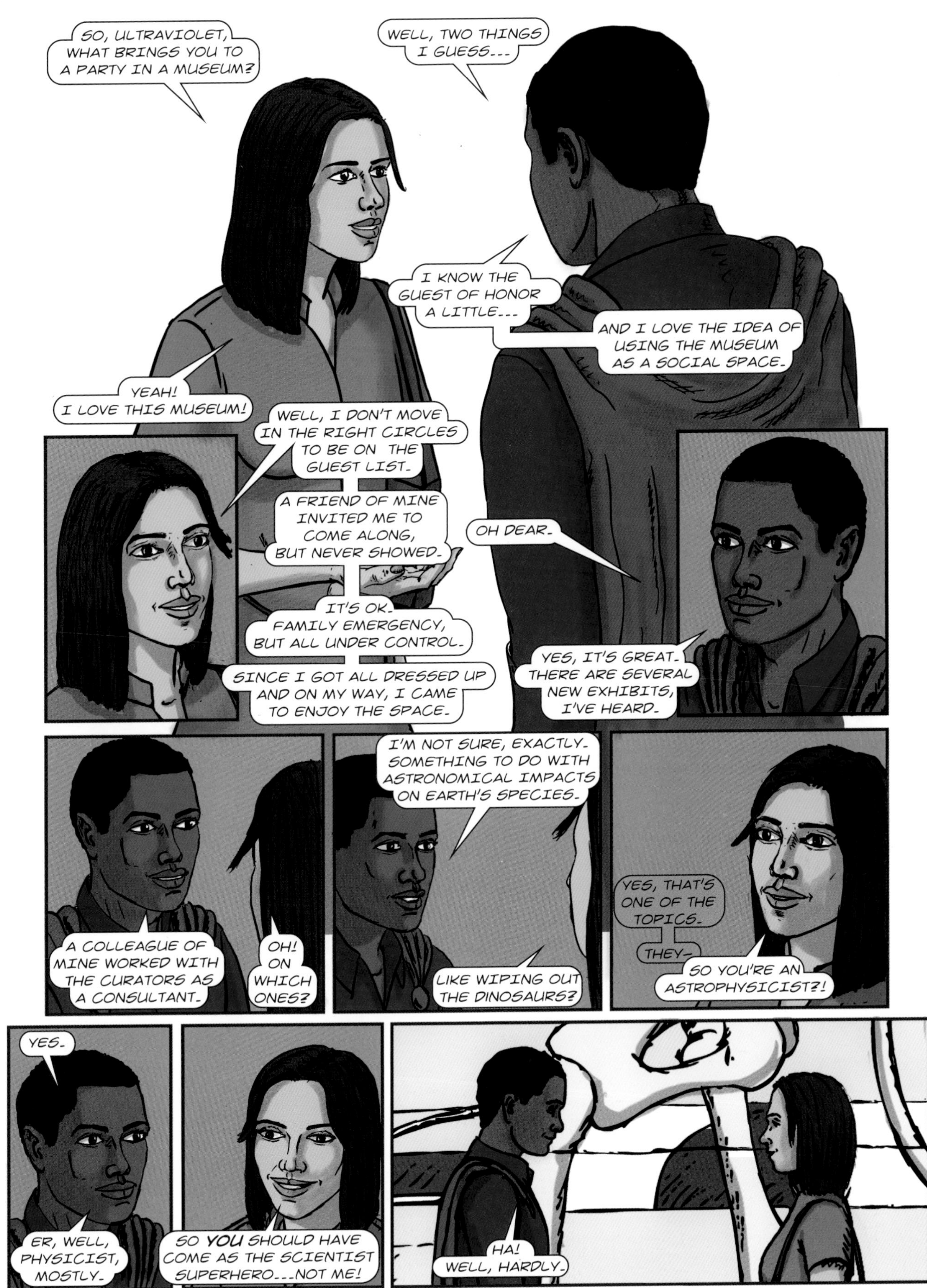
SO, ULTRAVIOLET, WHAT BRINGS YOU TO A PARTY IN A MUSEUM?
WELL, TWO THINGS I GUESS...
I KNOW THE GUEST OF HONOR A LITTLE...
AND I LOVE THE IDEA OF USING THE MUSEUM AS A SOCIAL SPACE.
YEAH! I LOVE THIS MUSEUM!
WELL, I DON'T MOVE IN THE RIGHT CIRCLES TO BE ON THE GUEST LIST.
A FRIEND OF MINE INVITED ME TO COME ALONG, BUT NEVER SHOWED.
OH DEAR.
IT'S OK. FAMILY EMERGENCY, BUT ALL UNDER CONTROL.
SINCE I GOT ALL DRESSED UP AND ON MY WAY, I CAME TO ENJOY THE SPACE.
YES, IT'S GREAT. THERE ARE SEVERAL NEW EXHIBITS, I'VE HEARD.
A COLLEAGUE OF MINE WORKED WITH THE CURATORS AS A CONSULTANT.
OH! ON WHICH ONES?
I'M NOT SURE, EXACTLY. SOMETHING TO DO WITH ASTRONOMICAL IMPACTS ON EARTH'S SPECIES.
LIKE WIPING OUT THE DINOSAURS?
YES, THAT'S ONE OF THE TOPICS.
THEY–
SO YOU'RE AN ASTROPHYSICIST?!
YES.
ER, WELL, PHYSICIST, MOSTLY.
SO **YOU** SHOULD HAVE COME AS THE SCIENTIST SUPERHERO...NOT ME!
HA! WELL, HARDLY.

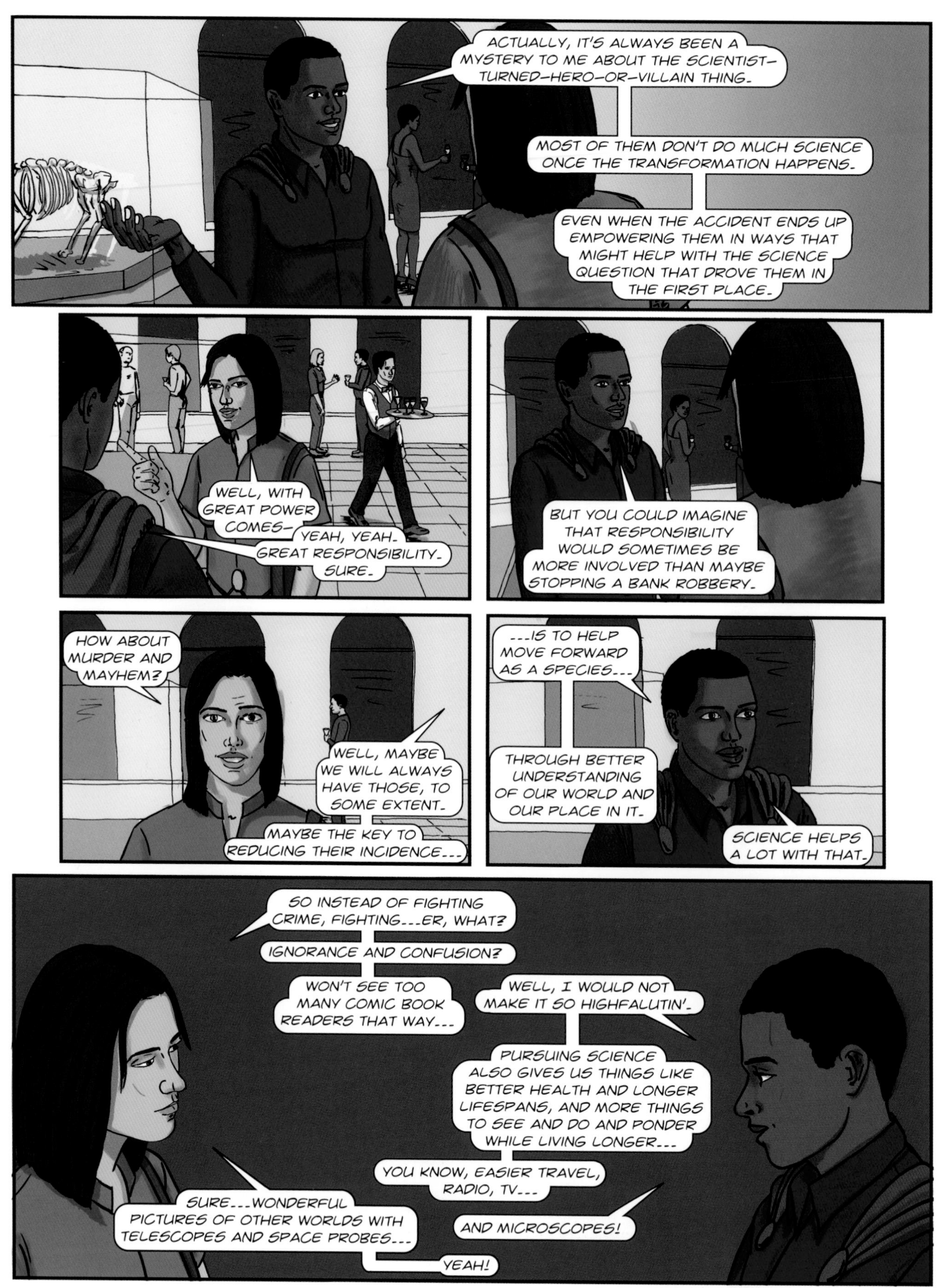
ACTUALLY, IT'S ALWAYS BEEN A MYSTERY TO ME ABOUT THE SCIENTIST-TURNED-HERO-OR-VILLAIN THING.
MOST OF THEM DON'T DO MUCH SCIENCE ONCE THE TRANSFORMATION HAPPENS.
EVEN WHEN THE ACCIDENT ENDS UP EMPOWERING THEM IN WAYS THAT MIGHT HELP WITH THE SCIENCE QUESTION THAT DROVE THEM IN THE FIRST PLACE.
WELL, WITH GREAT POWER COMES-
YEAH, YEAH. GREAT RESPONSIBILITY. SURE.
BUT YOU COULD IMAGINE THAT RESPONSIBILITY WOULD SOMETIMES BE MORE INVOLVED THAN MAYBE STOPPING A BANK ROBBERY.
HOW ABOUT MURDER AND MAYHEM?
WELL, MAYBE WE WILL ALWAYS HAVE THOSE, TO SOME EXTENT.
MAYBE THE KEY TO REDUCING THEIR INCIDENCE...
...IS TO HELP MOVE FORWARD AS A SPECIES...
THROUGH BETTER UNDERSTANDING OF OUR WORLD AND OUR PLACE IN IT.
SCIENCE HELPS A LOT WITH THAT.
SO INSTEAD OF FIGHTING CRIME, FIGHTING...ER, WHAT?
IGNORANCE AND CONFUSION?
WON'T SEE TOO MANY COMIC BOOK READERS THAT WAY...
WELL, I WOULD NOT MAKE IT SO HIGHFALUTIN'.
PURSUING SCIENCE ALSO GIVES US THINGS LIKE BETTER HEALTH AND LONGER LIFESPANS, AND MORE THINGS TO SEE AND DO AND PONDER WHILE LIVING LONGER...
YOU KNOW, EASIER TRAVEL, RADIO, TV...
AND MICROSCOPES!
SURE...WONDERFUL PICTURES OF OTHER WORLDS WITH TELESCOPES AND SPACE PROBES...
YEAH!

SO I SEE A GREAT STORY IN WHICH A SCIENTIST GETS SPECIAL ABILITIES...
BUT SHE TURNS AROUND AND GOES OFF AND DOES MORE SCIENCE WITH IT.
THE WORLD PLEADS WITH HER TO HELP FIGHT CRIME, ETC....
BUT SHE REFUSES.
AND SHE IS DESPISED FOR IT, EVEN THOUGH SHE'S MOTIVATED BY THE GREATER GOOD AS MUCH AS THE SUPERHERO WHO FLIES AROUND ENDLESSLY TRYING TO STOP CRIME, DIVERT TIDAL WAVES, ETC....
THAT COULD BE A GREAT STORY, YES...
BUT I GUESS IT WOULD DEPEND A LOT UPON WHAT SPECIAL ABILITIES THE SCIENTIST GETS.
MAYBE.
BUT ACTUALLY HUGE ADVANCES IN SCIENCE HAVE BEEN MADE BY, IN ONE WAY OR ANOTHER, ENHANCING EVEN A SINGLE ONE OF OUR ABILITIES TO EXPERIENCE THE WORLD.
REALLY?

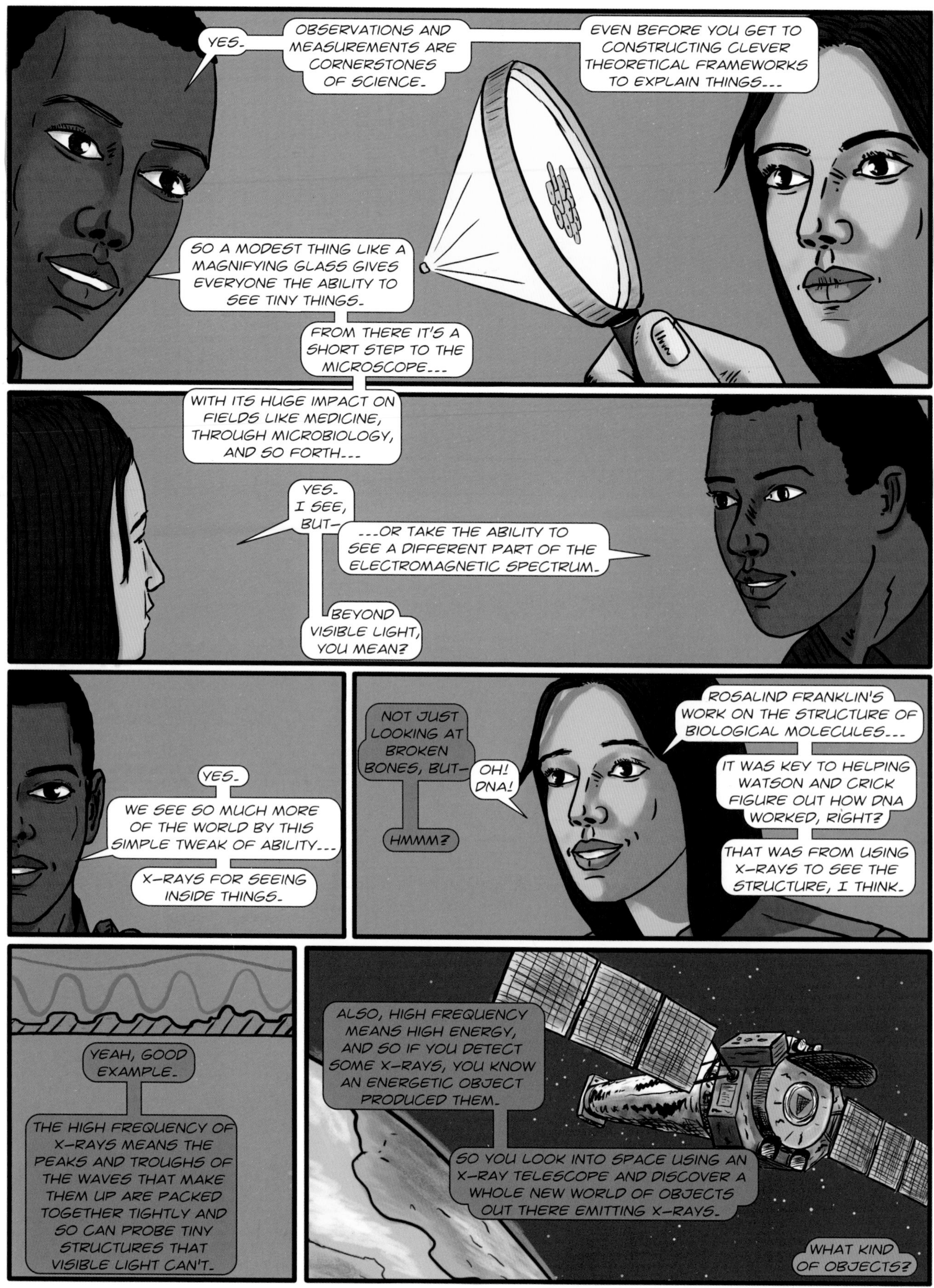
YES.
OBSERVATIONS AND MEASUREMENTS ARE CORNERSTONES OF SCIENCE.
EVEN BEFORE YOU GET TO CONSTRUCTING CLEVER THEORETICAL FRAMEWORKS TO EXPLAIN THINGS...
SO A MODEST THING LIKE A MAGNIFYING GLASS GIVES EVERYONE THE ABILITY TO SEE TINY THINGS.
FROM THERE IT'S A SHORT STEP TO THE MICROSCOPE...
WITH ITS HUGE IMPACT ON FIELDS LIKE MEDICINE, THROUGH MICROBIOLOGY, AND SO FORTH...
YES. I SEE, BUT-
...OR TAKE THE ABILITY TO SEE A DIFFERENT PART OF THE ELECTROMAGNETIC SPECTRUM.
BEYOND VISIBLE LIGHT, YOU MEAN?
YES.
WE SEE SO MUCH MORE OF THE WORLD BY THIS SIMPLE TWEAK OF ABILITY...
X-RAYS FOR SEEING INSIDE THINGS.
NOT JUST LOOKING AT BROKEN BONES, BUT-
OH! DNA!
HMMM?
ROSALIND FRANKLIN'S WORK ON THE STRUCTURE OF BIOLOGICAL MOLECULES...
IT WAS KEY TO HELPING WATSON AND CRICK FIGURE OUT HOW DNA WORKED, RIGHT?
THAT WAS FROM USING X-RAYS TO SEE THE STRUCTURE, I THINK.
YEAH, GOOD EXAMPLE.
THE HIGH FREQUENCY OF X-RAYS MEANS THE PEAKS AND TROUGHS OF THE WAVES THAT MAKE THEM UP ARE PACKED TOGETHER TIGHTLY AND SO CAN PROBE TINY STRUCTURES THAT VISIBLE LIGHT CAN'T.
ALSO, HIGH FREQUENCY MEANS HIGH ENERGY, AND SO IF YOU DETECT SOME X-RAYS, YOU KNOW AN ENERGETIC OBJECT PRODUCED THEM.
SO YOU LOOK INTO SPACE USING AN X-RAY TELESCOPE AND DISCOVER A WHOLE NEW WORLD OF OBJECTS OUT THERE EMITTING X-RAYS.
WHAT KIND OF OBJECTS?

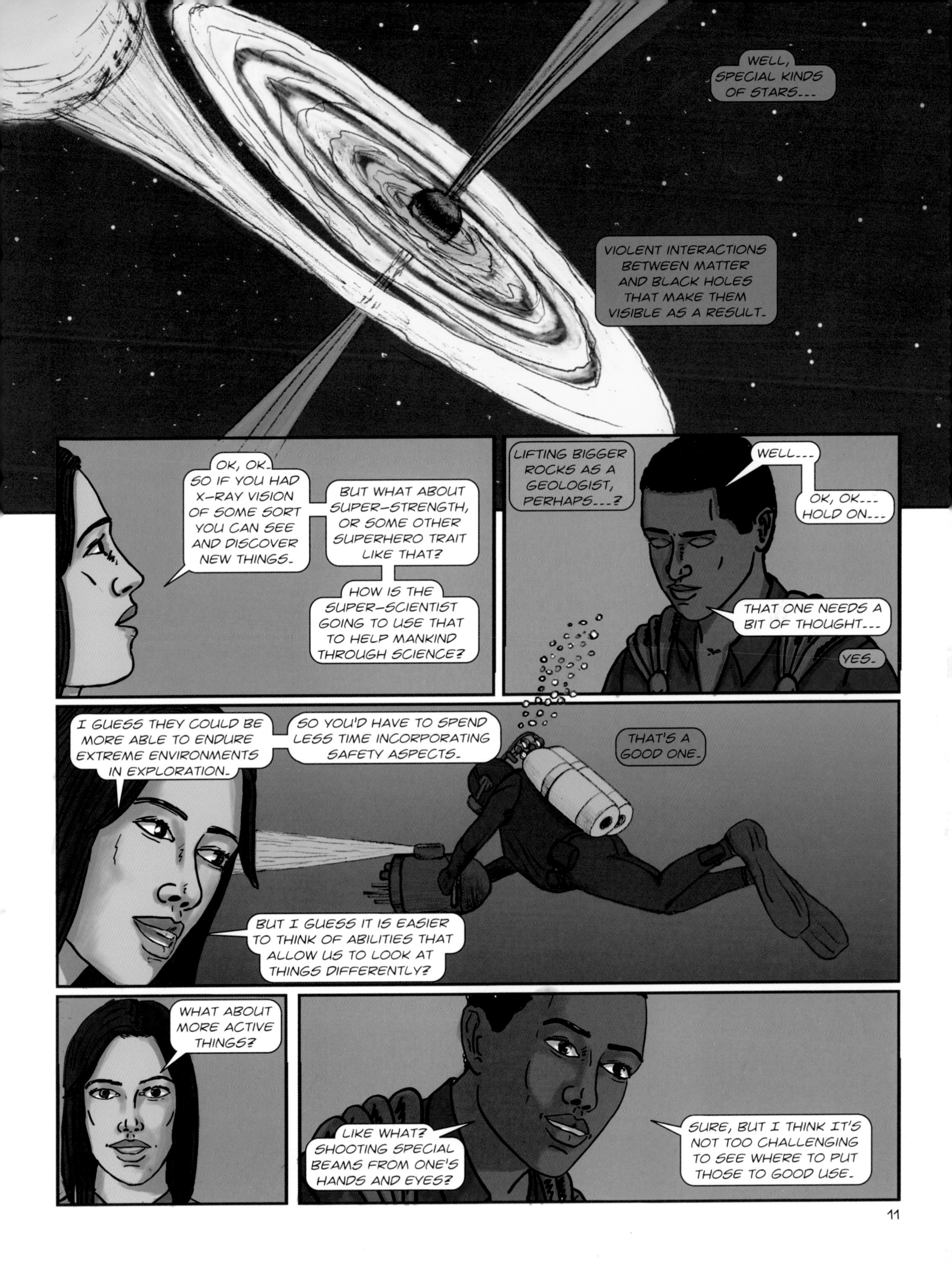
WELL, SPECIAL KINDS OF STARS...
VIOLENT INTERACTIONS BETWEEN MATTER AND BLACK HOLES THAT MAKE THEM VISIBLE AS A RESULT.
OK, OK. SO IF YOU HAD X-RAY VISION OF SOME SORT YOU CAN SEE AND DISCOVER NEW THINGS.
BUT WHAT ABOUT SUPER-STRENGTH, OR SOME OTHER SUPERHERO TRAIT LIKE THAT?
HOW IS THE SUPER-SCIENTIST GOING TO USE THAT TO HELP MANKIND THROUGH SCIENCE?
LIFTING BIGGER ROCKS AS A GEOLOGIST, PERHAPS...?
WELL...
OK, OK... HOLD ON...
THAT ONE NEEDS A BIT OF THOUGHT...
YES.
I GUESS THEY COULD BE MORE ABLE TO ENDURE EXTREME ENVIRONMENTS IN EXPLORATION.
SO YOU'D HAVE TO SPEND LESS TIME INCORPORATING SAFETY ASPECTS.
THAT'S A GOOD ONE.
BUT I GUESS IT IS EASIER TO THINK OF ABILITIES THAT ALLOW US TO LOOK AT THINGS DIFFERENTLY?
WHAT ABOUT MORE ACTIVE THINGS?
LIKE WHAT? SHOOTING SPECIAL BEAMS FROM ONE'S HANDS AND EYES?
SURE, BUT I THINK IT'S NOT TOO CHALLENGING TO SEE WHERE TO PUT THOSE TO GOOD USE.

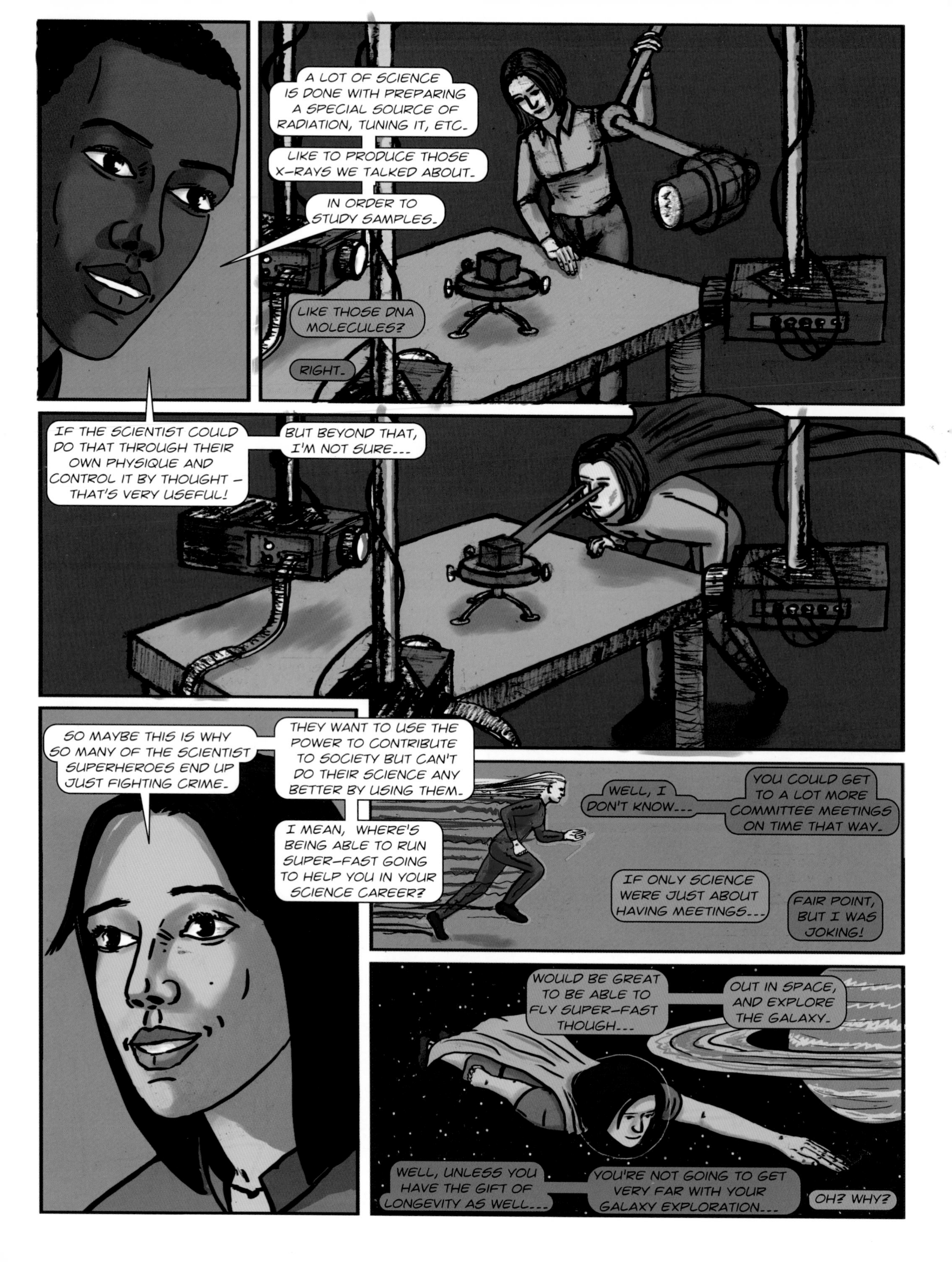
A LOT OF SCIENCE IS DONE WITH PREPARING A SPECIAL SOURCE OF RADIATION, TUNING IT, ETC.
LIKE TO PRODUCE THOSE X-RAYS WE TALKED ABOUT.
IN ORDER TO STUDY SAMPLES.
LIKE THOSE DNA MOLECULES?
RIGHT.
IF THE SCIENTIST COULD DO THAT THROUGH THEIR OWN PHYSIQUE AND CONTROL IT BY THOUGHT – THAT'S VERY USEFUL!
BUT BEYOND THAT, I'M NOT SURE...
SO MAYBE THIS IS WHY SO MANY OF THE SCIENTIST SUPERHEROES END UP JUST FIGHTING CRIME.
THEY WANT TO USE THE POWER TO CONTRIBUTE TO SOCIETY BUT CAN'T DO THEIR SCIENCE ANY BETTER BY USING THEM.
I MEAN, WHERE'S BEING ABLE TO RUN SUPER-FAST GOING TO HELP YOU IN YOUR SCIENCE CAREER?
WELL, I DON'T KNOW...
YOU COULD GET TO A LOT MORE COMMITTEE MEETINGS ON TIME THAT WAY.
IF ONLY SCIENCE WERE JUST ABOUT HAVING MEETINGS...
FAIR POINT, BUT I WAS JOKING!
WOULD BE GREAT TO BE ABLE TO FLY SUPER-FAST THOUGH...
OUT IN SPACE, AND EXPLORE THE GALAXY.
WELL, UNLESS YOU HAVE THE GIFT OF LONGEVITY AS WELL...
YOU'RE NOT GOING TO GET VERY FAR WITH YOUR GALAXY EXPLORATION...
OH? WHY?

BEST YOU CAN DO IS THE SPEED OF LIGHT. THAT'S 300 MILLION METERS PER SECOND.
NOW THAT'S REALLY FAST HERE ON EARTH, WHERE YOU CAN GO FROM ONE END TO THE OTHER IN A FEW HUNDREDTHS OF A SECOND...
...BUT EVEN FOR EXPLORING THE SOLAR SYSTEM, IT'S A PRETTY MODEST SPEED COMPARED TO THE DISTANCES WE ARE TALKING ABOUT.
REALLY? HOW COME?
WELL, IT TAKES ALMOST THREE HOURS TO GET OUT TO NEPTUNE, FOR EXAMPLE...
AND IT WOULD TAKE A LITTLE OVER FOUR YEARS TO GET TO THE NEAREST STAR, PROXIMA CENTAURI.

BUT WHY IS THE SPEED OF LIGHT THE SPEED LIMIT, ANYWAY? THIS IS TO DO WITH EINSTEIN AND SPECIAL RELATIVITY, RIGHT?
HMMM...
I'D PUT IT THE OTHER WAY.
IN A SENSE THE THINGS YOU MAY HAVE HEARD ABOUT FROM SPECIAL RELATIVITY FOLLOW FROM IT BEING A SPEED LIMIT...
HOW SO? I MEAN... SPECIAL RELATIVITY FOLLOWING FROM SPEED LIMIT?

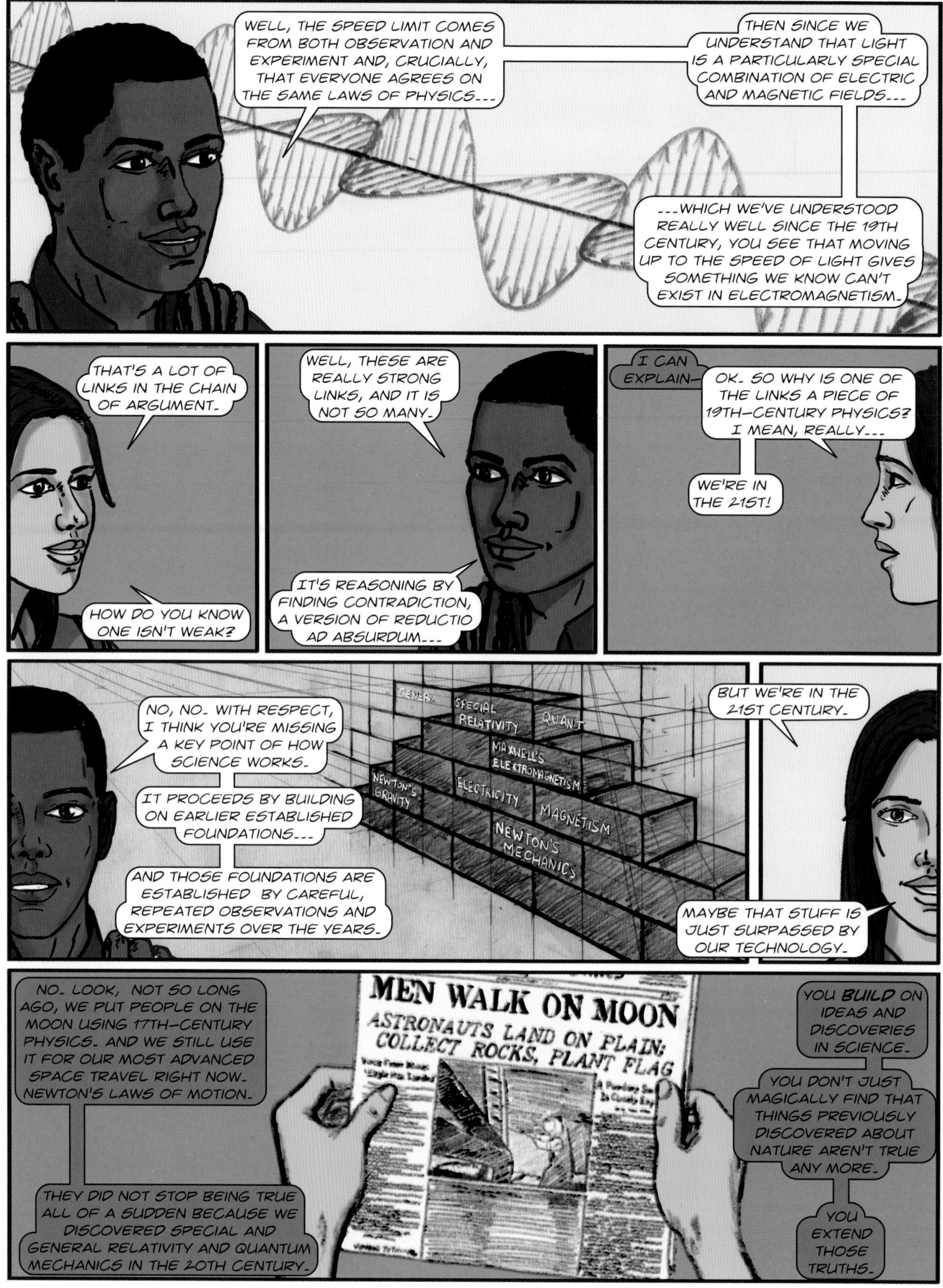

WELL, THE SPEED LIMIT COMES FROM BOTH OBSERVATION AND EXPERIMENT AND, CRUCIALLY, THAT EVERYONE AGREES ON THE SAME LAWS OF PHYSICS...
THEN SINCE WE UNDERSTAND THAT LIGHT IS A PARTICULARLY SPECIAL COMBINATION OF ELECTRIC AND MAGNETIC FIELDS...
...WHICH WE'VE UNDERSTOOD REALLY WELL SINCE THE 19TH CENTURY, YOU SEE THAT MOVING UP TO THE SPEED OF LIGHT GIVES SOMETHING WE KNOW CAN'T EXIST IN ELECTROMAGNETISM.
THAT'S A LOT OF LINKS IN THE CHAIN OF ARGUMENT.
HOW DO YOU KNOW ONE ISN'T WEAK?
WELL, THESE ARE REALLY STRONG LINKS, AND IT IS NOT SO MANY.
IT'S REASONING BY FINDING CONTRADICTION, A VERSION OF REDUCTIO AD ABSURDUM...
I CAN EXPLAIN—
OK. SO WHY IS ONE OF THE LINKS A PIECE OF 19TH-CENTURY PHYSICS? I MEAN, REALLY...
WE'RE IN THE 21ST!
NO, NO. WITH RESPECT, I THINK YOU'RE MISSING A KEY POINT OF HOW SCIENCE WORKS.
IT PROCEEDS BY BUILDING ON EARLIER ESTABLISHED FOUNDATIONS...
AND THOSE FOUNDATIONS ARE ESTABLISHED BY CAREFUL, REPEATED OBSERVATIONS AND EXPERIMENTS OVER THE YEARS.
SPECIAL RELATIVITY
MAXWELL'S ELECTROMAGNETISM
NEWTON'S GRAVITY
ELECTRICITY
MAGNETISM
NEWTON'S MECHANICS
BUT WE'RE IN THE 21ST CENTURY.
MAYBE THAT STUFF IS JUST SURPASSED BY OUR TECHNOLOGY.
NO. LOOK, NOT SO LONG AGO, WE PUT PEOPLE ON THE MOON USING 17TH-CENTURY PHYSICS. AND WE STILL USE IT FOR OUR MOST ADVANCED SPACE TRAVEL RIGHT NOW. NEWTON'S LAWS OF MOTION.
THEY DID NOT STOP BEING TRUE ALL OF A SUDDEN BECAUSE WE DISCOVERED SPECIAL AND GENERAL RELATIVITY AND QUANTUM MECHANICS IN THE 20TH CENTURY.
MEN WALK ON MOON
ASTRONAUTS LAND ON PLAIN; COLLECT ROCKS, PLANT FLAG
YOU BUILD ON IDEAS AND DISCOVERIES IN SCIENCE.
YOU DON'T JUST MAGICALLY FIND THAT THINGS PREVIOUSLY DISCOVERED ABOUT NATURE AREN'T TRUE ANY MORE.
YOU EXTEND THOSE TRUTHS.

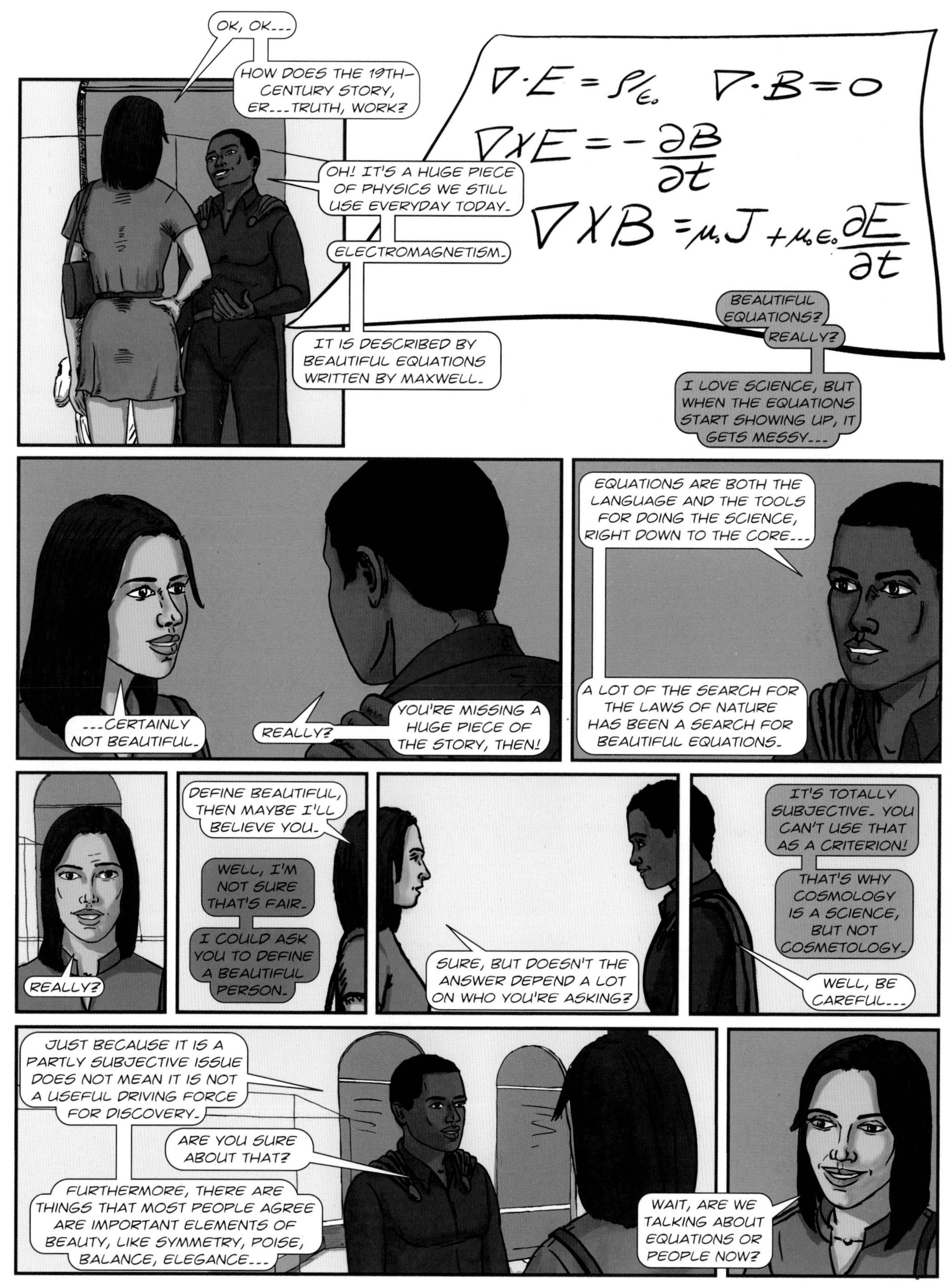
OK, OK...
HOW DOES THE 19TH-CENTURY STORY, ER...TRUTH, WORK?
OH! IT'S A HUGE PIECE OF PHYSICS WE STILL USE EVERYDAY TODAY.
ELECTROMAGNETISM.
IT IS DESCRIBED BY BEAUTIFUL EQUATIONS WRITTEN BY MAXWELL.
$\nabla \cdot E = \rho/\epsilon_0$ $\nabla \cdot B = 0$
$\nabla \times E = -\frac{\partial B}{\partial t}$
$\nabla \times B = \mu_0 J + \mu_0 \epsilon_0 \frac{\partial E}{\partial t}$
BEAUTIFUL EQUATIONS?
REALLY?
I LOVE SCIENCE, BUT WHEN THE EQUATIONS START SHOWING UP, IT GETS MESSY...
...CERTAINLY NOT BEAUTIFUL.
REALLY?
YOU'RE MISSING A HUGE PIECE OF THE STORY, THEN!
EQUATIONS ARE BOTH THE LANGUAGE AND THE TOOLS FOR DOING THE SCIENCE, RIGHT DOWN TO THE CORE...
A LOT OF THE SEARCH FOR THE LAWS OF NATURE HAS BEEN A SEARCH FOR BEAUTIFUL EQUATIONS.
REALLY?
DEFINE BEAUTIFUL, THEN MAYBE I'LL BELIEVE YOU.
WELL, I'M NOT SURE THAT'S FAIR.
I COULD ASK YOU TO DEFINE A BEAUTIFUL PERSON.
SURE, BUT DOESN'T THE ANSWER DEPEND A LOT ON WHO YOU'RE ASKING?
IT'S TOTALLY SUBJECTIVE. YOU CAN'T USE THAT AS A CRITERION!
THAT'S WHY COSMOLOGY IS A SCIENCE, BUT NOT COSMETOLOGY.
WELL, BE CAREFUL...
JUST BECAUSE IT IS A PARTLY SUBJECTIVE ISSUE DOES NOT MEAN IT IS NOT A USEFUL DRIVING FORCE FOR DISCOVERY.
ARE YOU SURE ABOUT THAT?
FURTHERMORE, THERE ARE THINGS THAT MOST PEOPLE AGREE ARE IMPORTANT ELEMENTS OF BEAUTY, LIKE SYMMETRY, POISE, BALANCE, ELEGANCE...
WAIT, ARE WE TALKING ABOUT EQUATIONS OR PEOPLE NOW?

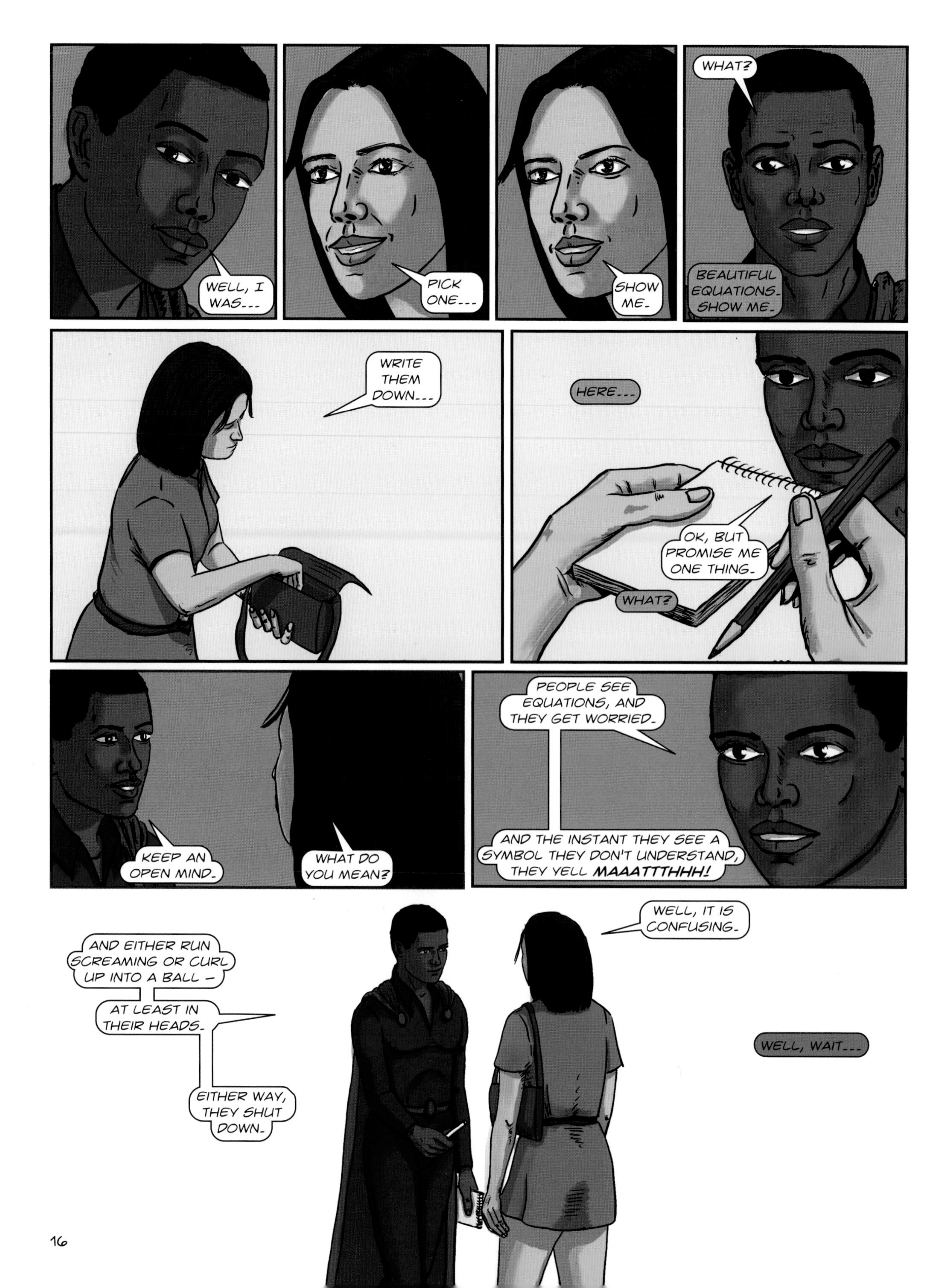
WELL, I WAS...
PICK ONE...
SHOW ME.
WHAT?
BEAUTIFUL EQUATIONS. SHOW ME.
WRITE THEM DOWN...
HERE...
OK, BUT PROMISE ME ONE THING.
WHAT?
KEEP AN OPEN MIND.
WHAT DO YOU MEAN?
PEOPLE SEE EQUATIONS, AND THEY GET WORRIED.
AND THE INSTANT THEY SEE A SYMBOL THEY DON'T UNDERSTAND, THEY YELL **MAAATTTHHH!**
AND EITHER RUN SCREAMING OR CURL UP INTO A BALL –
AT LEAST IN THEIR HEADS.
EITHER WAY, THEY SHUT DOWN.
WELL, IT IS CONFUSING.
WELL, WAIT...

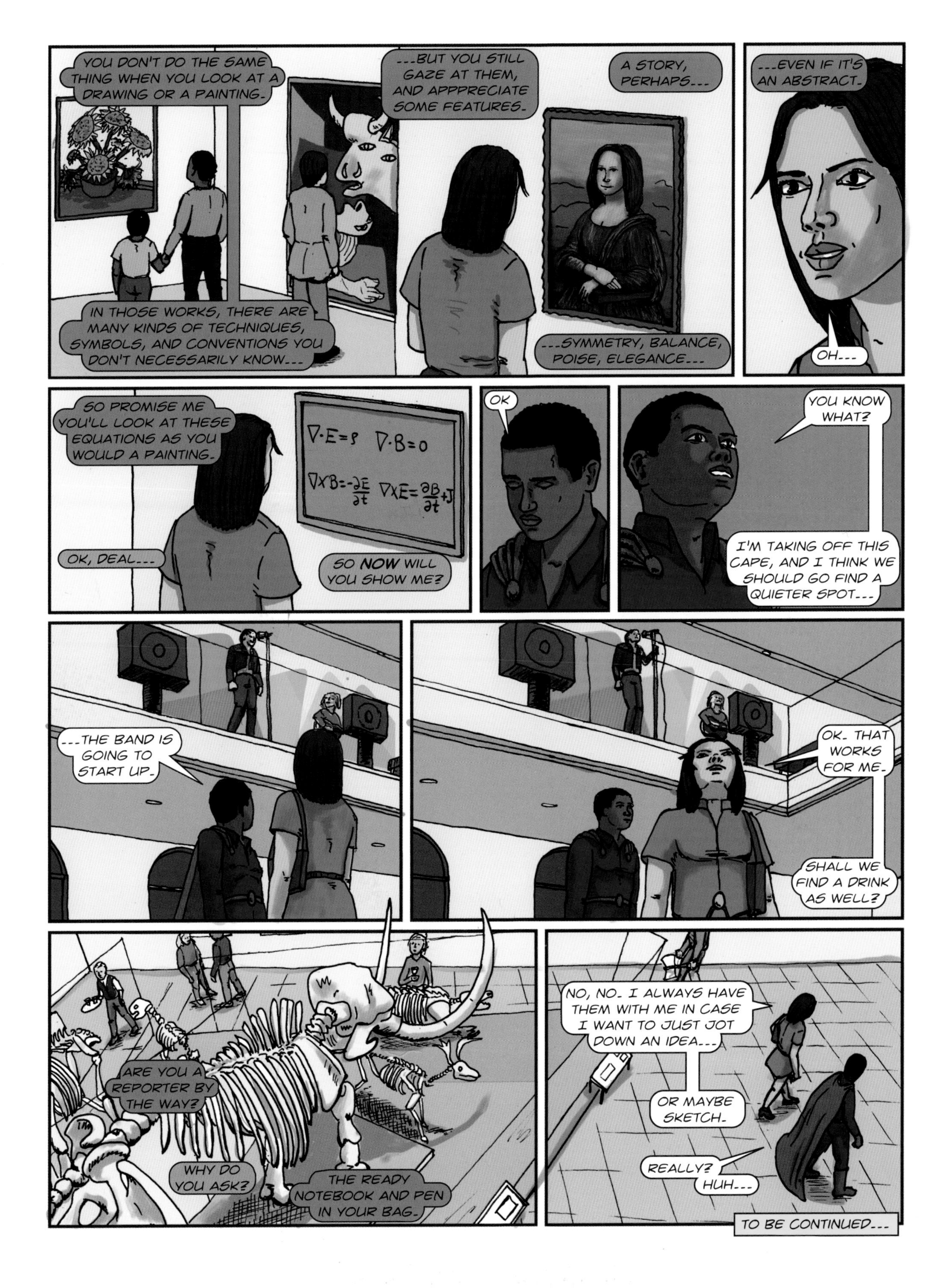
YOU DON'T DO THE SAME THING WHEN YOU LOOK AT A DRAWING OR A PAINTING.
IN THOSE WORKS, THERE ARE MANY KINDS OF TECHNIQUES, SYMBOLS, AND CONVENTIONS YOU DON'T NECESSARILY KNOW...
...BUT YOU STILL GAZE AT THEM, AND APPPRECIATE SOME FEATURES.
A STORY, PERHAPS...
...SYMMETRY, BALANCE, POISE, ELEGANCE...
...EVEN IF IT'S AN ABSTRACT.
OH...
SO PROMISE ME YOU'LL LOOK AT THESE EQUATIONS AS YOU WOULD A PAINTING.
∇·E=ρ ∇·B=0
∇×B=-∂E/∂t ∇×E=∂B/∂t+J
OK, DEAL...
SO **NOW** WILL YOU SHOW ME?
OK
YOU KNOW WHAT?
I'M TAKING OFF THIS CAPE, AND I THINK WE SHOULD GO FIND A QUIETER SPOT...
...THE BAND IS GOING TO START UP.
OK. THAT WORKS FOR ME.
SHALL WE FIND A DRINK AS WELL?
ARE YOU A REPORTER BY THE WAY?
WHY DO YOU ASK?
THE READY NOTEBOOK AND PEN IN YOUR BAG.
NO, NO. I ALWAYS HAVE THEM WITH ME IN CASE I WANT TO JUST JOT DOWN AN IDEA...
OR MAYBE SKETCH.
REALLY? HUH...
TO BE CONTINUED...

# Notes

***Pages 8–9*** – The list of scientist comic book heroes and villains is very long. Examples from the two biggest publishers of the hero genre of comics include Reed Richards (Mr. Fantastic), Susan Storm Richards (Invisible Woman), Bruce Banner (Hulk), T'Challa (Black Panther), Janet Van Dyne (Wasp), Henry Philip "Hank" McCoy (Beast), Pamela Lillian Isley (Poison Ivy), Ray Palmer (Atom), and so on. (They can all be looked up in the superhero database online: http://www.superherodb.com).

Some counterexamples to the usual model of giving up science after getting abilities: At least during the so-called Silver Age of comics,* Mr. Fantastic often seemed to busy himself with entirely curiosity-driven scientific research even after getting his abilities, although being able to stretch his limbs was presumably not much specific help beyond long reaches for all the numerous knobs and levers on his 1960s machinery (always depicted splendidly by Jack Kirby in that period). Similarly, Poison Ivy's activities as a villain were in fact driven by an urge to protect the environment, and she used her botanical knowledge and powers to help her.

*For more on that term, see Jim Casey's essay "Silver Age Comics" in Mark Bould, ed., *The Routledge Companion to Science Fiction* (London: Routledge, 2009).

***Page 10*** (panel 2) – Terrence Allen, *Microscopy: A Very Short Introduction* (Oxford: Oxford University Press, 2015).

***Page 10*** (panel 4) – An excellent account of this work can be found in this biography: Brenda Maddox, *Rosalind Franklin: The Dark Lady of DNA* (New York: HarperCollins, 2002).

***Page 10*** – The website for NASA's Chandra X-ray Observatory has lots of useful information: http://nasa.gov/chandra/.

***Page 12*** – For an introduction to the size of our solar system, galaxy, and beyond, the opening sections of these books are very good: Neil deGrasse Tyson, Michael A. Strauss, and J. Richard Gott, *Welcome to the Universe: An Astrophysical Tour* (Princeton, NJ: Princeton University Press, 2016); Adam Frank, *Astronomy: At Play in the Cosmos* (New York: W. W. Norton & Co., 2016).

***Page 12 and beyond*** – Here are two books about aspects of the history of our understanding of light and the electromagnetic spectrum: Sidney Perkowitz, *Empire of Light: A History of Discovery in Science and Art* (New York: Henry Holt, 1996); Ian Walmsley, *Light: A Very Short Introduction* (Oxford: Oxford University Press, 2015). The first one covers light in both science and the visual arts.

***Page 13*** – Here's a short and accessible account of aspects of special relativity: Brian Cox and Jeff R. Forshaw, *Why Does $E=mc^2$? (and Why Should We Care?)* (Cambridge, MA: Da Capo Press, 2009). Two unusual (and refreshing) diagrammatic introductions to special relativity can be found in Sander Bais, *Very Special Relativity: An Illustrated Guide* (Cambridge, MA: Harvard University Press, 2007); Tatsu Takeuchi, *An Illustrated Guide to Relativity* (Cambridge: Cambridge University Press, 2010).

***Page 14 and beyond*** - For a biography discussing the work of Faraday and Maxwell (and others) on electromagnetism, see Nancy Forbes and Basil Mahon, *Faraday, Maxwell, and the Electromagnetic Field: How Two Men Revolutionized Physics* (Amherst, NY: Prometheus Books, 2014).

***Page 15*** - Maxwell's equations are shown in "SI units" of measurement here. This means that two constants of nature, $\varepsilon_0$ and $\mu_0$, appear in them. Later on, they'll be written without those two constants, for simplicity. This will not change physics, but instead amounts to using different units of measurement in which those constants each have the value 1.

Further discussion of the symmetry and beauty of equations in physics (including Maxwell's) can be found in this delightful book: Frank Wilczek, *A Beautiful Question: Finding Nature's Deep Design* (New York: Penguin Press, 2015).

A collection of essays about the beauty, history, and other aspects of several important equations in science can be found in Graham Farmelo, ed., *It Must Be Beautiful: Great Equations of Modern Science* (London: Granta, 2002).

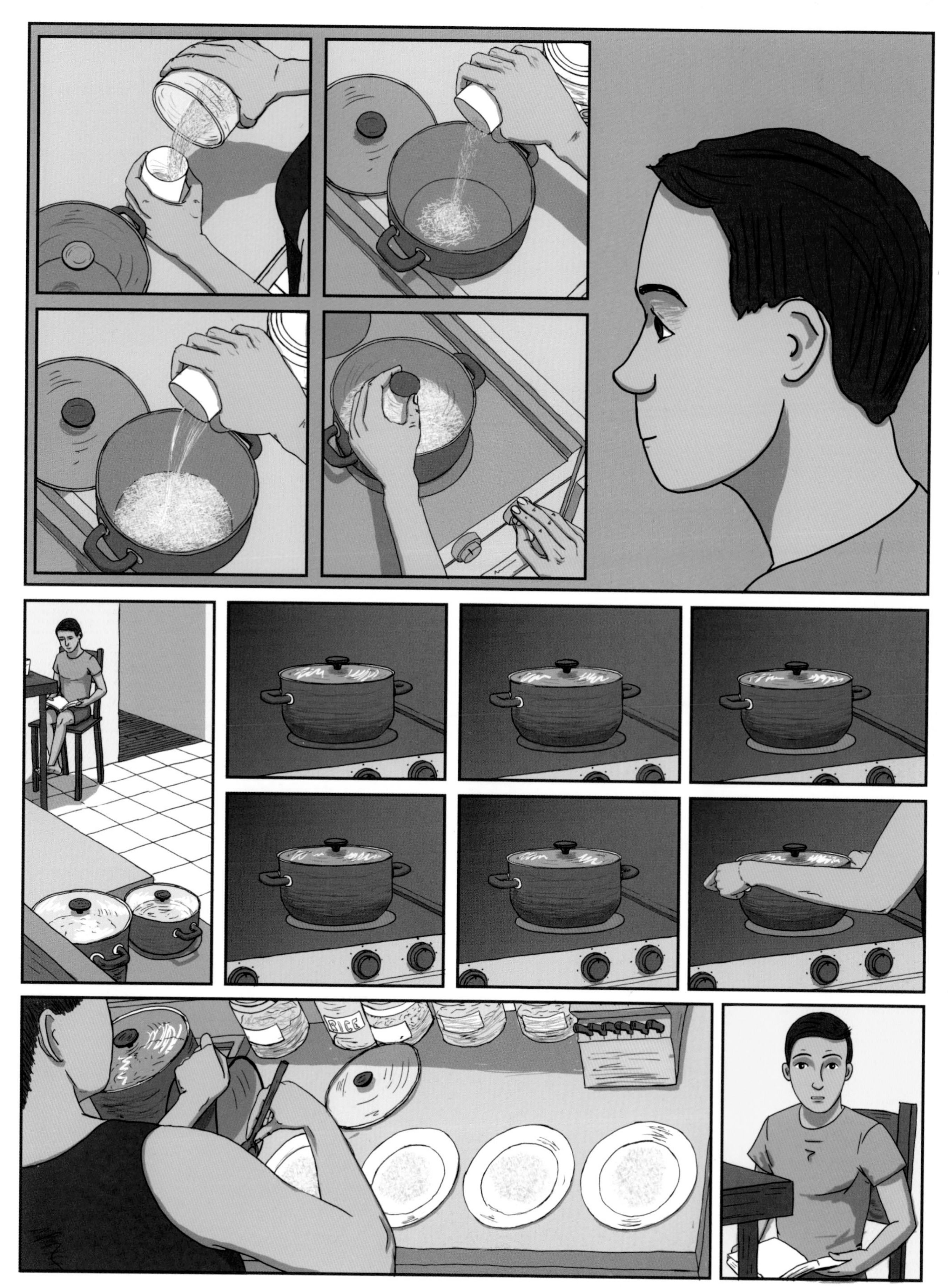
RICE

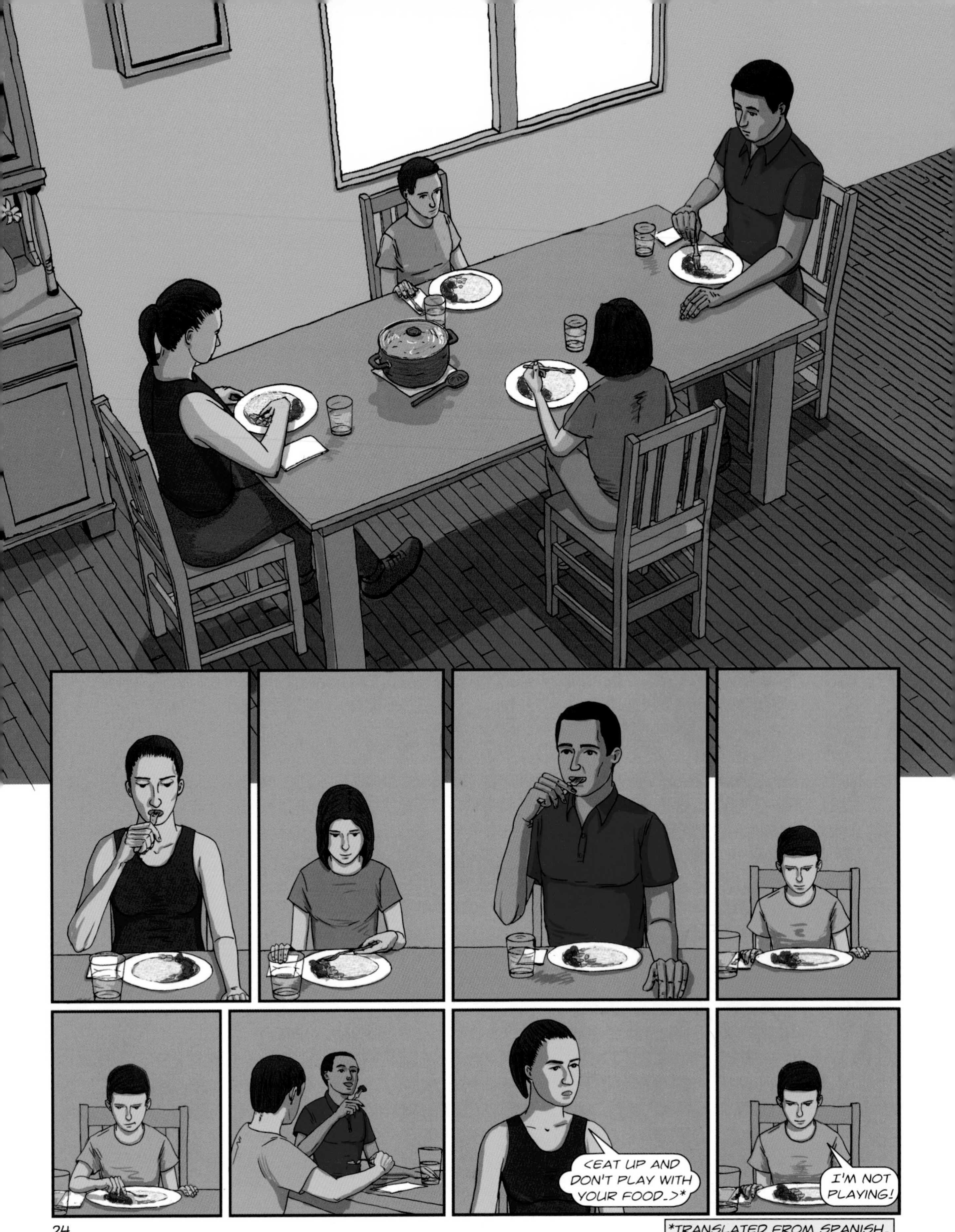




*TRANSLATED FROM SPANISH.

WHERE DOES ALL THE EXTRA RICE COME FROM?
WHAT?
MOM PUTS JUST A CUP OF RICE INTO THE PAN...
AND WHEN IT IS DONE, THERE'S A TON MORE!
SO WHAT?
WELL, YOU GET A LOT MORE RICE THAN YOU STARTED WITH – HOW?
ARE YOU GONNA LET ME WATCH MY SHOW? OR KEEP ASKING SILLY QUESTIONS?

I'M SORRY I SAID YOU WERE SILLY.
I'VE NEVER THOUGHT ABOUT THAT.
I GUESS I DON'T KNOW THE ANSWER EITHER.
THAT'S OK.
SO WHAT DO YOU THINK THE ANSWER IS?
I FIGURE YOU GET MORE BY THEM ALL BREAKING UP INTO BITS...
MAYBE...?
OR THEY MAKE BABY RICE SOMEHOW?
NO... WHERE WOULD ALL THE EXTRA STUFF COME FROM?

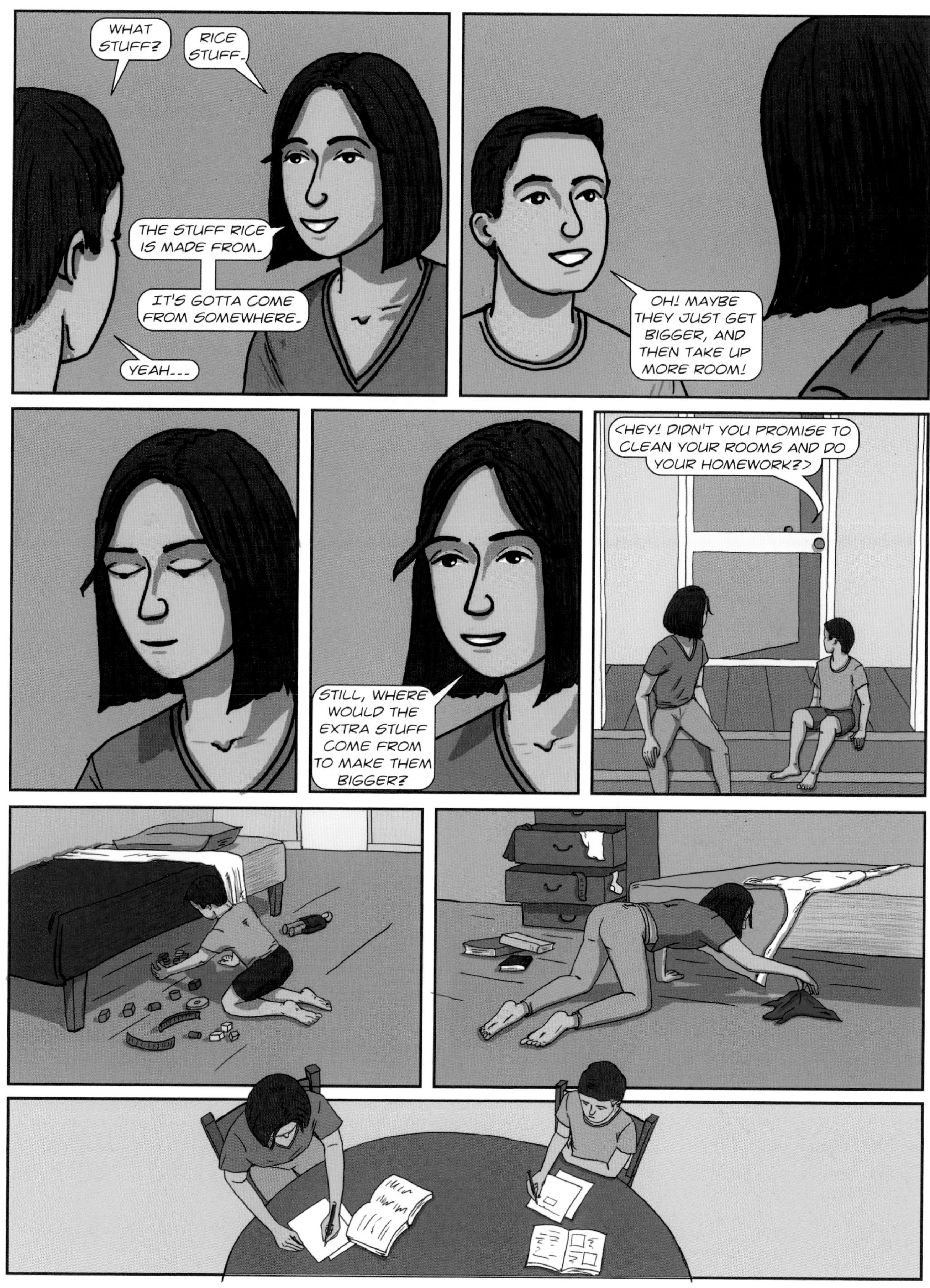

WHAT STUFF?
RICE STUFF.
THE STUFF RICE IS MADE FROM.
IT'S GOTTA COME FROM SOMEWHERE.
YEAH...
OH! MAYBE THEY JUST GET BIGGER, AND THEN TAKE UP MORE ROOM!
STILL, WHERE WOULD THE EXTRA STUFF COME FROM TO MAKE THEM BIGGER?
<HEY! DIDN'T YOU PROMISE TO CLEAN YOUR ROOMS AND DO YOUR HOMEWORK?>

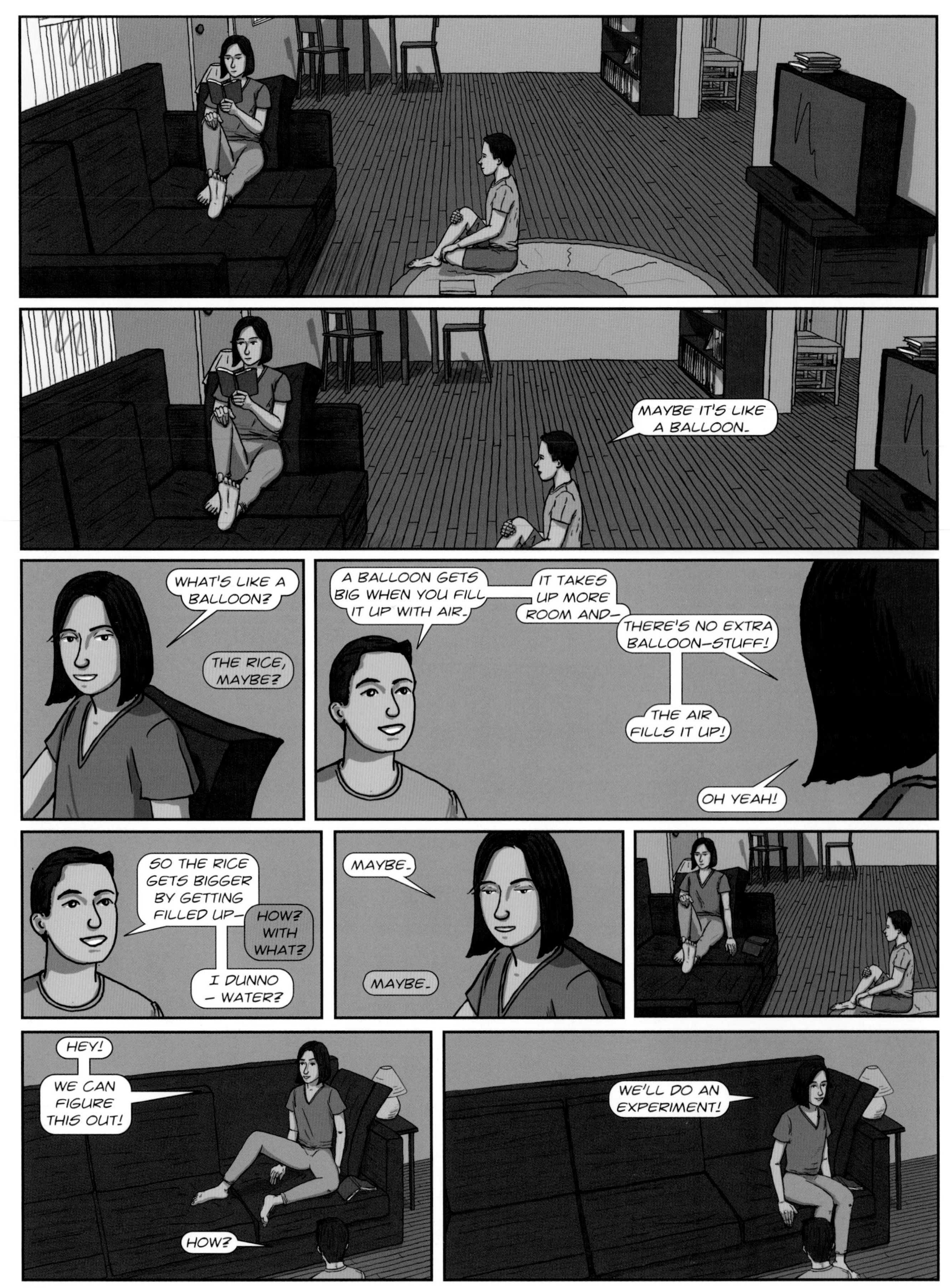
MAYBE IT'S LIKE A BALLOON.
WHAT'S LIKE A BALLOON?
THE RICE, MAYBE?
A BALLOON GETS BIG WHEN YOU FILL IT UP WITH AIR.
IT TAKES UP MORE ROOM AND—
THERE'S NO EXTRA BALLOON—STUFF!
THE AIR FILLS IT UP!
OH YEAH!
SO THE RICE GETS BIGGER BY GETTING FILLED UP—
HOW? WITH WHAT?
I DUNNO — WATER?
MAYBE.
MAYBE.
HEY!
WE CAN FIGURE THIS OUT!
HOW?
WE'LL DO AN EXPERIMENT!

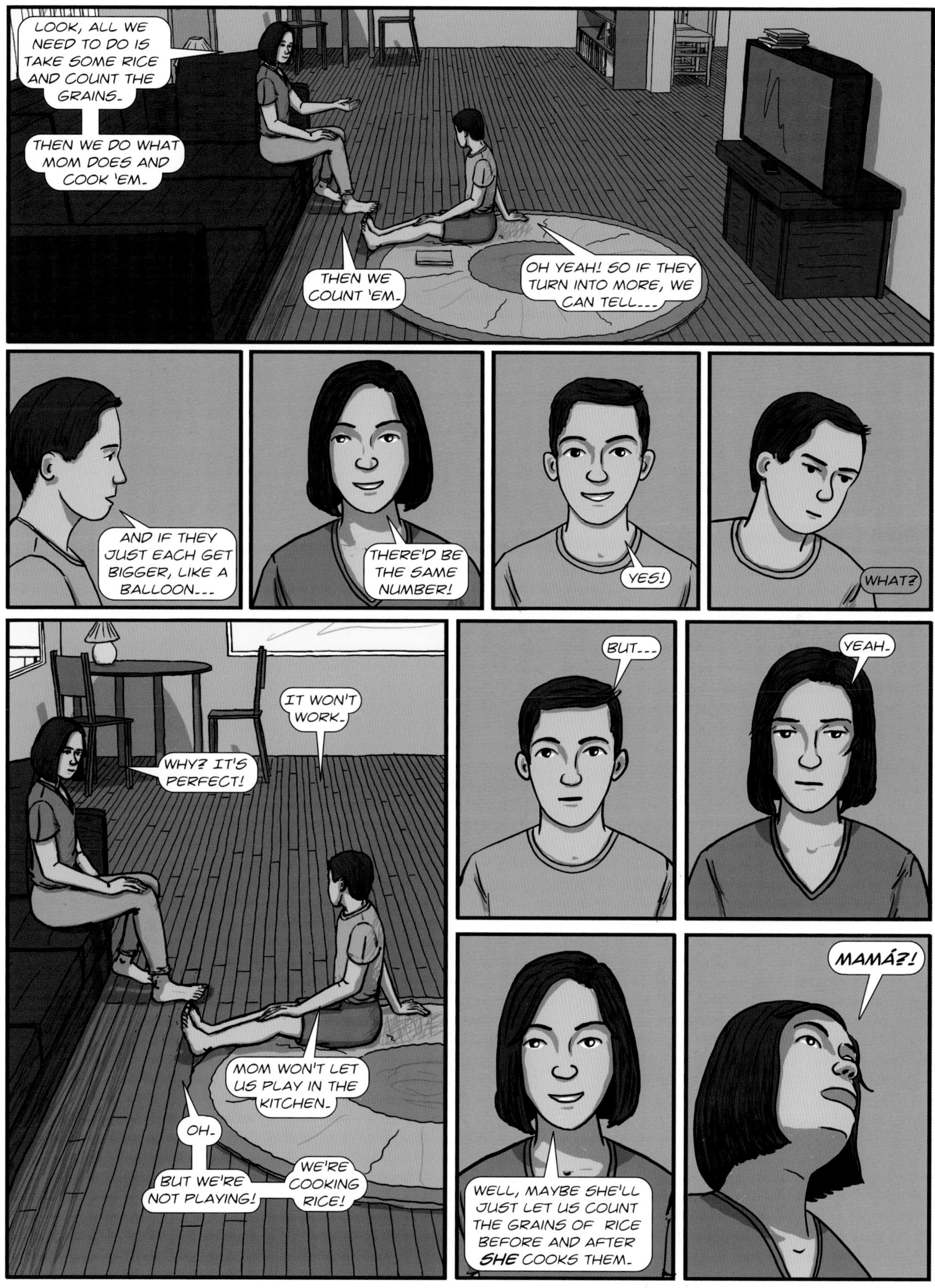
LOOK, ALL WE NEED TO DO IS TAKE SOME RICE AND COUNT THE GRAINS.
THEN WE DO WHAT MOM DOES AND COOK 'EM.
THEN WE COUNT 'EM.
OH YEAH! SO IF THEY TURN INTO MORE, WE CAN TELL...
AND IF THEY JUST EACH GET BIGGER, LIKE A BALLOON...
THERE'D BE THE SAME NUMBER!
YES!
WHAT?
WHY? IT'S PERFECT!
IT WON'T WORK.
MOM WON'T LET US PLAY IN THE KITCHEN.
OH.
BUT WE'RE NOT PLAYING!
WE'RE COOKING RICE!
BUT...
YEAH.
WELL, MAYBE SHE'LL JUST LET US COUNT THE GRAINS OF RICE BEFORE AND AFTER **SHE** COOKS THEM.
MAMÁ?!

MAMÁ!
<YES?>
ARE WE HAVING RICE FOR DINNER?
NO.
CAN WE?
<NO, WE ARE HAVING POTATOES.>
...AND WE DON'T HAVE ENOUGH RICE ANYWAY.
HOW ABOUT TOMORROW?
NO, I WON'T GO TO THE GROCERY STORE AGAIN UNTIL SATURDAY.
WHAT'S GOING ON?
NOTHING.
WE JUST WANTED TO SEE WHAT HAPPENS WHEN WE COOK IT...
WHAT HAPPENS?
IT COOKS.
IT'S AN EXPERIMENT!
YOU'RE NOT DOING EXPERIMENTS IN THE KITCHEN.
IT IS DANGEROUS TO PLAY IN HERE.
WE'RE NOT PLAYING!
NO! AND WE DON'T HAVE MONEY TO WASTE FOOD.
NOW GO AND PLAY...

OH!
WHAT?
WE DON'T NEED ALL THAT RICE.
WHAT DO YOU MEAN?
HOW DO WE DO THE EXPERIMENT?
WE ONLY NEED TO COOK ONE GRAIN.
OH YES!
MOM?
<YOU STILL HERE IN THE KITCHEN?>
MOM?
YES?
CAN WE HAVE JUST ONE GRAIN OF RICE?
ONE?
YES.
OK.

HAVE TWO RICE GRAINS THEN.
ONE EACH, SO YOU DON'T FIGHT OVER THEM.
WE WON'T FIGHT. WE'RE WORKING TOGETHER.
OK.
WHAT DO YOU SAY?
THANK YOU!
OK, WHAT IS IT NOW?
CAN YOU PUT MY GRAIN IN WITH THE POTATOES WHEN YOU COOK THEM?
THEN WHAT?
THAT'S IT.
WHEN THE POTATOES ARE DONE, PLEASE CAN YOU GIVE US THE RICE THAT IS LEFT IN THE PAN?
OK.
IF THAT WILL HELP YOU LEAVE ME IN PEACE TO MAKE DINNER.
THANK YOU, MAMÁ!

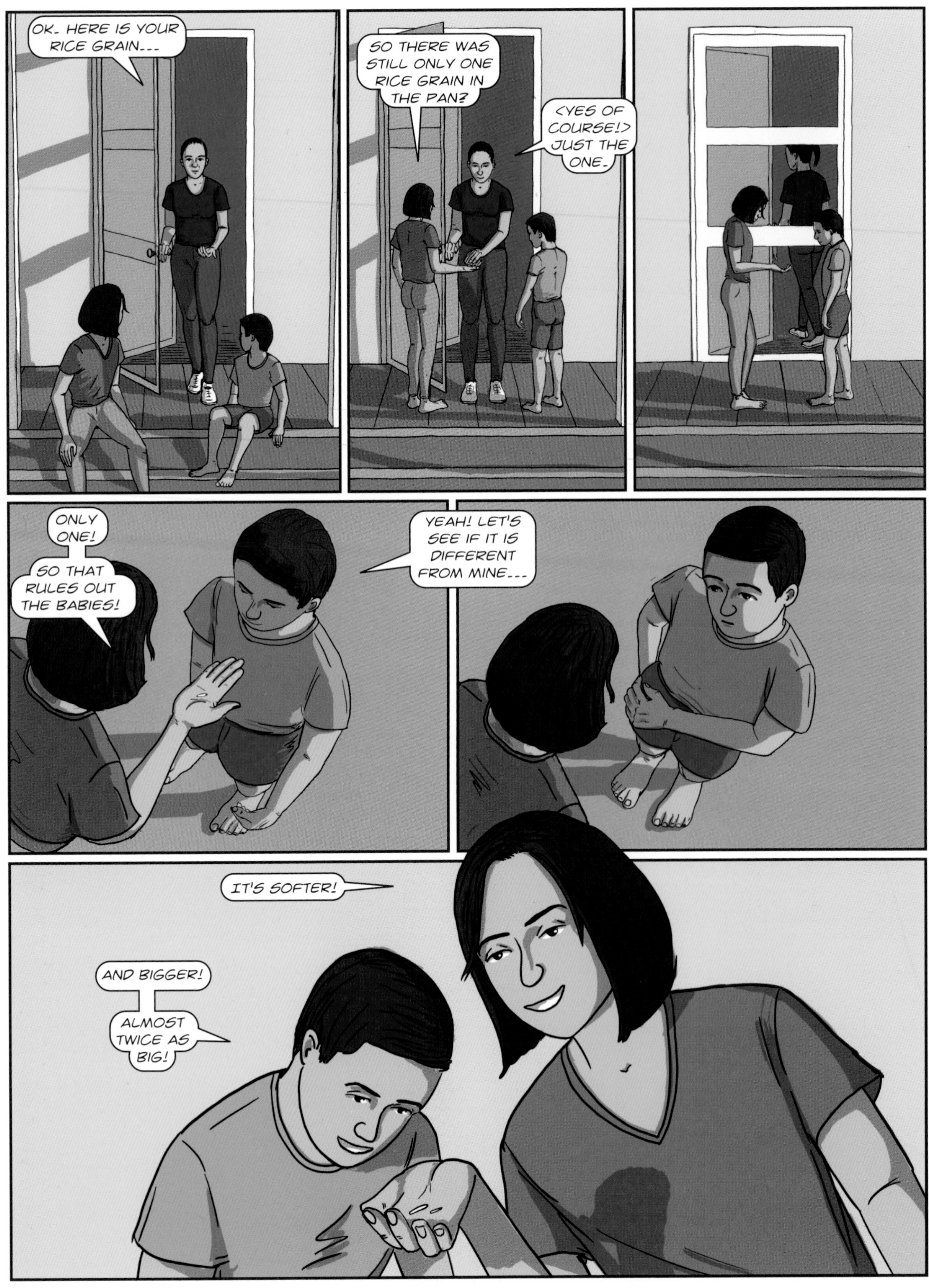
OK. HERE IS YOUR RICE GRAIN...
SO THERE WAS STILL ONLY ONE RICE GRAIN IN THE PAN?
<YES OF COURSE!> JUST THE ONE.
ONLY ONE!
SO THAT RULES OUT THE BABIES!
YEAH! LET'S SEE IF IT IS DIFFERENT FROM MINE...
IT'S SOFTER!
AND BIGGER!
ALMOST TWICE AS BIG!

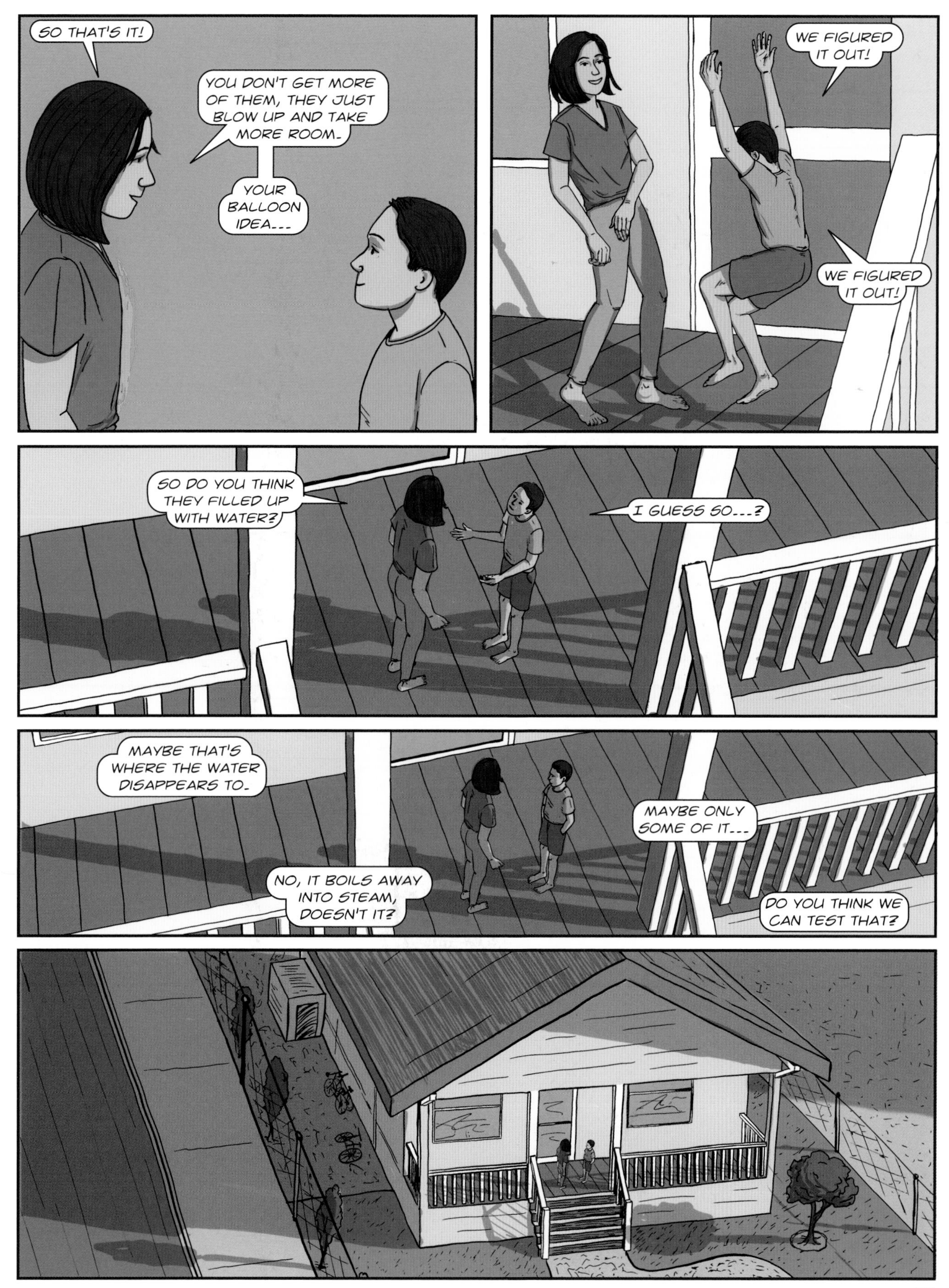
SO THAT'S IT!
YOU DON'T GET MORE OF THEM, THEY JUST BLOW UP AND TAKE MORE ROOM.
YOUR BALLOON IDEA...
WE FIGURED IT OUT!
WE FIGURED IT OUT!
SO DO YOU THINK THEY FILLED UP WITH WATER?
I GUESS SO...?
MAYBE THAT'S WHERE THE WATER DISAPPEARS TO.
NO, IT BOILS AWAY INTO STEAM, DOESN'T IT?
MAYBE ONLY SOME OF IT...
DO YOU THINK WE CAN TEST THAT?

## Notes

Cooking is a marvelous context in which to explore science and put the scientific method into action. The results are often immediately rewarding too!

Perhaps the most well-known (and well-loved) book on science and cooking is Harold McGee, *On Food and Cooking: The Science and Lore of the Kitchen* (New York: Scribner, 2004).

Much delight can be gained by reading the writings of Hervé This, one of the founders of modern "molecular gastronomy." Hervé's work contains descriptions of lots of fascinating experiments and processes exploring traditional methods and trying to understand how (and if) they work, and why. It also includes lots of interesting chemistry and physics along the way, and sometimes strong opinions. Two of his works are Hervé This, *Molecular Gastronomy: Exploring the Science of Flavor*, trans. M. B. DeBevoise (New York: Columbia University Press, 2006), and Hervé This, *The Science of the Oven*, trans. Jody Gladding (New York: Columbia University Press, 2009). (The title of the latter might be a bit confusing since the whole kitchen is explored.)

A more recent book, probably destined to become a classic too, is J. Kenji López-Alt, *The Food Lab: Better Home Cooking through Science* (New York: W. W. Norton & Co., 2015), which accompanies the column of the same name from the excellent Serious Eats website: http://www.seriouseats.com/the-food-lab/.

Speaking of websites, Professor Amy Rowat of the UCLA Division of Life Sciences and Department of Integrative Biology and Physiology has a project called Science and Food to which many scientists and professional chefs contribute. It has an excellent website: http://scienceandfood.org/.

Another great online resource very much in the spirit of the curiosity-driven exploration seen in the conversation is the website of the Exploratorium. It has a section devoted to food and cooking: http://www.exploratorium.edu/cooking/.

The Exploratorium was founded by Frank Oppenheimer, and you can learn more about him and his drive to illuminate science through exploration in this excellent biography: K. C. Cole, *Something Incredibly Wonderful Happens: Frank Oppenheimer and the World He Made Up* (Boston: Houghton Mifflin Harcourt, 2009).

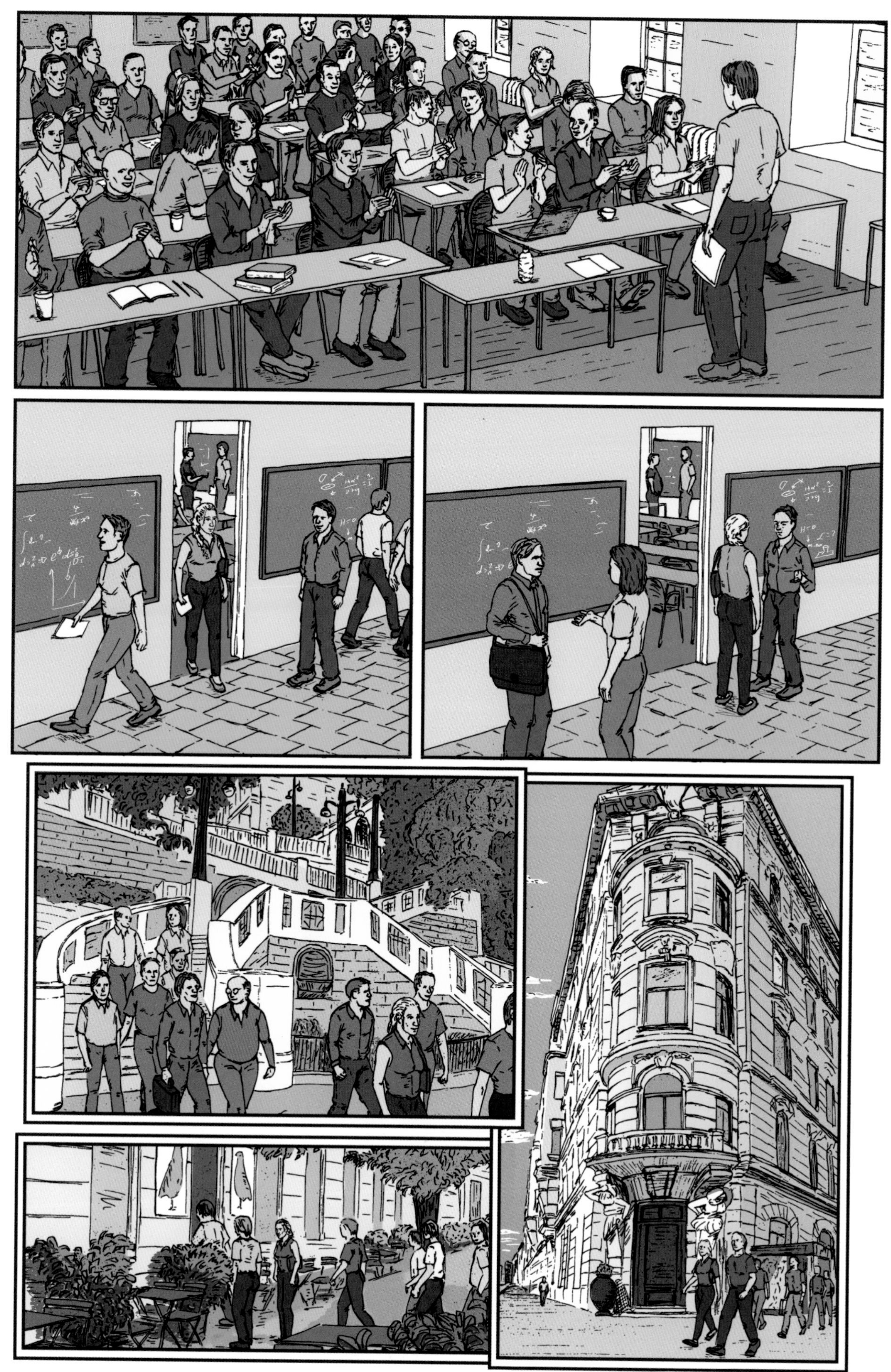

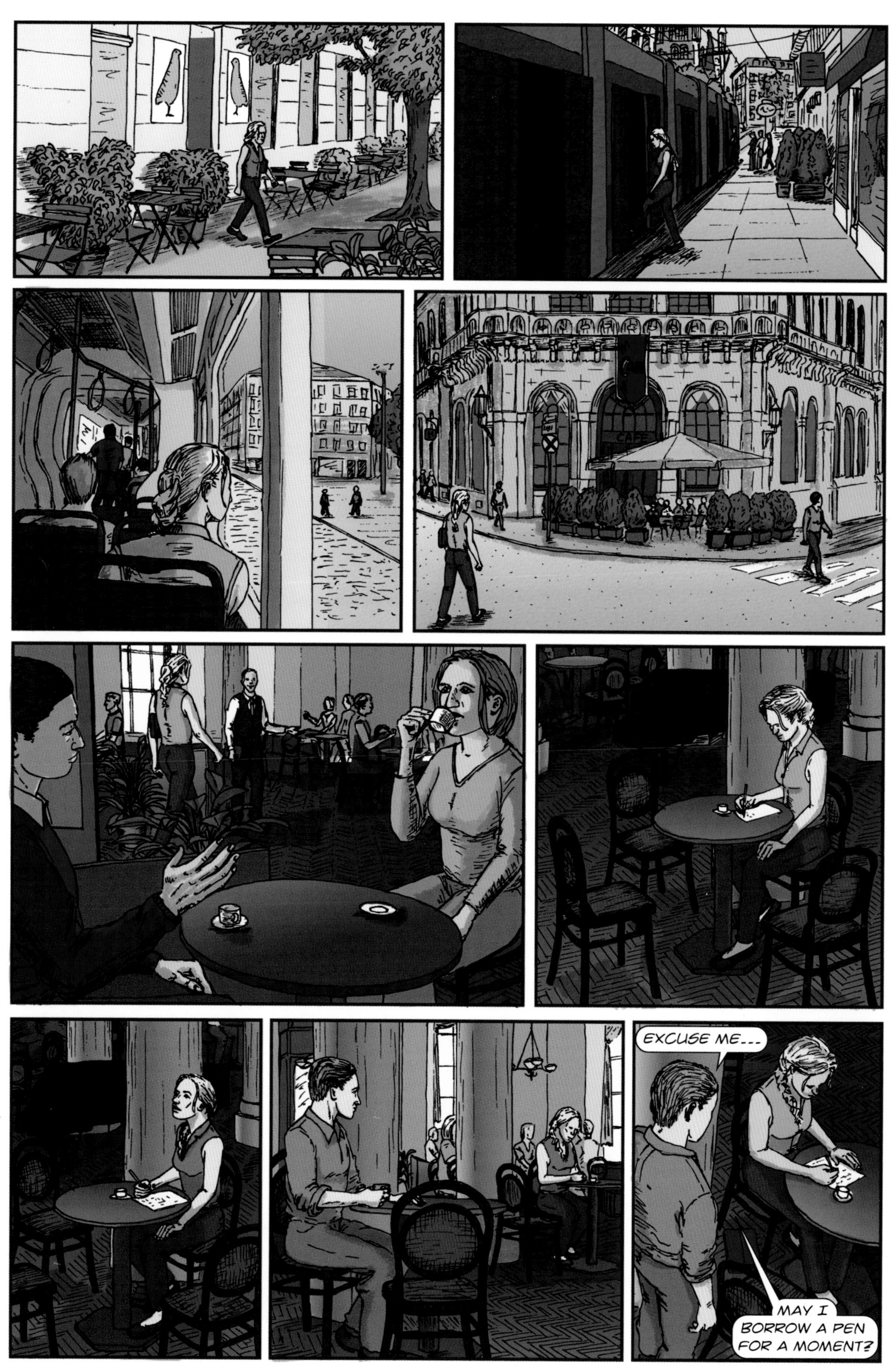

EXCUSE ME...
MAY I BORROW A PEN FOR A MOMENT?

JUST A SEC...
THAT'S THE BEST YOU CAN DO?
EXCUSE ME?
IT'S OK.
JUST A LITTLE DISAPPOINTING S'ALL.
UH...
I'D HAVE GONE WITH "DID YOU KNOW THIS WAS FREUD'S FAVORITE SPOT?" OR SOMETHING SIMILAR.
WOW, WAS IT?!
I'VE NO IDEA, REALLY. I MAY HAVE READ THAT ONLINE.
SO WHAT'S YOUR STORY?
GET YOUR COFFEE AND GRAB A SEAT.
I REALLY DID WANT TO BORROW A PEN...
OK, BUT WHAT ARE YOU UP TO IN THE CITY?
I'M HERE FOR A BUSINESS MEETING...
YOU KNOW, DISCUSSIONS ABOUT NEW MARKETS TO SELL WIDGETS TO.
WHAT KIND OF WIDGETS?
OH, THE GOOD KIND, OF COURSE.
HOW ABOUT YOU?
CONFERENCE.
OH? WHAT ABOUT?
PHYSICS.
REALLY? WHAT KIND?
OH, THE GOOD KIND, OF COURSE.

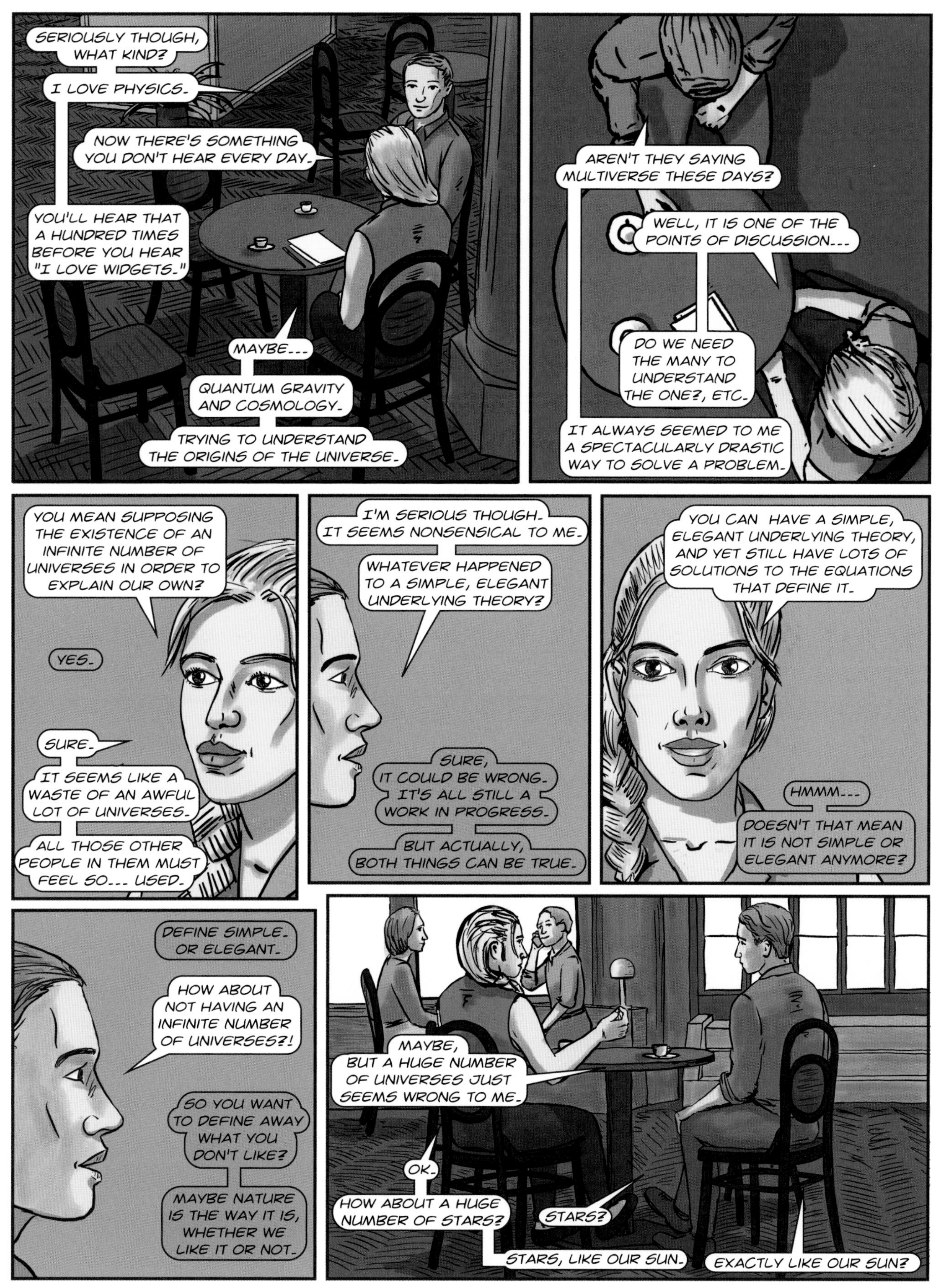
SERIOUSLY THOUGH, WHAT KIND?
I LOVE PHYSICS.
NOW THERE'S SOMETHING YOU DON'T HEAR EVERY DAY.
YOU'LL HEAR THAT A HUNDRED TIMES BEFORE YOU HEAR "I LOVE WIDGETS."
MAYBE...
QUANTUM GRAVITY AND COSMOLOGY.
TRYING TO UNDERSTAND THE ORIGINS OF THE UNIVERSE.
AREN'T THEY SAYING MULTIVERSE THESE DAYS?
WELL, IT IS ONE OF THE POINTS OF DISCUSSION...
DO WE NEED THE MANY TO UNDERSTAND THE ONE?, ETC.
IT ALWAYS SEEMED TO ME A SPECTACULARLY DRASTIC WAY TO SOLVE A PROBLEM.
YOU MEAN SUPPOSING THE EXISTENCE OF AN INFINITE NUMBER OF UNIVERSES IN ORDER TO EXPLAIN OUR OWN?
YES.
SURE.
IT SEEMS LIKE A WASTE OF AN AWFUL LOT OF UNIVERSES.
ALL THOSE OTHER PEOPLE IN THEM MUST FEEL SO... USED.
I'M SERIOUS THOUGH. IT SEEMS NONSENSICAL TO ME.
WHATEVER HAPPENED TO A SIMPLE, ELEGANT UNDERLYING THEORY?
SURE, IT COULD BE WRONG. IT'S ALL STILL A WORK IN PROGRESS.
BUT ACTUALLY, BOTH THINGS CAN BE TRUE.
YOU CAN HAVE A SIMPLE, ELEGANT UNDERLYING THEORY, AND YET STILL HAVE LOTS OF SOLUTIONS TO THE EQUATIONS THAT DEFINE IT.
HMMM...
DOESN'T THAT MEAN IT IS NOT SIMPLE OR ELEGANT ANYMORE?
DEFINE SIMPLE. OR ELEGANT.
HOW ABOUT NOT HAVING AN INFINITE NUMBER OF UNIVERSES?!
SO YOU WANT TO DEFINE AWAY WHAT YOU DON'T LIKE?
MAYBE NATURE IS THE WAY IT IS, WHETHER WE LIKE IT OR NOT.
MAYBE, BUT A HUGE NUMBER OF UNIVERSES JUST SEEMS WRONG TO ME.
OK.
HOW ABOUT A HUGE NUMBER OF STARS?
STARS?
STARS, LIKE OUR SUN.
EXACTLY LIKE OUR SUN?

SOME BIGGER, SOME SMALLER.
SOME COLDER, SOME HOTTER.
SOME REDDER, SOME BLUER.
I'M DR. SEUSS, BY THE WAY.
ER...
ONE STAR, TWO STAR.
RED STAR, BLUE STAR?
I GUESS I'M THE GRINCH?
HEH...
STARS?
SURE, I CAN SEE LOTS OF THEM AT NIGHT...
WHAT IS YOUR POINT?
WELL, WE PRETTY MUCH UNDERSTAND THE PHYSICS NEEDED TO STUDY STARS.
THERE'S A LOT OF EQUATIONS YOU CAN WRITE DOWN FOR THEM,
FROM GRAVITY, NUCLEAR PHYSICS, AND THERMODYNAMICS.
THERE'S A HUGE NUMBER OF SOLUTIONS TO THE EQUATIONS,
DEPENDING UPON WHAT YOU START WITH – HOW MUCH AND WHAT KIND OF RAW MATERIALS... HYDROGEN, HELIUM, ETC.
THE EQUATIONS DON'T TELL YOU THAT THERE'S ONLY ONE KIND OF STAR.

FAIR POINT.
BUT WE CAN **SEE** THOSE OTHER STARS.
AND THERE'S A FINITE NUMBER OF THEM, NOT INFINITE.
OK, LET'S BE CAREFUL HERE.
FIRST, THE EQUATIONS GIVE AN INFINITE SET OF POSSIBILITIES FOR THE PROPERTIES OF A STAR.
MAYBE A STAR JUST LIKE OUR SUN BUT JUST A FEW DEGREES DIFFERENT IN TEMPERATURE...
OR A BLUE SUPER GIANT STAR MANY TIMES BIGGER THAN OUR SUN.
I CAN AT LEAST LOOK UP AND SEE.
SECOND, THOSE STARS CAN EXIST WHETHER WE ARE LOOKING AT THEM OR NOT.
THEY DON'T NEED US TO BE LOOKING AT THEM.
THEY DON'T NEED US TO EXIST AT ALL.
FOR EXAMPLE, THE TEMPERATURE AND MASS OF OUR SUN IS NOT DETERMINED BY THE NUMBER AND VARIETY OF ALL THE STARS IN THE UNIVERSE.
HMMM...
SO MAYBE YOUR THEORY OF STARS IS INCOMPLETE.
A BETTER THEORY WOULD TELL YOU HOW MUCH MATERIAL IS IN THE WHOLE UNIVERSE, AND ALSO HOW IT IS DISTRIBUTED.
COMBINE THAT INFORMATION WITH YOUR STAR EQUATIONS AND THEN YOU'D DESCRIBE ALL STARS, EVERYWHERE.
ISN'T THAT WHAT THE THEORY OF EVERYTHING IS SUPPOSED TO DO, BUT FOR UNIVERSES?
REALLY?
I DON'T REALLY KNOW WHAT A THEORY OF EVERYTHING IS SUPPOSED TO DO.
DOES ANYONE?

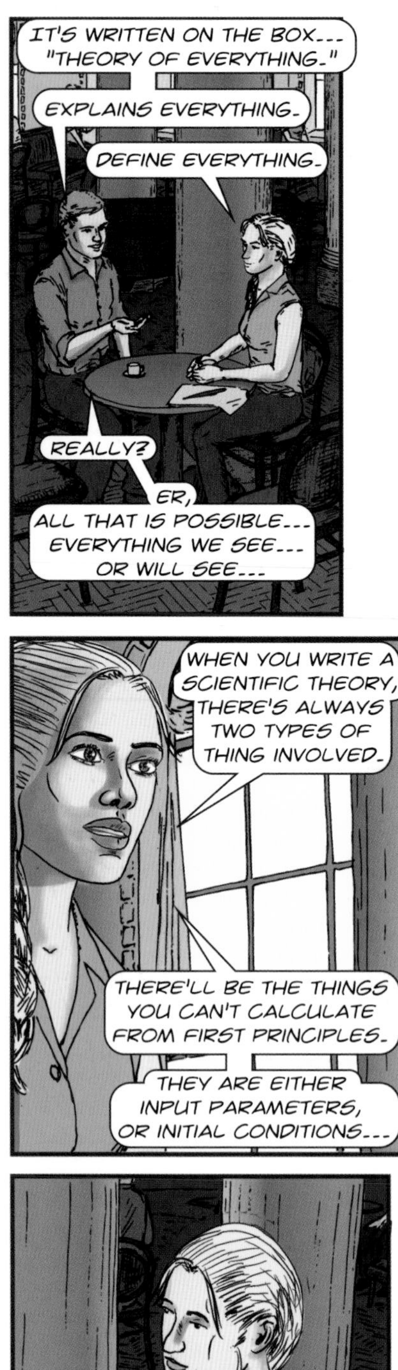
IT'S WRITTEN ON THE BOX...
"THEORY OF EVERYTHING."
EXPLAINS EVERYTHING.
DEFINE EVERYTHING.
REALLY?
ER,
ALL THAT IS POSSIBLE... EVERYTHING WE SEE... OR WILL SEE...
WHEN YOU WRITE A SCIENTIFIC THEORY, THERE'S ALWAYS TWO TYPES OF THING INVOLVED.
THERE'LL BE THE THINGS YOU CAN'T CALCULATE FROM FIRST PRINCIPLES.
THEY ARE EITHER INPUT PARAMETERS, OR INITIAL CONDITIONS...

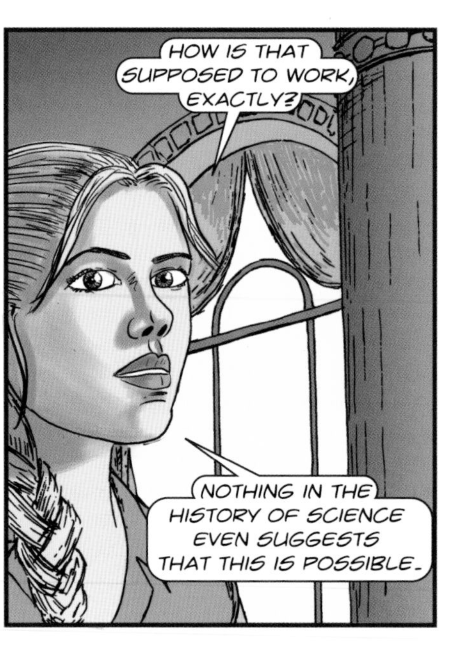
HOW IS THAT SUPPOSED TO WORK, EXACTLY?
NOTHING IN THE HISTORY OF SCIENCE EVEN SUGGESTS THAT THIS IS POSSIBLE.

BUT THAT'S THE DREAM, ISN'T IT?
WHOSE DREAM?
ALSO, WHO SAYS WE HAVE TO ALL DREAM THE SAME?
WELL THAT'S WHAT THEY SAY YOU GUYS ARE TRYING TO FIND.
AND THEN THERE'S EVERYTHING ELSE...
STUFF YOU CAN COMPUTE WITH YOUR EQUATIONS.
YOU HAVE A LIMITED DOMAIN WHERE THE THEORY WORKS...

THEORY OF EVERYTHING

ULTIMATELY, YOU'LL GET TO A POINT WHERE YOU MIGHT ASK WHERE THOSE INPUTS COME FROM.
BUT THE EQUATIONS WILL NEVER TELL YOU THAT.

YOU'VE GOT TO FIND WHERE YOUR THEORY HANDS OVER TO ANOTHER THEORY.
IN THAT ONE, SOME OF THOSE INPUT PARAMETERS MIGHT BE THINGS YOU CAN COMPUTE.
BUT THERE'LL STILL BE UNCOMPUTABLE PARAMETERS IN THAT LARGER THEORY.

THEN YOU'VE GOTTA FIND A BETTER THEORY.
SURE, BUT THE PROCESS NEVER STOPS, YOU SEE!
THAT'S MY POINT!

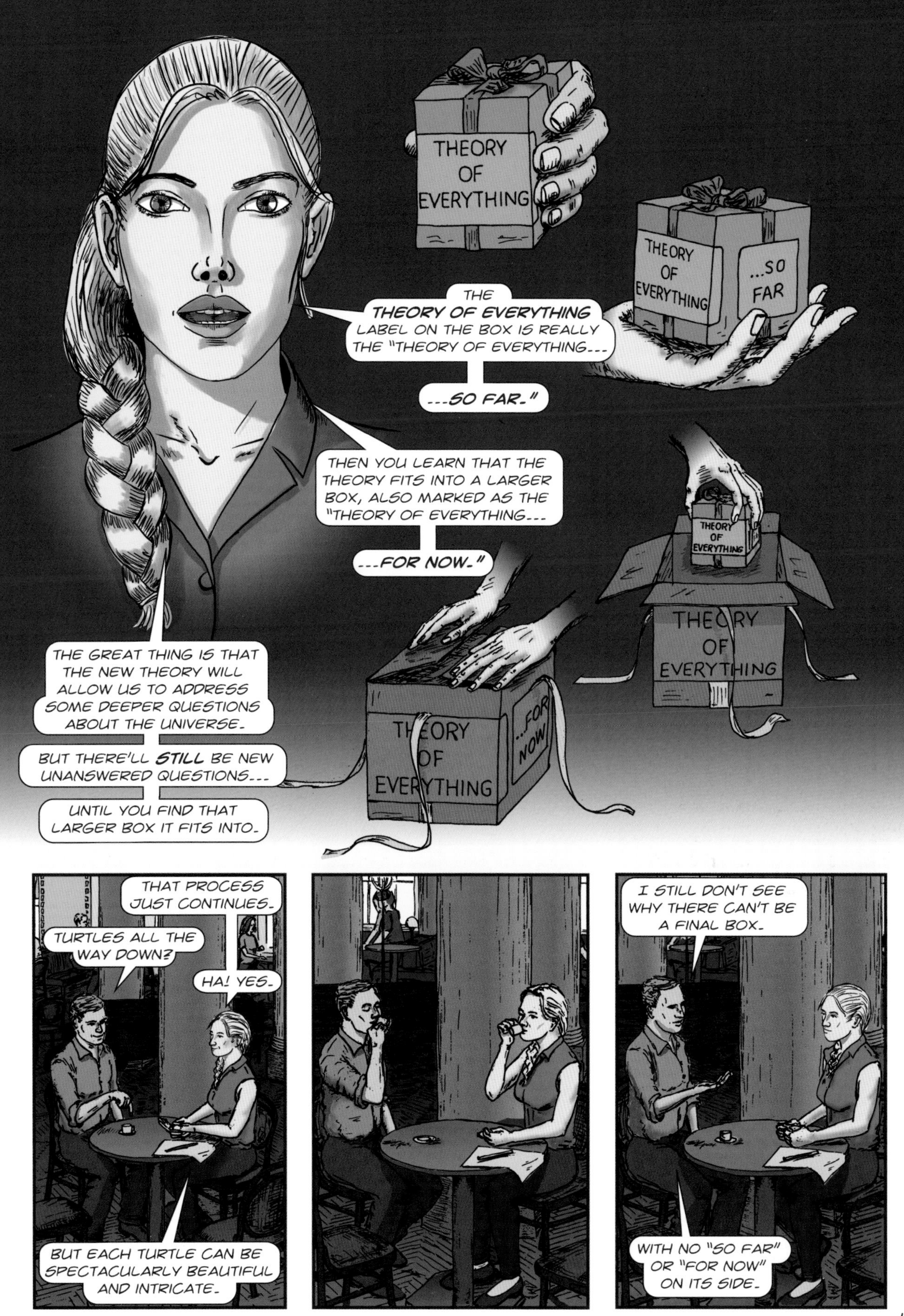
THEORY OF EVERYTHING
THEORY OF EVERYTHING
...SO FAR
THE THEORY OF EVERYTHING LABEL ON THE BOX IS REALLY THE "THEORY OF EVERYTHING...
...SO FAR."
THEN YOU LEARN THAT THE THEORY FITS INTO A LARGER BOX, ALSO MARKED AS THE "THEORY OF EVERYTHING...
...FOR NOW."
THEORY OF EVERYTHING
THEORY OF EVERYTHING
THEORY OF EVERYTHING
...FOR NOW
THE GREAT THING IS THAT THE NEW THEORY WILL ALLOW US TO ADDRESS SOME DEEPER QUESTIONS ABOUT THE UNIVERSE.
BUT THERE'LL STILL BE NEW UNANSWERED QUESTIONS...
UNTIL YOU FIND THAT LARGER BOX IT FITS INTO.
THAT PROCESS JUST CONTINUES.
TURTLES ALL THE WAY DOWN?
HA! YES.
BUT EACH TURTLE CAN BE SPECTACULARLY BEAUTIFUL AND INTRICATE.
I STILL DON'T SEE WHY THERE CAN'T BE A FINAL BOX.
WITH NO "SO FAR" OR "FOR NOW" ON ITS SIDE.

WELL I CAN'T PROVE THAT THERE ISN'T ONE.
JUST DOES NOT SEEM VERY PLAUSIBLE...
AND IT DOES NOT FIT WITH THE WAY SCIENCE ORGANIZES ITSELF.
WHAT DO YOU MEAN?
WELL, ISN'T IT AMAZING THAT NATURE IS UNDERSTANDABLE AT ALL?
SURE...
I THINK THAT'S POSSIBLE BECAUSE OF THE WAY THIS IS ALL ORGANIZED.
YOU MEAN...
I MEAN IT IS A BASIC PART OF THE DIALOGUE WE HAVE WITH NATURE.
WHY DOES THAT SAY THAT THERE'S AN INFINITE SET OF BOXES?
A THEORY IS NECESSARILY LIMITED TO HAVE ONLY A CERTAIN AMOUNT IT CAN COMPUTE.
THE REST HAD **BETTER** BE INPUTS...
OTHERWISE, WE'D HAVE TO EXPLAIN EVERYTHING AT ONCE
REALLY?
WELL, BACK IN THE 17TH CENTURY, WE'D NEVER HAVE UNDERSTOOD THE SIMPLE MOTION OF A THROWN BALL...
IF WE WERE NOT ABLE TO IGNORE A LOT OF STUFF.
YOU MEAN, LIKE AIR RESISTANCE?
OH, MUCH MORE THAN THAT!
NEWTON'S SECOND LAW OF MOTION, F=MA, IS KIND OF LIKE THE THEORY OF EVERYTHING FOR PROJECTILE MOTION HERE ON EARTH.
THE ACCELERATION 'A' COMES FROM THE EARTH'S GRAVITY, WHICH YOU MEASURE.
NOTICE THAT THE THEORY DOESN'T TELL YOU THE STRENGTH OF GRAVITY. IT'S AN INPUT!
YOU'D NEED NEWTON'S LAW OF GRAVITY TO COMPUTE IT FROM SCRATCH, INPUTTING EARTH'S MASS AND RADIUS.
THAT'S THE BIGGER BOX?
YES.
AND NEWTON'S LAW IN TURN COMES FROM EINSTEIN'S MORE FUNDAMENTAL GENERAL RELATIVITY — AN EVEN BIGGER BOX.

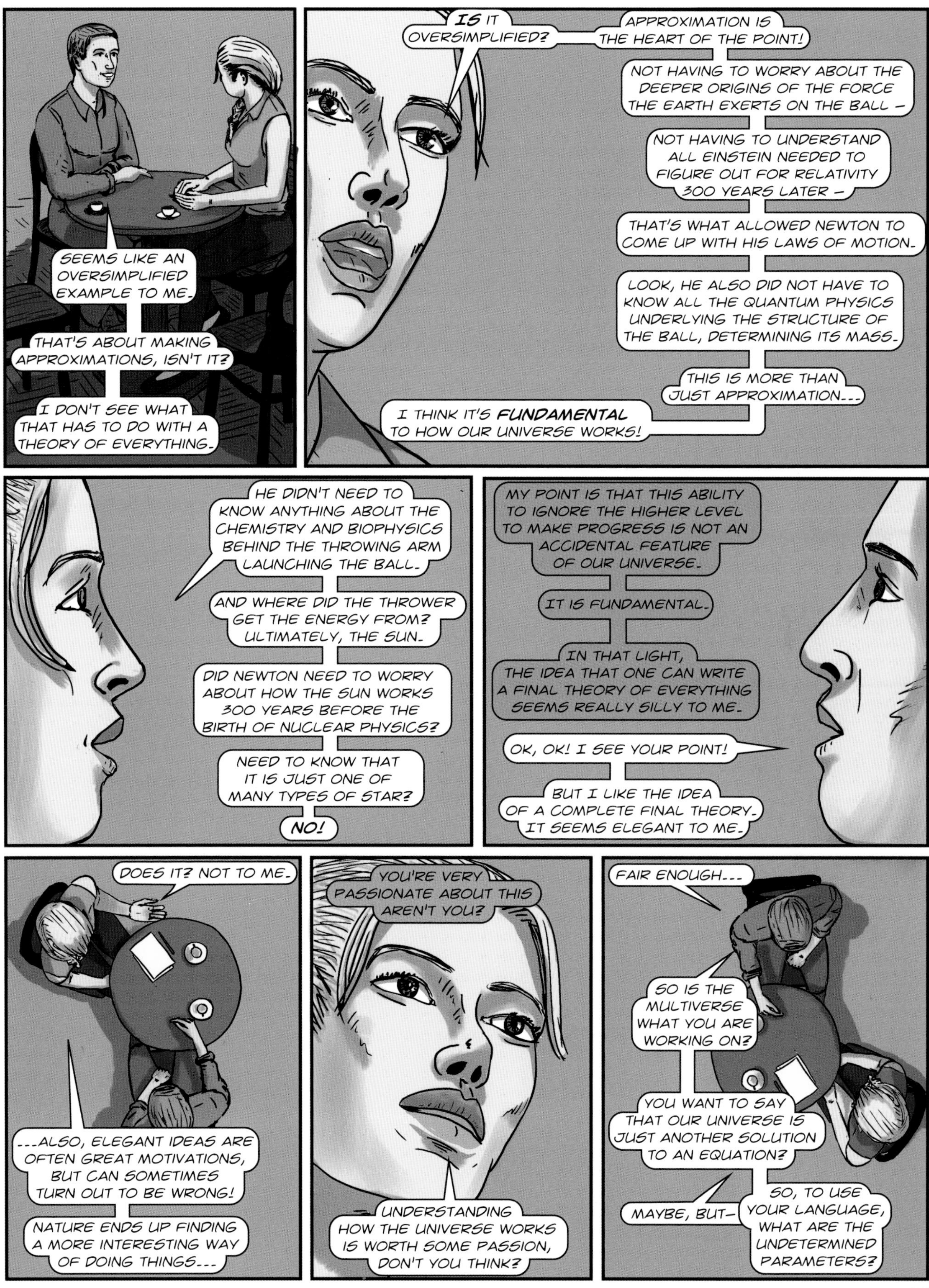

SEEMS LIKE AN OVERSIMPLIFIED EXAMPLE TO ME.
THAT'S ABOUT MAKING APPROXIMATIONS, ISN'T IT?
I DON'T SEE WHAT THAT HAS TO DO WITH A THEORY OF EVERYTHING.
IS IT OVERSIMPLIFIED?
APPROXIMATION IS THE HEART OF THE POINT!
NOT HAVING TO WORRY ABOUT THE DEEPER ORIGINS OF THE FORCE THE EARTH EXERTS ON THE BALL –
NOT HAVING TO UNDERSTAND ALL EINSTEIN NEEDED TO FIGURE OUT FOR RELATIVITY 300 YEARS LATER –
THAT'S WHAT ALLOWED NEWTON TO COME UP WITH HIS LAWS OF MOTION.
LOOK, HE ALSO DID NOT HAVE TO KNOW ALL THE QUANTUM PHYSICS UNDERLYING THE STRUCTURE OF THE BALL, DETERMINING ITS MASS.
THIS IS MORE THAN JUST APPROXIMATION...
I THINK IT'S FUNDAMENTAL TO HOW OUR UNIVERSE WORKS!
HE DIDN'T NEED TO KNOW ANYTHING ABOUT THE CHEMISTRY AND BIOPHYSICS BEHIND THE THROWING ARM LAUNCHING THE BALL.
AND WHERE DID THE THROWER GET THE ENERGY FROM? ULTIMATELY, THE SUN.
DID NEWTON NEED TO WORRY ABOUT HOW THE SUN WORKS 300 YEARS BEFORE THE BIRTH OF NUCLEAR PHYSICS?
NEED TO KNOW THAT IT IS JUST ONE OF MANY TYPES OF STAR?
NO!
MY POINT IS THAT THIS ABILITY TO IGNORE THE HIGHER LEVEL TO MAKE PROGRESS IS NOT AN ACCIDENTAL FEATURE OF OUR UNIVERSE.
IT IS FUNDAMENTAL.
IN THAT LIGHT, THE IDEA THAT ONE CAN WRITE A FINAL THEORY OF EVERYTHING SEEMS REALLY SILLY TO ME.
OK, OK! I SEE YOUR POINT!
BUT I LIKE THE IDEA OF A COMPLETE FINAL THEORY. IT SEEMS ELEGANT TO ME.
DOES IT? NOT TO ME.
...ALSO, ELEGANT IDEAS ARE OFTEN GREAT MOTIVATIONS, BUT CAN SOMETIMES TURN OUT TO BE WRONG!
NATURE ENDS UP FINDING A MORE INTERESTING WAY OF DOING THINGS...
YOU'RE VERY PASSIONATE ABOUT THIS AREN'T YOU?
UNDERSTANDING HOW THE UNIVERSE WORKS IS WORTH SOME PASSION, DON'T YOU THINK?
FAIR ENOUGH...
SO IS THE MULTIVERSE WHAT YOU ARE WORKING ON?
YOU WANT TO SAY THAT OUR UNIVERSE IS JUST ANOTHER SOLUTION TO AN EQUATION?
MAYBE, BUT–
SO, TO USE YOUR LANGUAGE, WHAT ARE THE UNDETERMINED PARAMETERS?

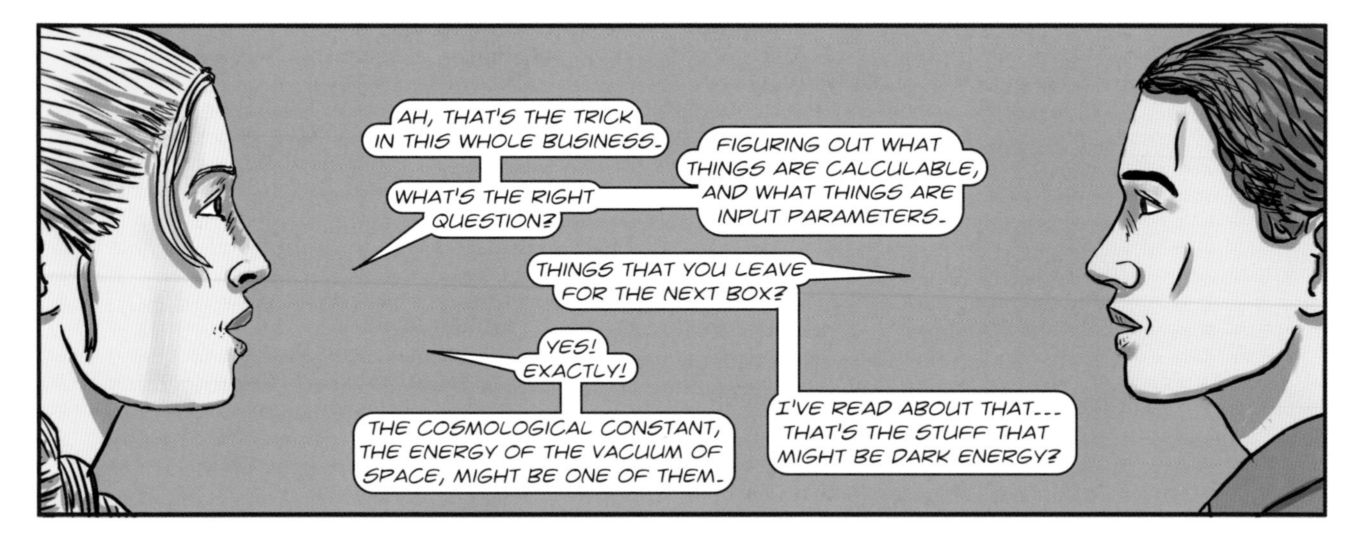
AH, THAT'S THE TRICK IN THIS WHOLE BUSINESS.
WHAT'S THE RIGHT QUESTION?
FIGURING OUT WHAT THINGS ARE CALCULABLE, AND WHAT THINGS ARE INPUT PARAMETERS.
THINGS THAT YOU LEAVE FOR THE NEXT BOX?
YES! EXACTLY!
THE COSMOLOGICAL CONSTANT, THE ENERGY OF THE VACUUM OF SPACE, MIGHT BE ONE OF THEM.
I'VE READ ABOUT THAT... THAT'S THE STUFF THAT MIGHT BE DARK ENERGY?

YES.
NOT EVERYONE AGREES THOUGH... THAT'S WHY WE HAVE CONFERENCES LIKE THE ONE I'M HERE FOR.
IT IS A **QUANTUM GRAVITY** QUESTION.
QUANTUM GRAVITY? YOU MEAN LIKE STRING THEORY?
ISN'T **THAT** SUPPOSED TO BE THE THEORY OF EVERYTHING?
HMM... YOU KNOW YOUR STUFF.

WELL, I READ AN ARTICLE ABOUT IT LAST YEAR.
NICE...
...AND ALSO ALL ABOUT THE WARRING SIDES.
WARRING SIDES?

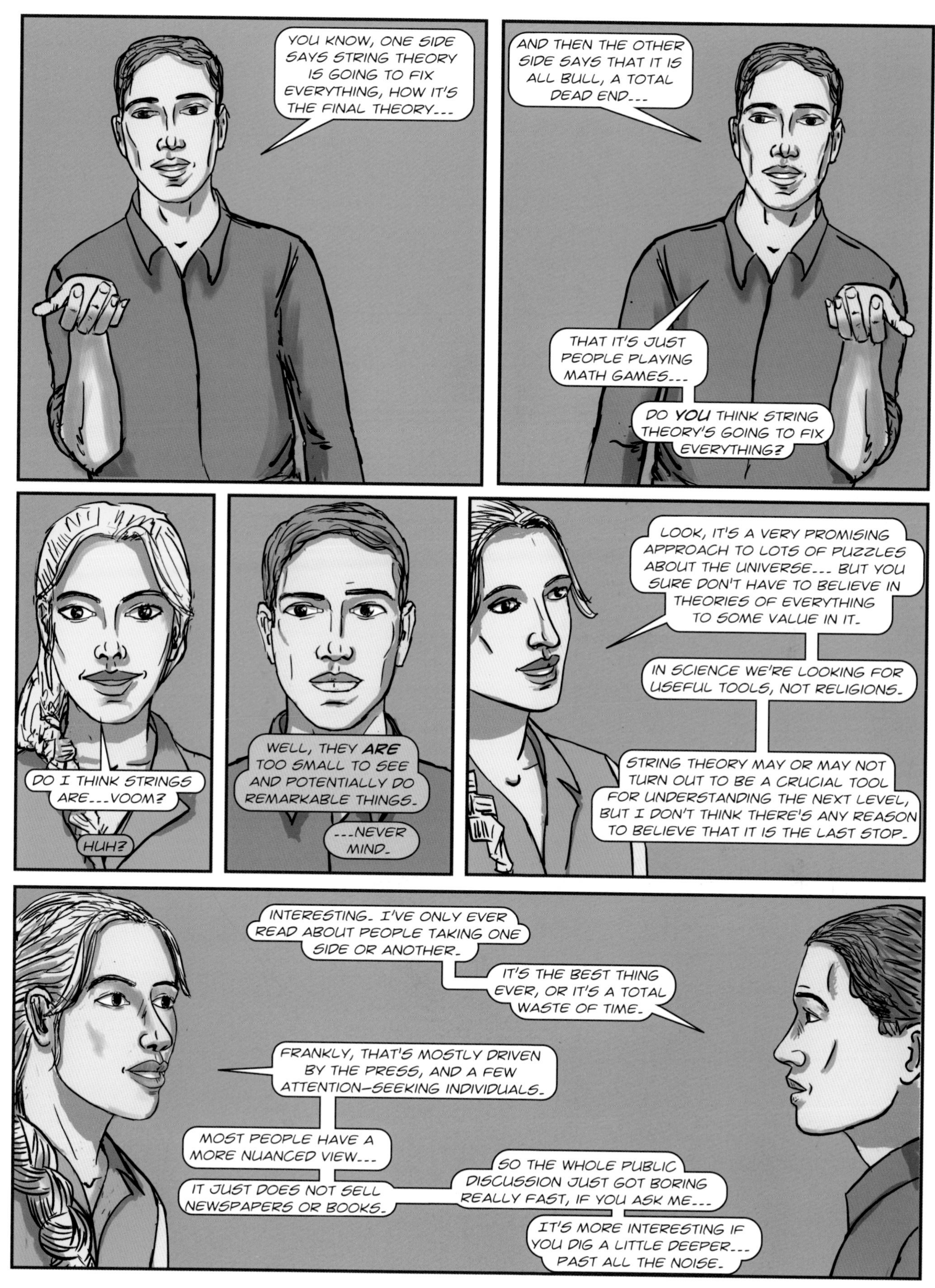

YOU KNOW, ONE SIDE SAYS STRING THEORY IS GOING TO FIX EVERYTHING, HOW IT'S THE FINAL THEORY...
AND THEN THE OTHER SIDE SAYS THAT IT IS ALL BULL, A TOTAL DEAD END...
THAT IT'S JUST PEOPLE PLAYING MATH GAMES...
DO **YOU** THINK STRING THEORY'S GOING TO FIX EVERYTHING?
DO I THINK STRINGS ARE...VOOM?
HUH?
WELL, THEY **ARE** TOO SMALL TO SEE AND POTENTIALLY DO REMARKABLE THINGS.
...NEVER MIND.
LOOK, IT'S A VERY PROMISING APPROACH TO LOTS OF PUZZLES ABOUT THE UNIVERSE... BUT YOU SURE DON'T HAVE TO BELIEVE IN THEORIES OF EVERYTHING TO SOME VALUE IN IT.
IN SCIENCE WE'RE LOOKING FOR USEFUL TOOLS, NOT RELIGIONS.
STRING THEORY MAY OR MAY NOT TURN OUT TO BE A CRUCIAL TOOL FOR UNDERSTANDING THE NEXT LEVEL, BUT I DON'T THINK THERE'S ANY REASON TO BELIEVE THAT IT IS THE LAST STOP.
INTERESTING. I'VE ONLY EVER READ ABOUT PEOPLE TAKING ONE SIDE OR ANOTHER.
IT'S THE BEST THING EVER, OR IT'S A TOTAL WASTE OF TIME.
FRANKLY, THAT'S MOSTLY DRIVEN BY THE PRESS, AND A FEW ATTENTION-SEEKING INDIVIDUALS.
MOST PEOPLE HAVE A MORE NUANCED VIEW...
IT JUST DOES NOT SELL NEWSPAPERS OR BOOKS.
SO THE WHOLE PUBLIC DISCUSSION JUST GOT BORING REALLY FAST, IF YOU ASK ME...
IT'S MORE INTERESTING IF YOU DIG A LITTLE DEEPER... PAST ALL THE NOISE.

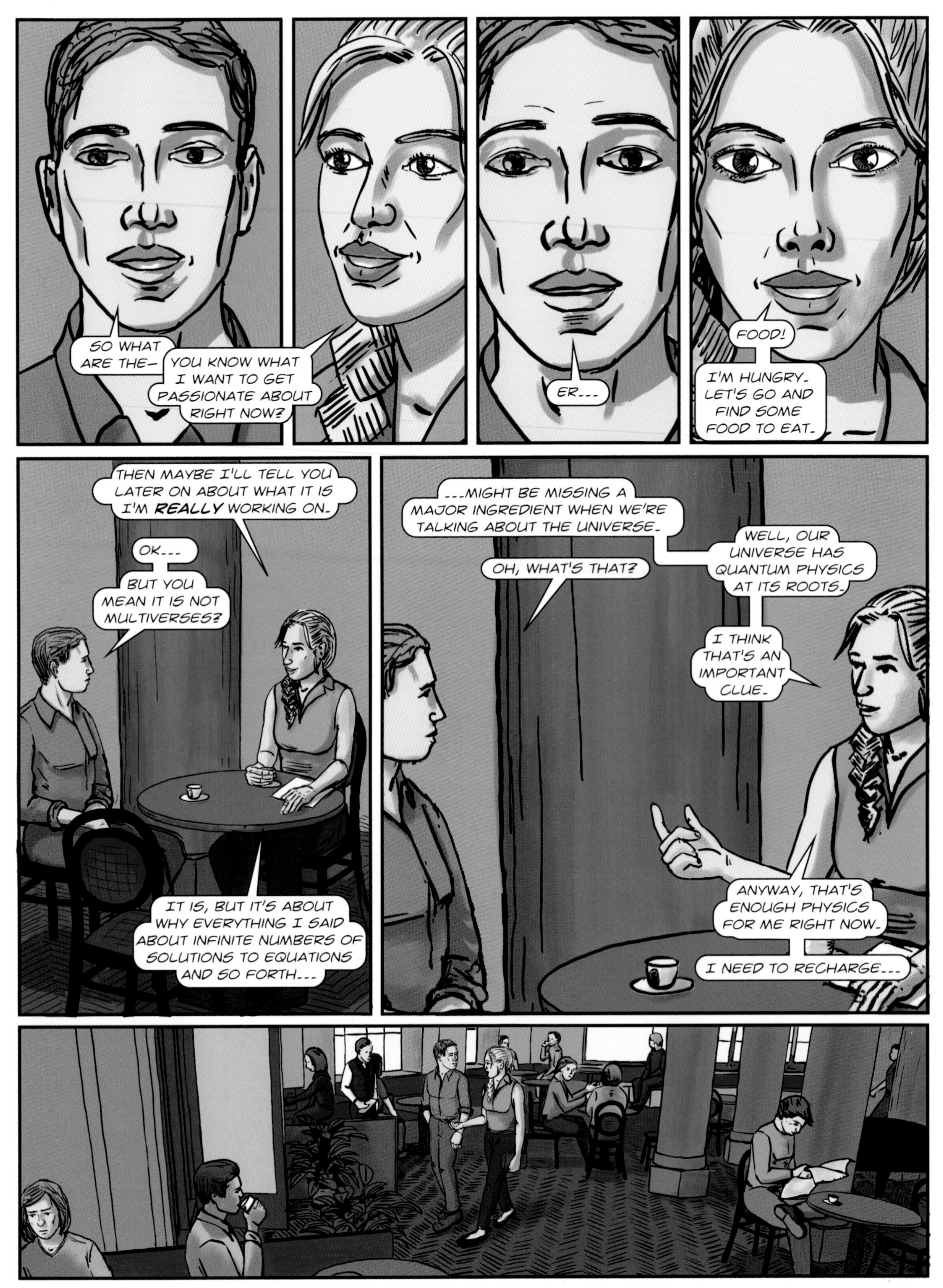
SO WHAT ARE THE–
YOU KNOW WHAT I WANT TO GET PASSIONATE ABOUT RIGHT NOW?
ER...
FOOD!
I'M HUNGRY. LET'S GO AND FIND SOME FOOD TO EAT.
THEN MAYBE I'LL TELL YOU LATER ON ABOUT WHAT IT IS I'M **REALLY** WORKING ON.
OK...
BUT YOU MEAN IT IS NOT MULTIVERSES?
IT IS, BUT IT'S ABOUT WHY EVERYTHING I SAID ABOUT INFINITE NUMBERS OF SOLUTIONS TO EQUATIONS AND SO FORTH...
...MIGHT BE MISSING A MAJOR INGREDIENT WHEN WE'RE TALKING ABOUT THE UNIVERSE.
OH, WHAT'S THAT?
WELL, OUR UNIVERSE HAS QUANTUM PHYSICS AT ITS ROOTS.
I THINK THAT'S AN IMPORTANT CLUE.
ANYWAY, THAT'S ENOUGH PHYSICS FOR ME RIGHT NOW.
I NEED TO RECHARGE...

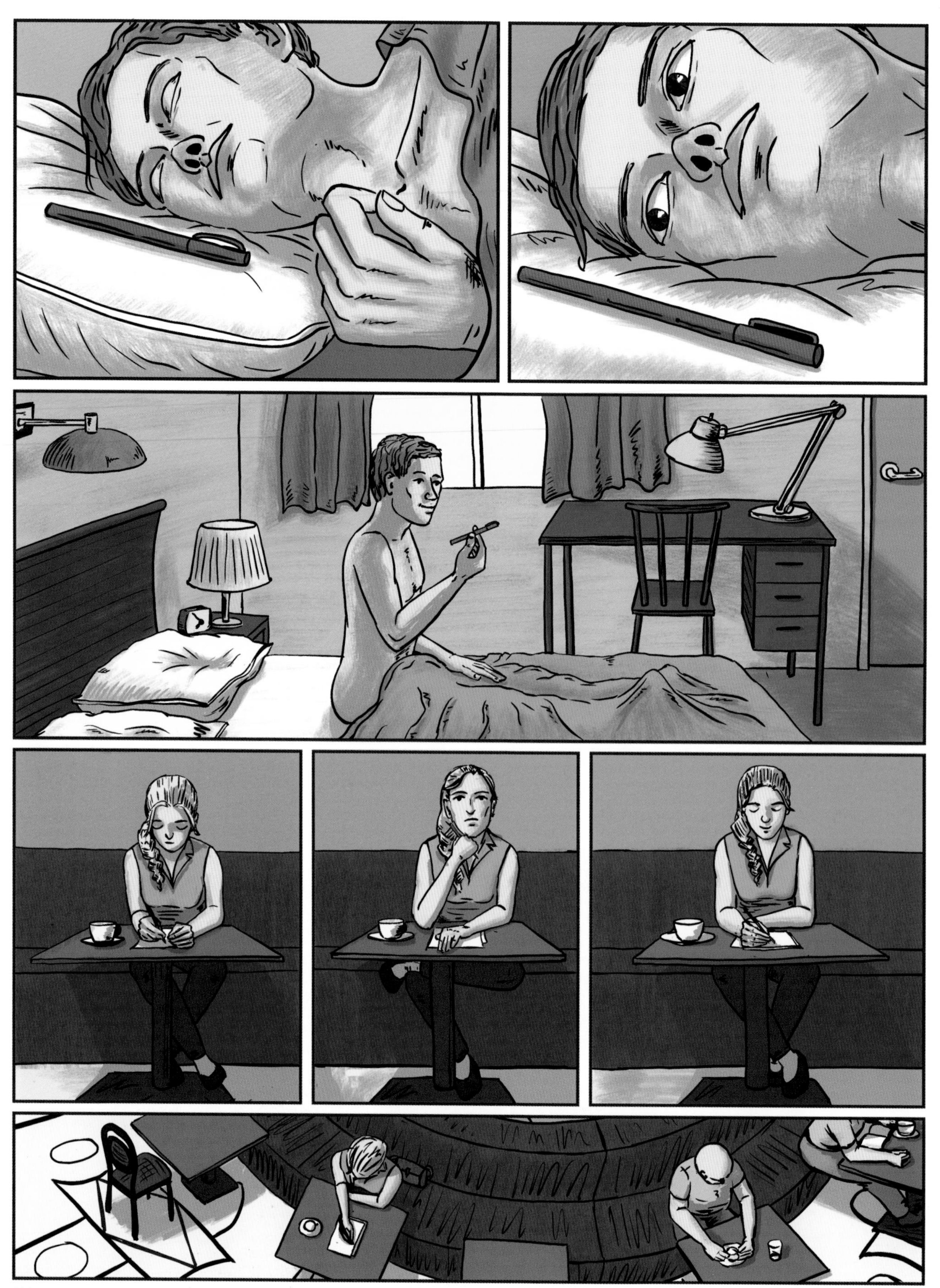

# Notes

***Page 45*** – Before going off to think about the multiverse, perhaps it's worth reading a good book devoted to the history of modern cosmology and its experimental and observational underpinnings: Simon Singh, *Big Bang: The Origin of the Universe* (New York: HarperCollins, 2005).

The multiverse remains a speculative idea, about which there's a lot of passionate discussion and argument among scientists and nonscientists alike. There are many ways of approaching the idea of a multiverse, and some are very distinct from each other. Unfortunately, there's often a lot of confusion about the various notions of the multiverse. Here are some sources that might help untangle things:

Multiverse ideas of numerous sorts are nicely described in the following books: Lisa Randall, *Warped Passages: Unraveling the Mysteries of the Universe's Hidden Dimensions* (New York: Ecco, 2005); John Gribbin, *In Search of the Multiverse: Parallel Worlds, Hidden Dimensions and the Ultimate Quest for the Frontiers of Reality* (Hoboken, NJ: Wiley, 2009); Brian Greene, *The Hidden Reality: Parallel Universes and the Deep Laws of the Cosmos* (New York: Alfred A. Knopf, 2011); Max Tegmark, *Our Mathematical Universe: My Quest for the Ultimate Nature of Reality* (New York: Alfred A. Knopf, 2014); David Wallace, *The Emergent Multiverse: Quantum Theory According to the Everett Interpretation* (Oxford: Oxford University Press, 2012); Alex Vilenkin, *Many Worlds in One: The Search for Other Universes* (New York: Hill and Wang, 2006). See also the short essays by Martin J. Rees, Andrei Linde, and Max Tegmark, among others, in *This Explains Everything*, ed. John Brockman (New York: Harper Perennial, 2013).

***Pages 46–47*** – For more about stars, galaxies, and beyond, including interviews with a variety of research scientists, see Adam Frank, *Astronomy: At Play in the Cosmos* (New York: W. W. Norton & Co., 2016). Part I of the Tyson, Strauss, and Gott book (mentioned in the chapter 1 note for page 18) also discusses the physics of stars, including their life cycle. David Garfinkle and Richard Garfinkle, *Three Steps to the Universe: From the Sun to Black Holes and the Mystery of Dark Matter* (Chicago: University of Chicago Press, 2008), has an excellent short summary of the life cycle of a star. It also will be useful as extra reading in support of other physics and ideas to come later on in this book. Here's a more advanced book entirely devoted to the physics of stars: Kenneth R. Lang, *The Life and Death of Stars* (Cambridge: Cambridge University Press, 2013).

***Page 46*** (panels 1–3) – See this celebration of the diversity of forms in the world: Dr. Seuss, *One Fish Two Fish, Red Fish, Blue Fish* (New York: Random House, 1960).

***Pages 48–50*** – So much is said these days about the idea of a search for a "theory of everything" that people might be excused for thinking that it is a given that such a thing exists. It has become such a mainstream idea (helped by vigorous press coverage of a narrow selection of research topics, and also by the decades of success of the particle physics enterprise) that it is not clear anymore whether people (including professional physicists) stop and think through what it would mean, or even if everyone agrees on what a TOE actually is. The fact is that nobody knows if such a thing exists. It has been a subject of some heated debate historically, and perhaps it will become so again—well beyond the café setting of this chapter. For an authoritative (still, after all these years) discussion of the idea of finding a TOE or a "final theory," and also of the idea of the multiverse, see Steven Weinberg, *Dreams of a Final Theory* (New York: Pantheon Books, 1992).

**Page 52** – Dark energy and the cosmological constant (or "vacuum energy") is discussed in some of the various multiverse references given above in the notes for page 45; see, for example, the works of Randall, Greene, and Vilenkin. It will come up again in chapter 9. Quantum gravity will come up more in chapters 6–10. Suggestions for reading given there start with the chapter 6 notes. String theory is one of a number of approaches to quantum gravity. It is currently unknown whether any of the approaches describe nature.

**Page 53** (panels 3 and 4) – Read what's said about Little Cat Z and what he has under his hat in Dr. Seuss, *The Cat in the Hat Comes Back* (New York: Random House, 1958).

**Page 53** (panel 6) – "dig a little deeper ... past all the noise." See, for example, all the work discussed in the conversation in chapter 9.

**Pages 52–53** – An excellent and nuanced discussion of research in string theory, including some of the arguments over various alternative approaches to the problems it is being used to address, appears in Joseph Conlon, *Why String Theory?* (London: CRC Press, 2015). Many accounts of string theory exist in book form, and Brian Greene, *The Elegant Universe* (New York: W. W. Norton & Co., 1999), remains an excellent introduction for nonexperts. Here's a short and clear book for nonexperts with some of the core modern discoveries (such as dualities, D-branes, etc.) and applications described: Steve S. Gubser, *The Little Book of String Theory* (Princeton, NJ: Princeton University Press, 2010). The following presents a nonphysicist's take on the subject, serving as another highly accessible introduction to some of the ideas of string theory: George Musser, *A Complete Idiot's Guide to String Theory* (London: Alpha, 2009). An interesting history of string theory, with several interviews and observations, is Dean Rickles, *A Brief History of String Theory: From Dual Models to M-theory* (Berlin: Springer-Verlag, 2014).

IT'S ALL ABOUT HOW ELECTRIC AND MAGNETIC FIELDS ARE PRODUCED...
WAIT, WAIT...
SO FIELDS ARE LIKE THOSE LINES PEOPLE DRAW TO SHOW YOU HOW ELECTRONS GET PUSHED AROUND, RIGHT?
I REMEMBER THEM IN SCHOOL.
YES. KIND OF.
NOT JUST ELECTRONS. ANY CHARGED PARTICLE, REALLY.
DO THOSE LINES REALLY EXIST?
WELL, FIELDS ARE THOUGHT OF AS REAL PHYSICAL THINGS.
THEY CAN MOVE AROUND, CHANGE, AND SO FORTH.
BUT THE LINES? ARE THEY REAL?
THE LINES ARE A BIT LIKE THE CONTOUR LINES ON A MAP...
THEY SHOW THE STRENGTH OF THE FIELD AND THE DIRECTION IT WILL PUSH ON A PARTICLE.

JUST LIKE ON A CONTOUR MAP OF A LANDSCAPE.
THOSE LINES AREN'T REAL...
BUT THEY HELP YOU DEDUCE THE LOCATION OF THE STEEPEST CLIFFS...
THE HIGHEST HILLS, THE DEEPEST VALLEY.
ALL REAL THINGS.
OH WOW.
I DIDN'T KNOW IT WAS LIKE A TOPO MAP...
I USE THEM FOR HIKING ALL THE TIME!
SO ELECTRICALLY CHARGED PARTICLES GET PUSHED ALONG ELECTRIC FIELD LINES...
AND THEY SPIRAL AROUND MAGNETIC FIELD LINES.
THE FASTER THEY MOVE, THE TIGHTER THE SPIRAL...
THAT'S A BIT DIFFERENT FROM WHAT YOU READ FOR GRAVITY ON A MAP,
BUT THE IDEA IS THE SAME:
THE LINES ARE A GUIDE TO A FIELD, WHICH IS VERY MUCH A REAL PHYSICAL THING.
YOU DON'T SEE DIAGRAMS OF ELECTRIC AND MAGNETIC FIELDS AS MUCH AS YOU DO TOPO MAPS THOUGH.
TRUE.
BUT YOU DO SEE THEM SOMETIMES.

A LOVELY GIANT EXAMPLE IS THE MAGNETIC FIELD OF THE EARTH.
IT PROTECTS US FROM SOLAR STORMS MADE OF SHOWERS OF CHARGED PARTICLES PRODUCED BY THE SUN.
PROTECTS?
LIKE A FORCE FIELD IN SCI-FI?
YES...
EXCEPT THAT NATURE INVENTED IT FIRST.
THE CHARGED PARTICLES WILL SPIRAL ALONG THOSE MAGNETIC FIELD LINES
AND ARE DIVERTED TO EITHER THE NORTH OR SOUTH POLES.
THEY CASCADE INTO OUR ATMOSPHERE THERE.
THE COLLISION MAKES THE AIR MOLECULES GLOW, GIVING WHAT'S CALLED...
OH!
AURORA BOREALIS! NORTHERN LIGHTS!
YES. EXACTLY.
THAT'S WHY THEY'RE MOSTLY AT THE NORTH AND SOUTH POLES –
THE MAGNETIC FIELD DIVERTS THEM THERE, AND IT DIVERTS MANY MORE OF THEM COMPLETELY AWAY FROM THE EARTH.
THAT'S AMAZING...
ANYWAY, WHAT I WAS GOING TO SAY IS THAT MAXWELL'S EQUATIONS TELL YOU PRECISELY HOW THESE FIELDS ARE MADE.
I SEE... SHOW ME!

OK, HERE GOES...
I'M GOING TO ARRANGE THINGS ON THE PAGE LIKE THIS...
∇.E
E STANDS FOR ELECTRIC FIELD...
∇.E =0 ∇·B=0
∇×E=-∂B
B STANDS FOR MAGNETIC FIELD...
∇·E =0 ∇·B=0
∇×E=-∂B/∂t ∇×B= ∂E/∂t
t IS TIME?
YES.
SO THERE YOU HAVE IT.
THESE TELL YOU HOW ELECTRIC AND MAGNETIC FIELDS ARE RELATED TO EACH OTHER.
REALLY?
HOW DO I READ THAT? WHAT'S THE BIG TRIANGLE THING?

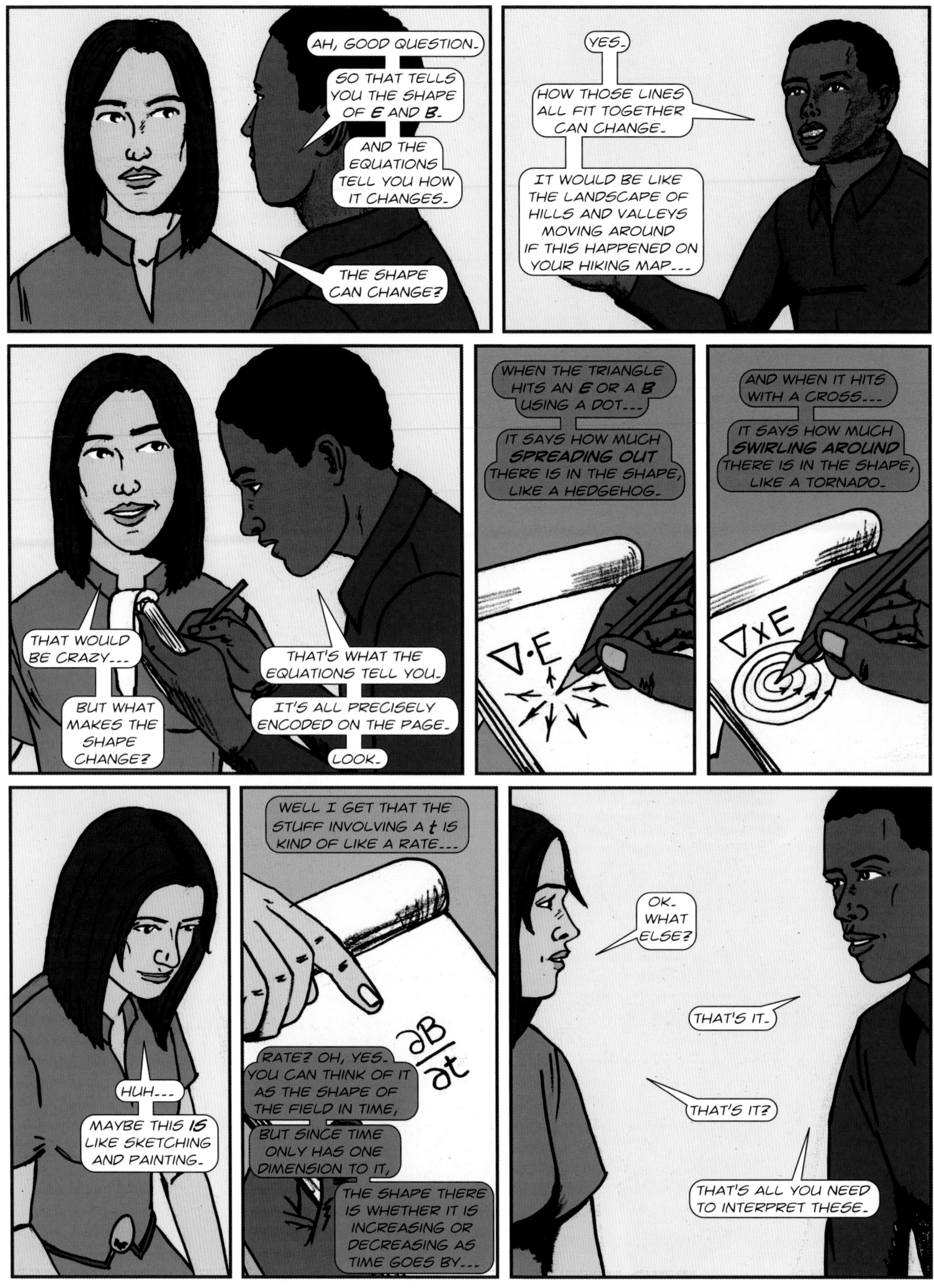

AH, GOOD QUESTION.
SO THAT TELLS YOU THE SHAPE OF E AND B.
AND THE EQUATIONS TELL YOU HOW IT CHANGES.
THE SHAPE CAN CHANGE?
YES.
HOW THOSE LINES ALL FIT TOGETHER CAN CHANGE.
IT WOULD BE LIKE THE LANDSCAPE OF HILLS AND VALLEYS MOVING AROUND IF THIS HAPPENED ON YOUR HIKING MAP...
THAT WOULD BE CRAZY...
BUT WHAT MAKES THE SHAPE CHANGE?
THAT'S WHAT THE EQUATIONS TELL YOU.
IT'S ALL PRECISELY ENCODED ON THE PAGE.
LOOK.
WHEN THE TRIANGLE HITS AN E OR A B USING A DOT...
IT SAYS HOW MUCH SPREADING OUT THERE IS IN THE SHAPE, LIKE A HEDGEHOG.
∇·E
AND WHEN IT HITS WITH A CROSS...
IT SAYS HOW MUCH SWIRLING AROUND THERE IS IN THE SHAPE, LIKE A TORNADO.
∇xE
HUH...
MAYBE THIS IS LIKE SKETCHING AND PAINTING.
WELL I GET THAT THE STUFF INVOLVING A t IS KIND OF LIKE A RATE...
∂B/∂t
RATE? OH, YES. YOU CAN THINK OF IT AS THE SHAPE OF THE FIELD IN TIME,
BUT SINCE TIME ONLY HAS ONE DIMENSION TO IT,
THE SHAPE THERE IS WHETHER IT IS INCREASING OR DECREASING AS TIME GOES BY...
OK. WHAT ELSE?
THAT'S IT.
THAT'S IT?
THAT'S ALL YOU NEED TO INTERPRET THESE.

REALLY?
WHAT'S THAT CIRCLE MEAN?
∇·E=0 ∇·B
OH, YOU KNOW THAT ONE...
IT'S ZERO.
HMMMPH.
OK!
SO WHERE'S THE BEAUTY?
AH. THAT CAN BE SEEN IN THE LAYOUT ON THE PAGE,
AND THE RELATIONSHIPS BETWEEN THINGS.
SO YOU'VE GOT THE SAME THING HAPPENING TO BOTH ELECTRIC AND MAGNETIC IN THIS TOP LINE...
∇·E=0 ∇·B=0
THAT'S ALREADY CUTE.
WAIT...
THAT'S...
NO.
GO ON!
WELL, FROM WHAT YOU SAID ABOUT THE TRIANGLE THING...
THAT'S SAYING THAT THERE'S NO HEDGEHOG-LIKE SPREADING OUT FOR E AND B...
THAT'S RIGHT!
BUT I REMEMBER FROM HIGH SCHOOL THAT THEY DO... DON'T THEY?
OK... WE'LL COME TO THAT.
OK...
YOU TRIED TO PULL A FAST ONE ON ME?
NO! NO! JUST GOING STEP BY STEP.

THE NEXT LINE SAYS THAT THE AMOUNT OF SWIRL IN E IS PRODUCED BY CHANGES IN B OVER TIME...
AND THAT THE OTHER WAY AROUND IS TRUE AS WELL..
$\nabla \cdot E = 0$ $\nabla \cdot B = 0$
$\nabla \times E = -\frac{\partial B}{\partial t}$ $\nabla \times B = \frac{\partial E}{\partial t}$
CHANGES IN E PRODUCE A SWIRL IN THE B FIELD?
YES!
NEAT AS HELL, RIGHT?
HOW SO?
WELL UNTIL THE LATE 19TH CENTURY, ELECTRIC AND MAGNETIC SEEMED VERY DIFFERENT.
THE EQUATIONS SHOW THAT THEY ARE ON THE SAME FOOTING,
AS DISCOVERED IN EXPERIMENTS!
ELECTRICITY AND MAGNETISM...
E
B
...BECAME ELECTROMAGNETISM.
UNIFIED!
ALL CAPTURED BY THE EQUATIONS!
I SEE.
THAT'S AMAZING!
YOU KNOW...
WHEN I THINK ABOUT SYMMETRY AS PART OF BEAUTY, I'M THINKING OF THIS...

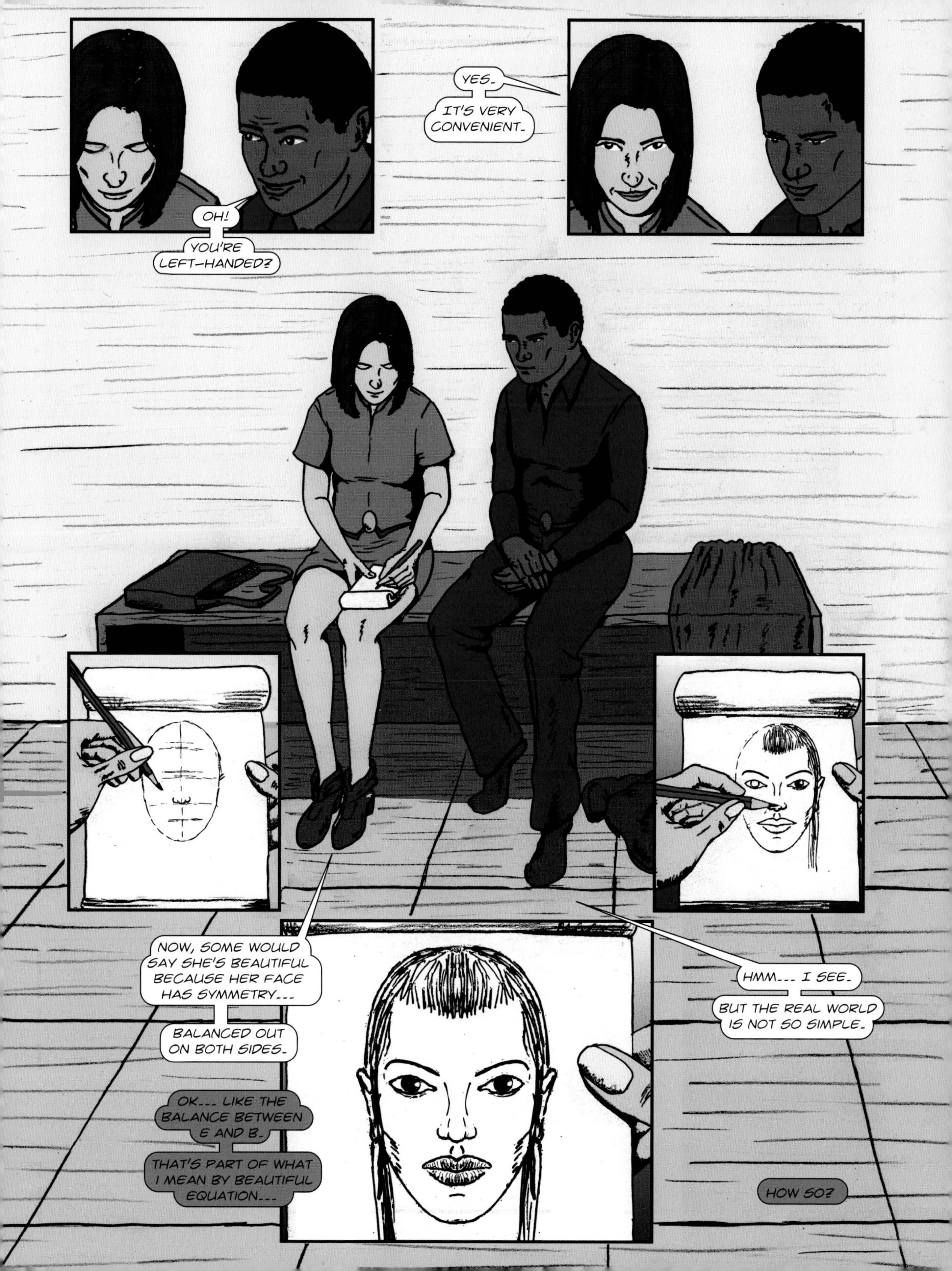
OH!
YOU'RE LEFT-HANDED?
YES.
IT'S VERY CONVENIENT.
NOW, SOME WOULD SAY SHE'S BEAUTIFUL BECAUSE HER FACE HAS SYMMETRY...
BALANCED OUT ON BOTH SIDES.
HMM... I SEE.
BUT THE REAL WORLD IS NOT SO SIMPLE.
OK... LIKE THE BALANCE BETWEEN E AND B.
THAT'S PART OF WHAT I MEAN BY BEAUTIFUL EQUATION...
HOW SO?

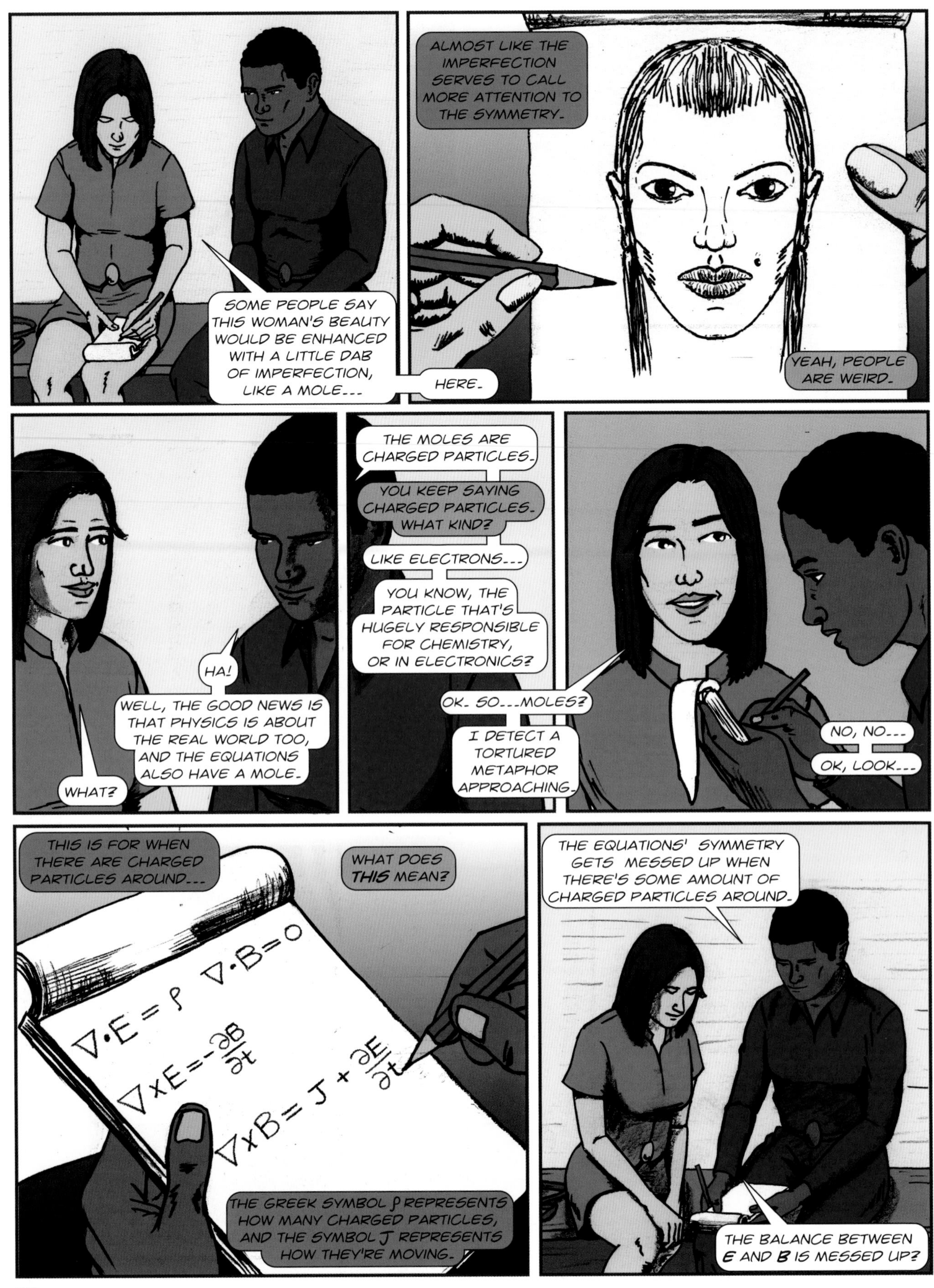
SOME PEOPLE SAY THIS WOMAN'S BEAUTY WOULD BE ENHANCED WITH A LITTLE DAB OF IMPERFECTION, LIKE A MOLE...
HERE.
ALMOST LIKE THE IMPERFECTION SERVES TO CALL MORE ATTENTION TO THE SYMMETRY.
YEAH, PEOPLE ARE WEIRD.
HA!
WELL, THE GOOD NEWS IS THAT PHYSICS IS ABOUT THE REAL WORLD TOO, AND THE EQUATIONS ALSO HAVE A MOLE.
WHAT?
THE MOLES ARE CHARGED PARTICLES.
YOU KEEP SAYING CHARGED PARTICLES. WHAT KIND?
LIKE ELECTRONS...
YOU KNOW, THE PARTICLE THAT'S HUGELY RESPONSIBLE FOR CHEMISTRY, OR IN ELECTRONICS?
OK. SO...MOLES?
I DETECT A TORTURED METAPHOR APPROACHING.
NO, NO...
OK, LOOK...
THIS IS FOR WHEN THERE ARE CHARGED PARTICLES AROUND...
WHAT DOES THIS MEAN?
$\nabla \cdot E = \rho$ $\nabla \cdot B = 0$
$\nabla \times E = -\frac{\partial B}{\partial t}$
$\nabla \times B = J + \frac{\partial E}{\partial t}$
THE GREEK SYMBOL ρ REPRESENTS HOW MANY CHARGED PARTICLES, AND THE SYMBOL J REPRESENTS HOW THEY'RE MOVING.
THE EQUATIONS' SYMMETRY GETS MESSED UP WHEN THERE'S SOME AMOUNT OF CHARGED PARTICLES AROUND.
THE BALANCE BETWEEN E AND B IS MESSED UP?

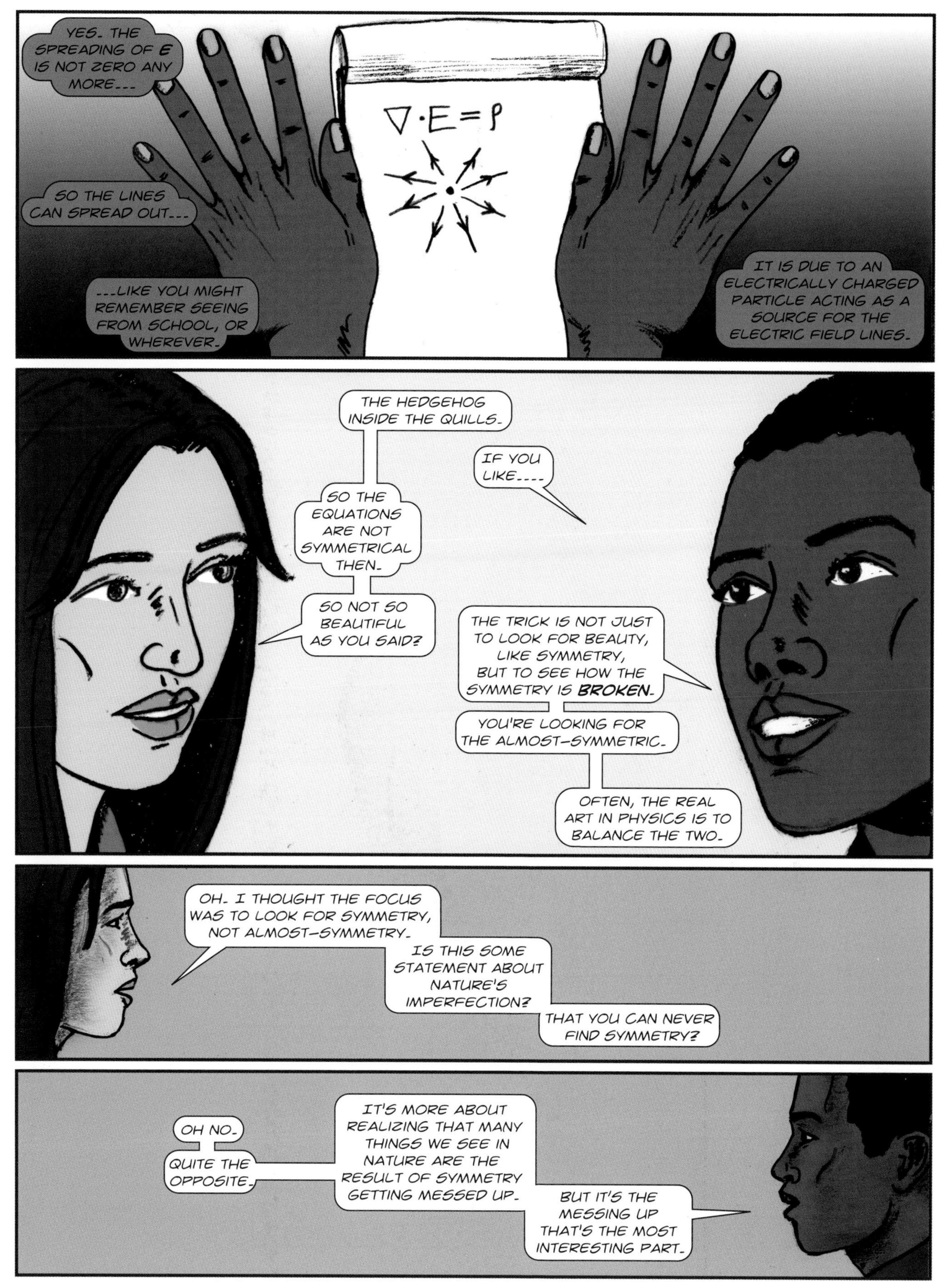
YES. THE SPREADING OF E IS NOT ZERO ANY MORE...
SO THE LINES CAN SPREAD OUT...
...LIKE YOU MIGHT REMEMBER SEEING FROM SCHOOL, OR WHEREVER.
∇·E = ρ
IT IS DUE TO AN ELECTRICALLY CHARGED PARTICLE ACTING AS A SOURCE FOR THE ELECTRIC FIELD LINES.
THE HEDGEHOG INSIDE THE QUILLS.
IF YOU LIKE....
SO THE EQUATIONS ARE NOT SYMMETRICAL THEN.
SO NOT SO BEAUTIFUL AS YOU SAID?
THE TRICK IS NOT JUST TO LOOK FOR BEAUTY, LIKE SYMMETRY, BUT TO SEE HOW THE SYMMETRY IS **BROKEN**.
YOU'RE LOOKING FOR THE ALMOST-SYMMETRIC.
OFTEN, THE REAL ART IN PHYSICS IS TO BALANCE THE TWO.
OH. I THOUGHT THE FOCUS WAS TO LOOK FOR SYMMETRY, NOT ALMOST-SYMMETRY.
IS THIS SOME STATEMENT ABOUT NATURE'S IMPERFECTION?
THAT YOU CAN NEVER FIND SYMMETRY?
OH NO.
QUITE THE OPPOSITE.
IT'S MORE ABOUT REALIZING THAT MANY THINGS WE SEE IN NATURE ARE THE RESULT OF SYMMETRY GETTING MESSED UP.
BUT IT'S THE MESSING UP THAT'S THE MOST INTERESTING PART.

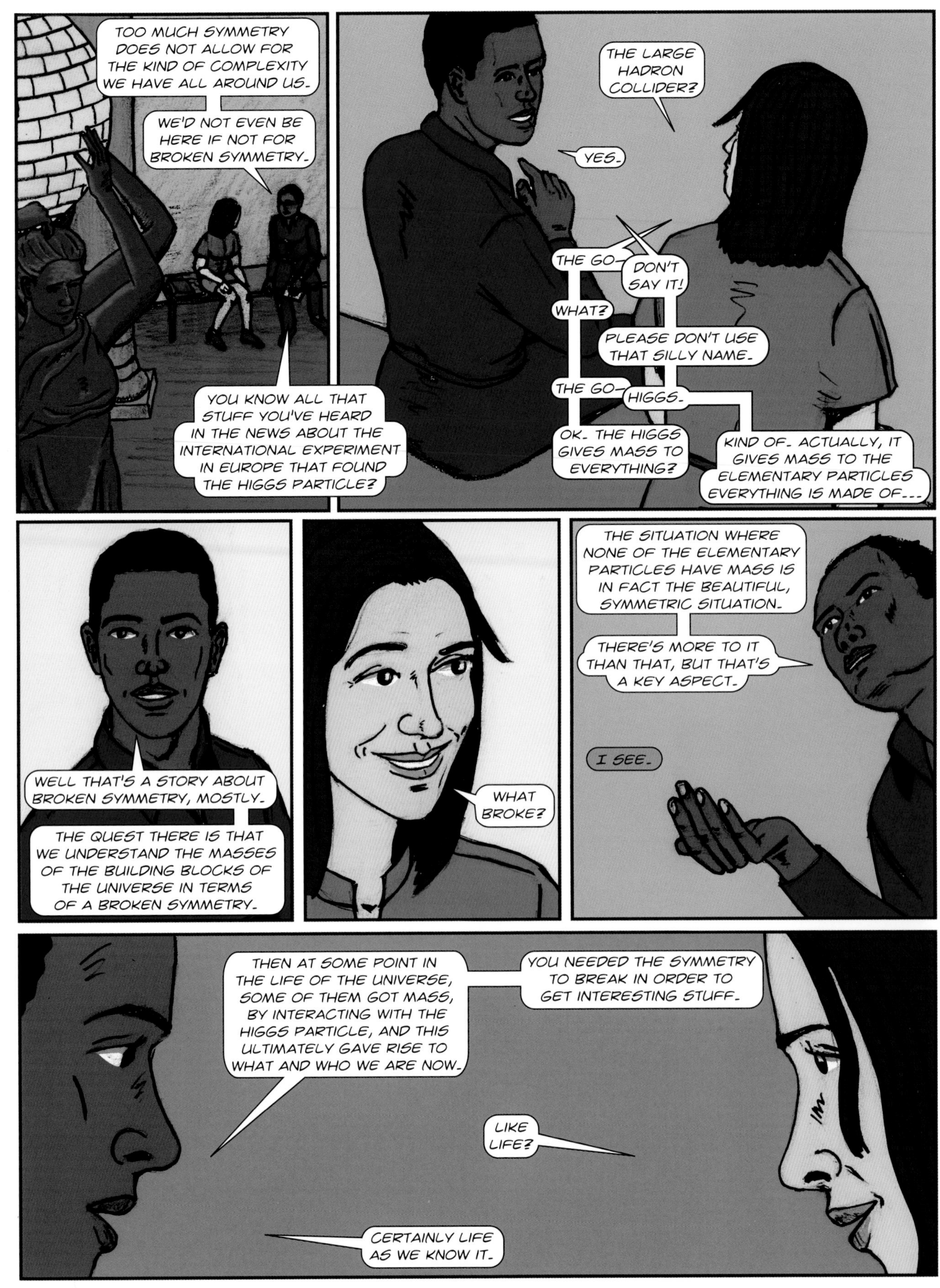
TOO MUCH SYMMETRY DOES NOT ALLOW FOR THE KIND OF COMPLEXITY WE HAVE ALL AROUND US.
WE'D NOT EVEN BE HERE IF NOT FOR BROKEN SYMMETRY.
YOU KNOW ALL THAT STUFF YOU'VE HEARD IN THE NEWS ABOUT THE INTERNATIONAL EXPERIMENT IN EUROPE THAT FOUND THE HIGGS PARTICLE?
THE LARGE HADRON COLLIDER?
YES.
THE GO-
DON'T SAY IT!
WHAT?
PLEASE DON'T USE THAT SILLY NAME.
THE GO-
HIGGS.
OK. THE HIGGS GIVES MASS TO EVERYTHING?
KIND OF. ACTUALLY, IT GIVES MASS TO THE ELEMENTARY PARTICLES EVERYTHING IS MADE OF...
WELL THAT'S A STORY ABOUT BROKEN SYMMETRY, MOSTLY.
THE QUEST THERE IS THAT WE UNDERSTAND THE MASSES OF THE BUILDING BLOCKS OF THE UNIVERSE IN TERMS OF A BROKEN SYMMETRY.
WHAT BROKE?
THE SITUATION WHERE NONE OF THE ELEMENTARY PARTICLES HAVE MASS IS IN FACT THE BEAUTIFUL, SYMMETRIC SITUATION.
THERE'S MORE TO IT THAN THAT, BUT THAT'S A KEY ASPECT.
I SEE.
THEN AT SOME POINT IN THE LIFE OF THE UNIVERSE, SOME OF THEM GOT MASS, BY INTERACTING WITH THE HIGGS PARTICLE, AND THIS ULTIMATELY GAVE RISE TO WHAT AND WHO WE ARE NOW.
YOU NEEDED THE SYMMETRY TO BREAK IN ORDER TO GET INTERESTING STUFF.
LIKE LIFE?
CERTAINLY LIFE AS WE KNOW IT.

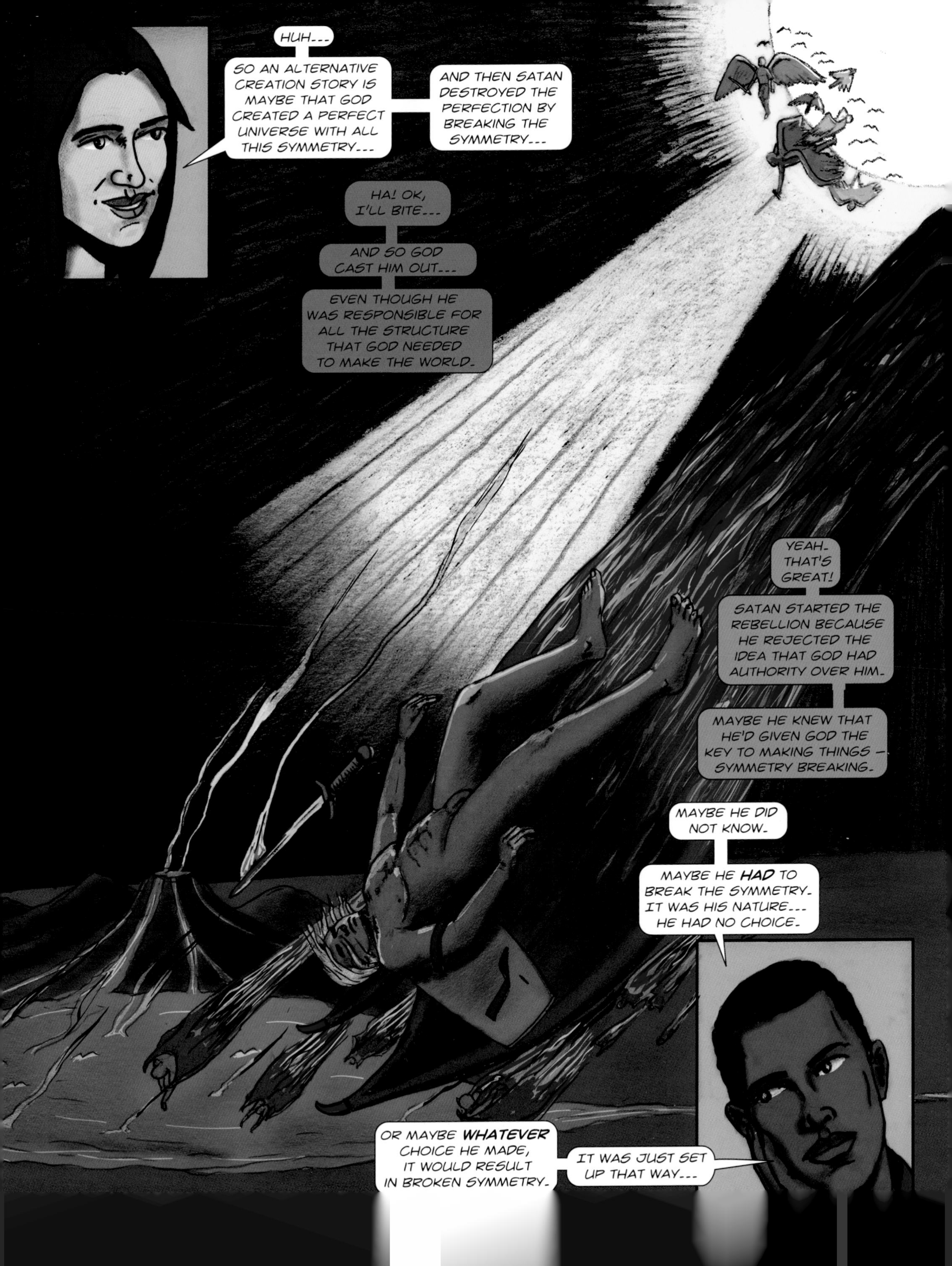
HUH...
SO AN ALTERNATIVE CREATION STORY IS MAYBE THAT GOD CREATED A PERFECT UNIVERSE WITH ALL THIS SYMMETRY...
AND THEN SATAN DESTROYED THE PERFECTION BY BREAKING THE SYMMETRY...
HA! OK, I'LL BITE...
AND SO GOD CAST HIM OUT...
EVEN THOUGH HE WAS RESPONSIBLE FOR ALL THE STRUCTURE THAT GOD NEEDED TO MAKE THE WORLD.
YEAH. THAT'S GREAT!
SATAN STARTED THE REBELLION BECAUSE HE REJECTED THE IDEA THAT GOD HAD AUTHORITY OVER HIM.
MAYBE HE KNEW THAT HE'D GIVEN GOD THE KEY TO MAKING THINGS — SYMMETRY BREAKING.
MAYBE HE DID NOT KNOW.
MAYBE HE **HAD** TO BREAK THE SYMMETRY. IT WAS HIS NATURE... HE HAD NO CHOICE.
OR MAYBE **WHATEVER** CHOICE HE MADE, IT WOULD RESULT IN BROKEN SYMMETRY.
IT WAS JUST SET UP THAT WAY...

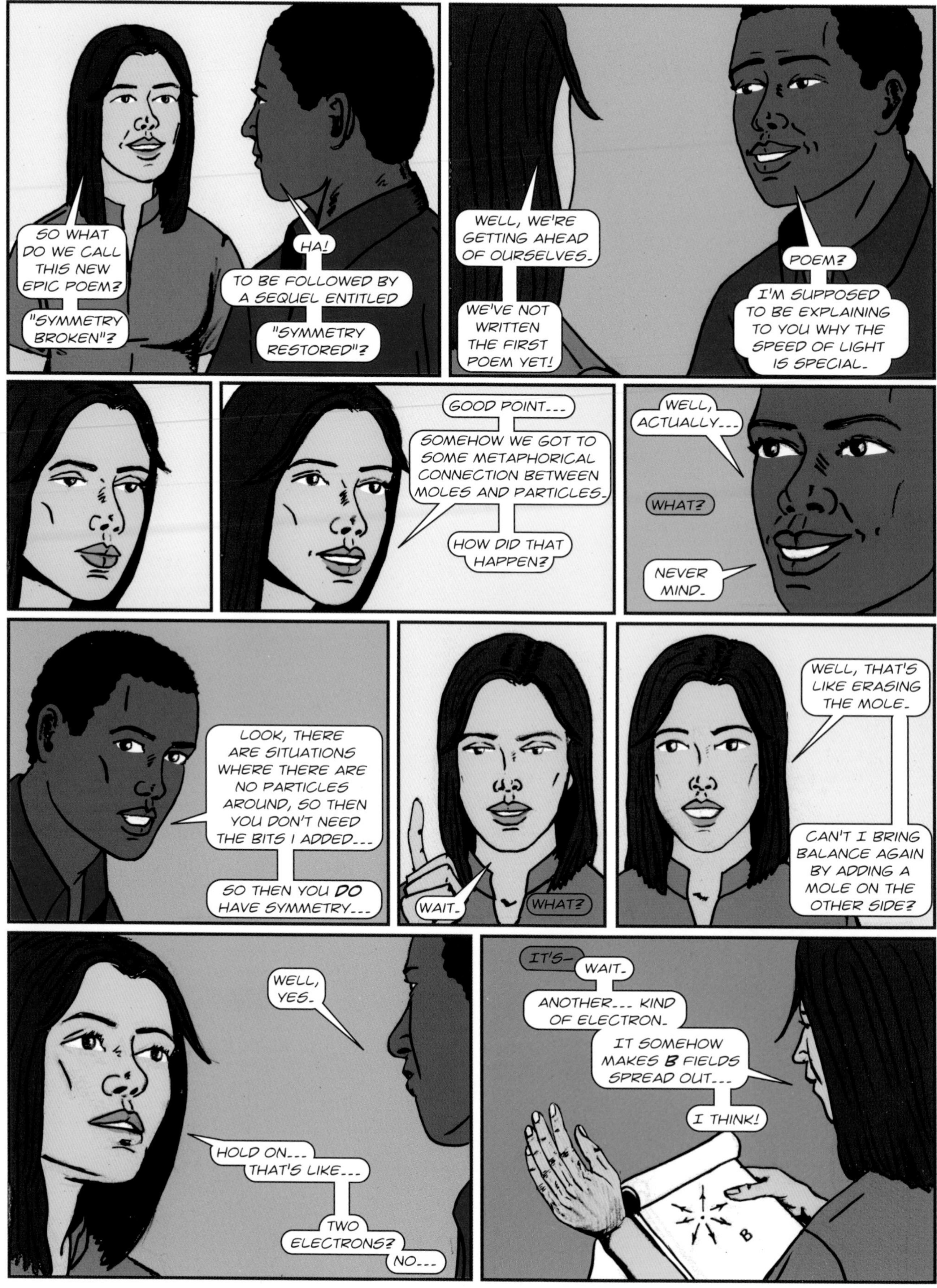

SO WHAT DO WE CALL THIS NEW EPIC POEM?
"SYMMETRY BROKEN"?
HA!
TO BE FOLLOWED BY A SEQUEL ENTITLED
"SYMMETRY RESTORED"?
WELL, WE'RE GETTING AHEAD OF OURSELVES.
WE'VE NOT WRITTEN THE FIRST POEM YET!
POEM?
I'M SUPPOSED TO BE EXPLAINING TO YOU WHY THE SPEED OF LIGHT IS SPECIAL.
GOOD POINT...
SOMEHOW WE GOT TO SOME METAPHORICAL CONNECTION BETWEEN MOLES AND PARTICLES.
HOW DID THAT HAPPEN?
WELL, ACTUALLY...
WHAT?
NEVER MIND.
LOOK, THERE ARE SITUATIONS WHERE THERE ARE NO PARTICLES AROUND, SO THEN YOU DON'T NEED THE BITS I ADDED...
SO THEN YOU **DO** HAVE SYMMETRY...
WAIT.
WHAT?
WELL, THAT'S LIKE ERASING THE MOLE.
CAN'T I BRING BALANCE AGAIN BY ADDING A MOLE ON THE OTHER SIDE?
WELL, YES.
HOLD ON... THAT'S LIKE...
TWO ELECTRONS?
NO...
IT'S—
WAIT.
ANOTHER... KIND OF ELECTRON.
IT SOMEHOW MAKES **B** FIELDS SPREAD OUT...
I THINK!
B

HEY, YOU'RE GOOD AT READING THE PATTERNS!
YOU'VE DISCOVERED–
ANTIMATTER!?
NO...
A PARTICLE CALLED A **MAGNETIC MONOPOLE.**
IT HAS MAGNETIC CHARGE INSTEAD OF ELECTRIC CHARGE.
OH.
BUT NOBODY'S EVER SEEN ONE,
SO FAR.
FOR SOME REASON NATURE HAS EITHER HIDDEN THEM AWAY, OR CHOSEN NOT TO USE THIS BEAUTIFUL POSSIBILITY AT ALL.
THERE'S AN INTERESTING STORY THERE.
BUT IT IS IMPRESSIVE THAT YOU DEDUCED THAT POSSIBILITY BY LOOKING AT THE EQUATIONS.
THAT'S OFTEN HOW REASONING USING SYMMETRY LEADS TO DISCOVERY.
THANKS.
YOU KNOW, IF I KNEW SO MUCH COULD BE FIGURED OUT BY REASONING IN PICTURES, I'D HAVE GONE INTO PHYSICS.
REASONING IN PICTURES IS ONE OF OUR MOST POWERFUL TOOLS OF DISCOVERY.
I HAD NO IDEA.

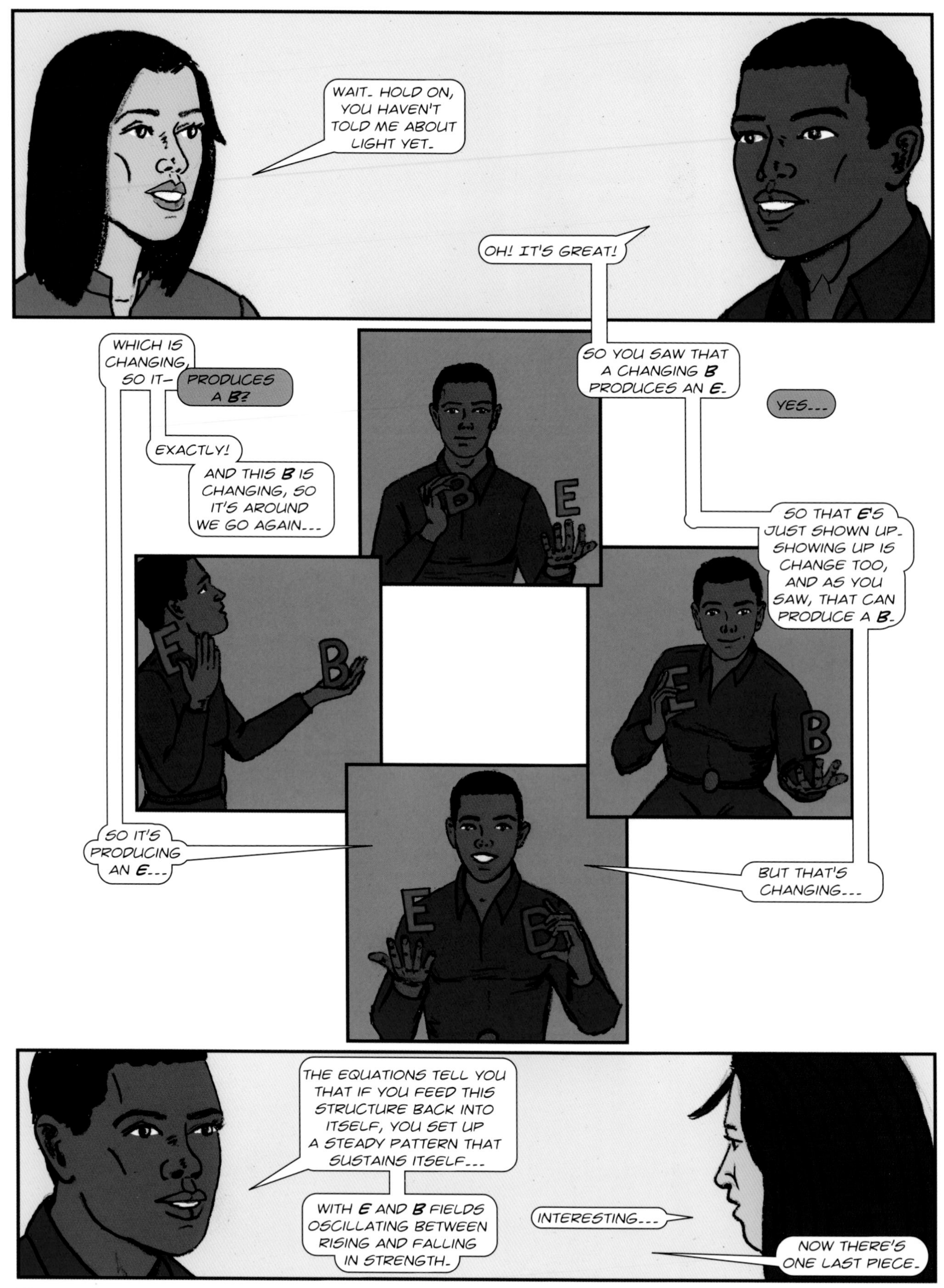

WAIT. HOLD ON, YOU HAVEN'T TOLD ME ABOUT LIGHT YET.
OH! IT'S GREAT!
SO YOU SAW THAT A CHANGING **B** PRODUCES AN **E**.
YES...
SO THAT **E**'S JUST SHOWN UP. SHOWING UP IS CHANGE TOO, AND AS YOU SAW, THAT CAN PRODUCE A **B**.
BUT THAT'S CHANGING...
SO IT'S PRODUCING AN **E**...
WHICH IS CHANGING, SO IT–
PRODUCES A **B**?
EXACTLY!
AND THIS **B** IS CHANGING, SO IT'S AROUND WE GO AGAIN...
THE EQUATIONS TELL YOU THAT IF YOU FEED THIS STRUCTURE BACK INTO ITSELF, YOU SET UP A STEADY PATTERN THAT SUSTAINS ITSELF...
WITH **E** AND **B** FIELDS OSCILLATING BETWEEN RISING AND FALLING IN STRENGTH.
INTERESTING...
NOW THERE'S ONE LAST PIECE.

E
B
WELL, IT TURNS OUT THAT THIS ONLY WORKS UNDER THE CONDITION THAT THE WHOLE CIRCUS I JUST DESCRIBED...
E
B
...MOVES FORWARD AS IT OSCILLATES BACK AND FORTH.
ISN'T THAT LIKE A WAVE?
YES! EXACTLY!
AND THE EQUATIONS SHOW THAT IT GOES ONLY AT ONE SPECIFIC SPEED:
THE SPEED OF LIGHT!
OH, RIGHT...LIGHT IS AN ELECTROMAGNETIC WAVE!
THAT'S WHAT THIS IS, RIGHT?
YES...
COOL.
IT'S A CONSEQUENCE OF THAT SYMMETRY BETWEEN ELECTRICITY AND MAGNETISM THAT THE EQUATIONS HAVE IN THIS SITUATION.
WHEN THE SYMMETRY BREAKS, IT GOES AWAY?
YES. IF THERE ARE LOTS OF CHARGED PARTICLES AROUND...
LIKE INSIDE A METAL...
OR IN THE EARLY YEARS OF OUR UNIVERSE, ACTUALLY...
REALLY?!
THE UNIVERSE WAS **DARK** BECAUSE OF A BROKEN SYMMETRY?!
FOR A WHILE, YES...
HUH.
THAT'S MORE MATERIAL FOR THE EPIC POEM...
OH! DRINKS!

SO WHY CAN'T I GET UP TO THE SPEED OF LIGHT?
AH.
ONE OF THE NICEST WAYS TO SEE IT IS TO ARGUE BY CONTRADICTION.
HOW?
IMAGINE CATCHING UP TO A LIGHT BEAM.
YOU KNOW, USE YOUR POWERS TO FLY OR RUN UP TO THE SPEED OF LIGHT.
OK...
NOW LOOK AT THE LIGHT WAVE, OR DO MEASUREMENTS ON IT, OR WHATEVER...
E
B
SO IF YOU CATCH UP TO A LIGHT BEAM, YOU'LL JUST SEE ELECTRIC AND MAGNETIC FIELDS OSCILLATING BACK AND FORTH...
...BUT JUST STAYING PUT, IN ONE SPOT.
SO WHAT?
THE EQUATIONS THAT TOLD YOU LIGHT IS A WAVE HAVE NO SOLUTION LIKE THAT.
THE WAVE IS ALWAYS MOVING, AND IT IS ALWAYS AT THE SPEED OF LIGHT.
OH...
NO OTHER SPEED, AND CERTAINLY NOT ZERO.
THAT'S IT?
NO TAKING INTO ACCOUNT THE METHOD I USE TO SPEED UP?
THAT'S IT.
NONE OF THE STATEMENTS I'VE HEARD ABOUT INFINITE ENERGY AND SO FORTH?
NOPE.
HMMM...
I THOUGHT IT WOULD BE MORE COMPLICATED.
WELL, ONE COULD MAKE IT MORE COMPLICATED, AND PEOPLE DO, BUT THERE'S REALLY NO NEED.
BUT IT ASSUMES THESE 19TH-CENTURY MAXWELL EQUATIONS ARE TRUE, RIGHT?
SURE.
ISN'T THAT A WEAK LINK?
WE'RE IN THE 21ST CENTURY, YOU MIGHT HAVE NOTICED.

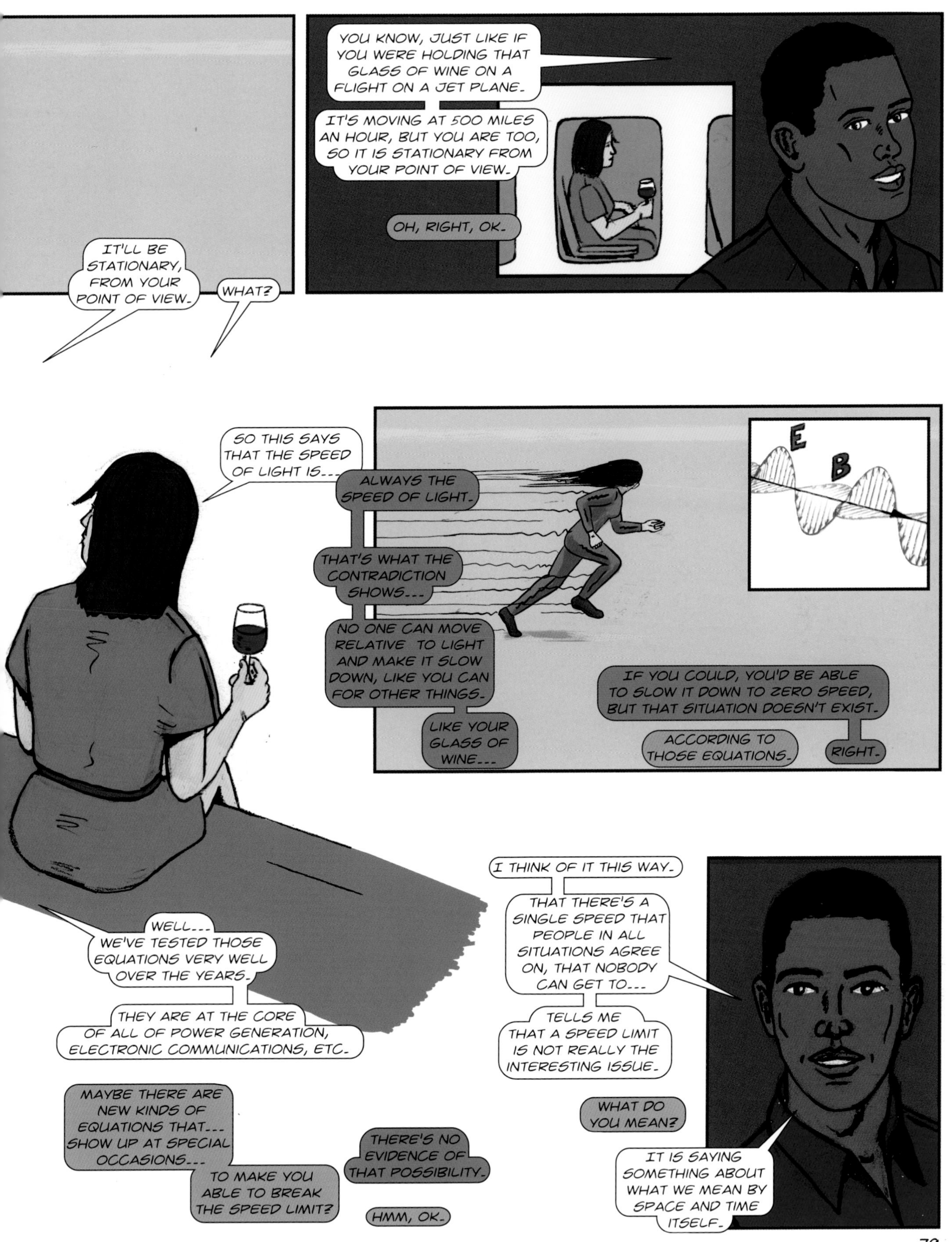

YOU KNOW, JUST LIKE IF YOU WERE HOLDING THAT GLASS OF WINE ON A FLIGHT ON A JET PLANE.
IT'S MOVING AT 500 MILES AN HOUR, BUT YOU ARE TOO, SO IT IS STATIONARY FROM YOUR POINT OF VIEW.
OH, RIGHT, OK.
IT'LL BE STATIONARY, FROM YOUR POINT OF VIEW.
WHAT?
SO THIS SAYS THAT THE SPEED OF LIGHT IS...
ALWAYS THE SPEED OF LIGHT.
THAT'S WHAT THE CONTRADICTION SHOWS...
NO ONE CAN MOVE RELATIVE TO LIGHT AND MAKE IT SLOW DOWN, LIKE YOU CAN FOR OTHER THINGS.
LIKE YOUR GLASS OF WINE...
E
B
IF YOU COULD, YOU'D BE ABLE TO SLOW IT DOWN TO ZERO SPEED, BUT THAT SITUATION DOESN'T EXIST.
ACCORDING TO THOSE EQUATIONS.
RIGHT.
WELL... WE'VE TESTED THOSE EQUATIONS VERY WELL OVER THE YEARS.
THEY ARE AT THE CORE OF ALL OF POWER GENERATION, ELECTRONIC COMMUNICATIONS, ETC.
MAYBE THERE ARE NEW KINDS OF EQUATIONS THAT... SHOW UP AT SPECIAL OCCASIONS...
TO MAKE YOU ABLE TO BREAK THE SPEED LIMIT?
THERE'S NO EVIDENCE OF THAT POSSIBILITY.
HMM, OK.
I THINK OF IT THIS WAY.
THAT THERE'S A SINGLE SPEED THAT PEOPLE IN ALL SITUATIONS AGREE ON, THAT NOBODY CAN GET TO...
TELLS ME THAT A SPEED LIMIT IS NOT REALLY THE INTERESTING ISSUE.
WHAT DO YOU MEAN?
IT IS SAYING SOMETHING ABOUT WHAT WE MEAN BY SPACE AND TIME ITSELF.

QUANTITIES LIKE SPEED ARE REALLY A RESULT OF MAKING SPACE MEASUREMENTS FOR POSITION, AND TIME MEASUREMENTS FOR DEPARTURE AND ARRIVAL.
AND SO DEDUCING MOTION.
SURE.
SO THE APPARENT SPEED LIMIT IS TELLING US SOMETHING ABOUT HOW SPACE AND TIME –
THE PLACE WHERE WE DO MEASUREMENTS AND HAVE SENSIBLE THINGS LIKE CAUSE AND EFFECT –
IS PUT TOGETHER.
CAUSE AND EFFECT ARE PART OF THIS STORY?
YES, YOU KNOW: **THIS** HAPPENED...
...THEN **THAT** HAPPENED AFTER BECAUSE OF THAT FIRST THING HAPPENING...
CAUSE AND EFFECT.
THAT'S ALL PART OF WHAT WE CALL SPACE AND TIME.
RIGHT, YES.
YOU'RE SAYING THAT THE SPEED OF LIGHT IS CONNECTED TO THAT SOMEHOW?
WHEN YOU TRACE IT ALL THROUGH, YOU END UP WITH THE STUFF YOU'VE HEARD OF FROM EINSTEIN: SPECIAL RELATIVITY.
SPACE AND TIME FOR AN OBSERVER DOING PHYSICS IS PART OF A BIGGER THING CALLED **SPACETIME**.
WHAT YOU'D CALL SPACE AND TIME ARE A PARTICULAR WAY OF SLICING IT.
I SEE.
SOMEONE WALKING BY, SINCE THEY ARE IN A DIFFERENT STATE OF MOTION, SLICES SPACETIME DIFFERENTLY, TO GET WHAT **THEY** CALL SPACE AND TIME.
THEY WILL BE DIFFERENT FROM YOURS.
OH, THIS IS WHERE YOU GET CLOCKS RUNNING AT DIFFERENT SPEEDS AND ALL THAT?
YES, IT'S ALL JUST DIFFERENT WAYS OF SLICING SPACETIME.
THE DIFFERENCES ARE TINY FOR THE EVERYDAY MOTION WE DO IN ORDINARY LIFE, BUT CAN BE HUGE IN OTHER SITUATIONS.

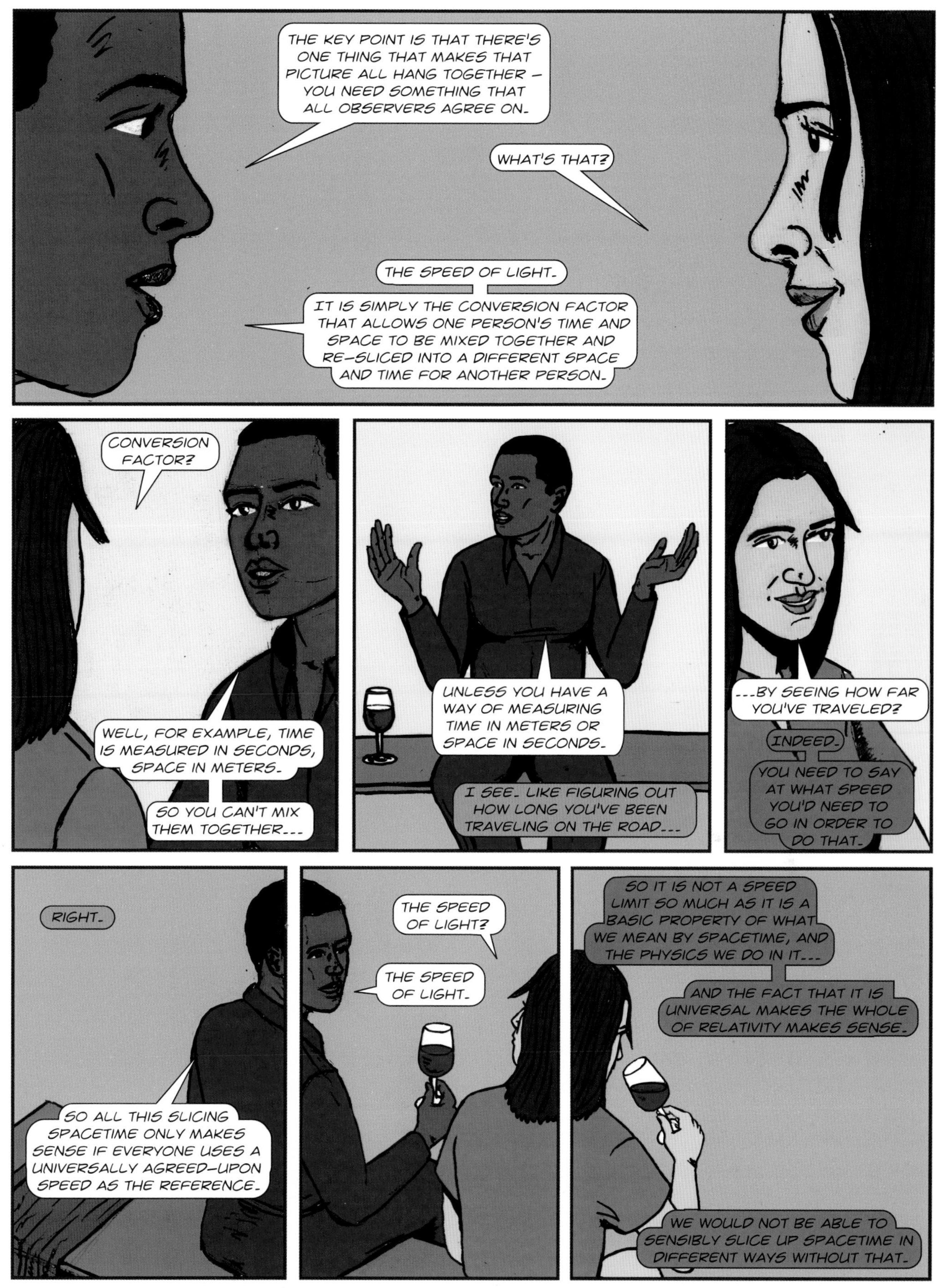
THE KEY POINT IS THAT THERE'S ONE THING THAT MAKES THAT PICTURE ALL HANG TOGETHER — YOU NEED SOMETHING THAT ALL OBSERVERS AGREE ON.
WHAT'S THAT?
THE SPEED OF LIGHT.
IT IS SIMPLY THE CONVERSION FACTOR THAT ALLOWS ONE PERSON'S TIME AND SPACE TO BE MIXED TOGETHER AND RE-SLICED INTO A DIFFERENT SPACE AND TIME FOR ANOTHER PERSON.
CONVERSION FACTOR?
WELL, FOR EXAMPLE, TIME IS MEASURED IN SECONDS, SPACE IN METERS.
SO YOU CAN'T MIX THEM TOGETHER...
UNLESS YOU HAVE A WAY OF MEASURING TIME IN METERS OR SPACE IN SECONDS.
I SEE. LIKE FIGURING OUT HOW LONG YOU'VE BEEN TRAVELING ON THE ROAD...
...BY SEEING HOW FAR YOU'VE TRAVELED?
INDEED.
YOU NEED TO SAY AT WHAT SPEED YOU'D NEED TO GO IN ORDER TO DO THAT.
RIGHT.
SO ALL THIS SLICING SPACETIME ONLY MAKES SENSE IF EVERYONE USES A UNIVERSALLY AGREED-UPON SPEED AS THE REFERENCE.
THE SPEED OF LIGHT?
THE SPEED OF LIGHT.
SO IT IS NOT A SPEED LIMIT SO MUCH AS IT IS A BASIC PROPERTY OF WHAT WE MEAN BY SPACETIME, AND THE PHYSICS WE DO IN IT...
AND THE FACT THAT IT IS UNIVERSAL MAKES THE WHOLE OF RELATIVITY MAKES SENSE.
WE WOULD NOT BE ABLE TO SENSIBLY SLICE UP SPACETIME IN DIFFERENT WAYS WITHOUT THAT.

GOING FASTER THAN LIGHT WOULD MESS THAT ALL UP?
WELL, IT WOULD CERTAINLY BREAK IT VERY BADLY IN WAYS WE'VE NEVER SEEN.
EVEN FUNDAMENTAL THINGS LIKE CAUSE AND EFFECT WOULD GET MESSED UP.
YOU'D BE ABLE TO MAKE TIME MACHINES, AND DO CRAZY THINGS THAT DON'T MAKE SENSE.
CRAZY THINGS?
WELL, THE TIME MACHINE COULD LET YOU GO BACK IN TIME AND STOP YOURSELF FROM MAKING THE TIME MACHINE....
IT'S NOT CLEAR AS TO WHAT LINE TO TREAD TO KEEP DOING SENSIBLE PHYSICS IF **THAT** IS POSSIBLE.
WAIT...
SO THE KEY TO MAKING TIME MACHINES IS TO MESS UP SPACETIME?
WELL, KIND OF. BUT IT'S GETTING PRETTY LATE...
THAT'S A CONVERSATION FOR ANOTHER TIME.
HA!
YES.
BUT WE COULD FIX THAT BY GOING BACK IN TIME...
HOW?
WE GO BACK TO WHEN WE FIRST MET AT THE MAMMOTH...
AND THEN HAVE THE CONVERSATION ABOUT TIME TRAVEL INSTEAD.
BUT THEN WE WOULD NOT BE MOTIVATED TO—
I KNOW, I KNOW...
THINGS WOULDN'T LIKELY GO THE SAME WAY ANYHOW...

WELL...
YES...
WELL, IT WAS GREAT TALKING TO YOU.
I HOPE YOU DID NOT MIND ALL THE QUESTIONS!
NOT AT ALL.
IT WAS A PLEASURE TO EXCHANGE IDEAS.
I ENJOYED IT! THE PARTY TURNED OUT NOT TO BE A BUST AFTER ALL.
I AGREE. BUT I STILL WANT TO GET OUT OF THIS SILLY COSTUME.
OH RIGHT!
I GUESS I SHOULD SAY "GOODBYE, ULTRAVIOLET."
HA! YES.
GOODBYE, CATALYST.
ART

## Notes

***Pages 62–64*** – The modern concept of a field is usually attributed to Maxwell, building on Faraday's idea of "lines of force." Maxwell's equations are a complete encapsulation of the experimental results of Faraday and many others, and also include phenomena Maxwell discovered, guided by the need to have the equations preserve an internal logic. They are explored in this chapter, and the book by Forbes and Mahon mentioned in the chapter 1 note for page 14 and beyond is a good historical resource here.

***Page 65 and beyond*** – In his enthusiastic scribbling in the notebook, our interlocutor has simplified the equations a bit by choosing units of measurement such that the two constants of nature that usually appear here (see the chapter 1 notes for page 15) are set to the value 1. The essential points about the symmetry of the equations remain true. This is a typical thing for physicists to do to remove unnecessary clutter from equations and see the structure more clearly.

***Page 72*** – The Higgs mechanism and the search for the Higgs particle (the direct evidence of an important kind of symmetry breaking) is described here: Jon Butterworth, *Most Wanted Particle: The Inside Story of the Hunt for the Higgs, the Heart of the Future of Physics* (New York: The Experiment, LLC, 2015).

***Pages 73–74*** – There are numerous editions of John Milton's *Paradise Lost* to choose from. This annotated edition is a great resource: John Milton, *Paradise Lost: An Authoritative Text, Backgrounds and Sources, Criticism*, ed. Gordon Teskey (New York: W. W. Norton & Co., 2005).

***Pages 74–77*** – Further reading is in the books about light mentioned in the chapter 1 note for page 12 and beyond.

***Page 77*** (panels 6 and 7) – The broken symmetry here is the fact that there are now point sources of the E field, but not the B field. Don't mix this up with other important broken symmetries in the early universe such as the one tied up with the Higgs mechanism described in the book by Butterworth (see above). The book by Singh on cosmology (see the chapter 3 notes for page 45) will illuminate the early universe remark. As for metals, it is hard to find a generally accessible book to point to. (Why are there so many books written for general audiences about black holes, exotic particles, strings, other universes, extra dimensions, etc., and so few about the fascinating physics of the things around us, like metals?) In both cases, for different reasons, the normal state of affairs is that there are lots of unbound electrons present. Those electrons interact strongly with the light, absorbing it as it tries to pass among them, making the situation opaque or dark. Andrew Zangwill, *Modern Electrodynamics* (Cambridge: Cambridge University Press, 2013) is a good (but advanced) source to explore.

***Pages 79–82*** – Further reading suggestions for special relativity were given in the chapter 1 note for page 13. More discussion of relativity, both special and general, can be found in the book by Randall, mentioned in the chapter 3 notes for page 45, and very thoroughly (but still at a nonspecialist level) in Kip Thorne, *Black Holes and Time Warps: Einstein's Outrageous Legacy* (New York: W. W. Norton & Co., 1994). The latter book will be an especially useful resource about black holes later on. Also, if you want to read about aspects of time travel (a topic that the characters in the chapter did not have time to get to), this book will help.

LOVE Never Dies
LOVE Never Dies
HA!
OH, PLEASE.
LOVE Never Dies
EXCUSE ME, I'M CURIOUS
WHY DO YOU SAY THAT?
I'M LAUGHING AT THE TITLE.
WHY IS THAT FUNNY?
WELL, EVERYTHING DIES.
EVERYTHING.
REALLY?
YES.
WE DIE, ALL ANIMALS DIE...
EVEN STARS DIE.

STARS? WELL-
LOOK, EVEN THE UNIVERSE PROBABLY DIES.
HOW CAN THE UNIVERSE DIE?
WHAT DOES THAT ACTUALLY MEAN?
WELL, IT WAS BORN.
IT WENT THROUGH A HOT, YOUNG, ENERGETIC PHASE FOR A WHILE...
IT'S EXPANDING AND COOLING DOWN...
IT'LL PROBABLY EVENTUALLY DIE IN ONE WAY OR ANOTHER.
DIE HOW?
WELL, PROBABLY BY EITHER BEING COLD AND LIFELESS...
OR BY BEING SO STRETCHED OUT AND SPREAD APART THAT THERE'S NOTHING INTERESTING GOING ON.
OR BOTH.
WELL, WHAT ABOUT GOD?
GODS DIE TOO.
NO WAY...
GOD'S IMMORTAL.
...
FIRST OF ALL, ATTRIBUTING IMMORTALITY TO GODS IS KIND OF JUST A NEW FAD.
THE TWO IDEAS ARE SEPARATE REALLY.
BUT IN ANY CASE, WHETHER PEOPLE SAY THEY ARE IMMORTAL OR NOT...
THE GODS DIE WHEN PEOPLE STOP BELIEVING IN THEM.

THEY'RE REALLY ALL ABOUT POWER IN ONE FORM OR ANOTHER.
THOSE POWERS LOSE THEIR POTENCY IF NOBODY BELIEVES IN THEIR EFFECTS...
THAT'S JUST DEPRESSING.
REALLY?
YES, BUT NOT THAT GODS AND EVERYTHING DIE –
I THINK YOU ARE WRONG ON THAT POINT –
BUT THAT YOU ACTUALLY BELIEVE THAT.
I DON'T THINK IT IS DEPRESSING AT ALL.
NEITHER OF THOSE THINGS, FOR THAT MATTER...
DEATH IS AN ENDING OF SORTS, BUT IT IS ALSO A BEGINNING.
I SUPPOSE SO.
YOUR BODY BECOMES FOOD FOR BUGS AND WORMS. LOTS OF WORM FAMILIES DO WELL OUT OF YOU.
UNLESS YOU'RE CREMATED, I GUESS.
OH, AND FLOWERS. PUSHING UP DAISIES AND ALL THAT.

WELL, IT RUNS A LOT DEEPER THAN THAT, TO MY MIND.
BUT YES, IT IS SORT OF ABOUT RECYCLING, OF ONE KIND OR ANOTHER.
SOME BASIC THINGS COME TOGETHER AND TAKE FORM FOR A WHILE... MAYBE A VERY COMPLEX FORM.
AND THERE'S POTENTIAL –
HOPE, MAYBE –
IN THAT FORM.
BUT EVENTUALLY THAT FORM GOES AWAY – AND THAT'S OK. IT GIVES WAY TO NEW FORMS.
ALSO, IT WAS ITSELF PART OF SOME MUCH LARGER FORM TOO.
THAT LITTLE FLOWER PLAYS A PART IN A LARGER DRAMA THAT GOES WELL BEYOND THE PLACE WHERE IT IS ROOTED...

A CLOUD OF MATTER COLLAPSES UNDER GRAVITY TO MAKE A STAR SYSTEM...
YOU KNOW, THAT MATTER WAS STUFF LEFT OVER FROM SOME PREVIOUS FORMS – HYDROGEN FROM THE BIG BANG, AND MORE COMPLEX ELEMENTS FROM STARS LONG DEAD.
THIS SUN MIGHT HAVE ORBITING PLANETS, PERHAPS HARBORING LIFE, INCLUDING THE FLOWER.
IT WILL FLOURISH FOR A WHILE...
BUT ONE DAY, THAT STAR SYSTEM WILL DIE TOO...
THE STAR WILL RUN OUT OF FUEL NEEDED TO STOP IT COLLAPSING UNDER GRAVITY, IT'LL COLLAPSE, PERHAPS GO SUPERNOVA,
...EXPLODING ALL THAT MATERIAL OUT INTO THE UNIVERSE FOR USE IN OTHER SYSTEMS.

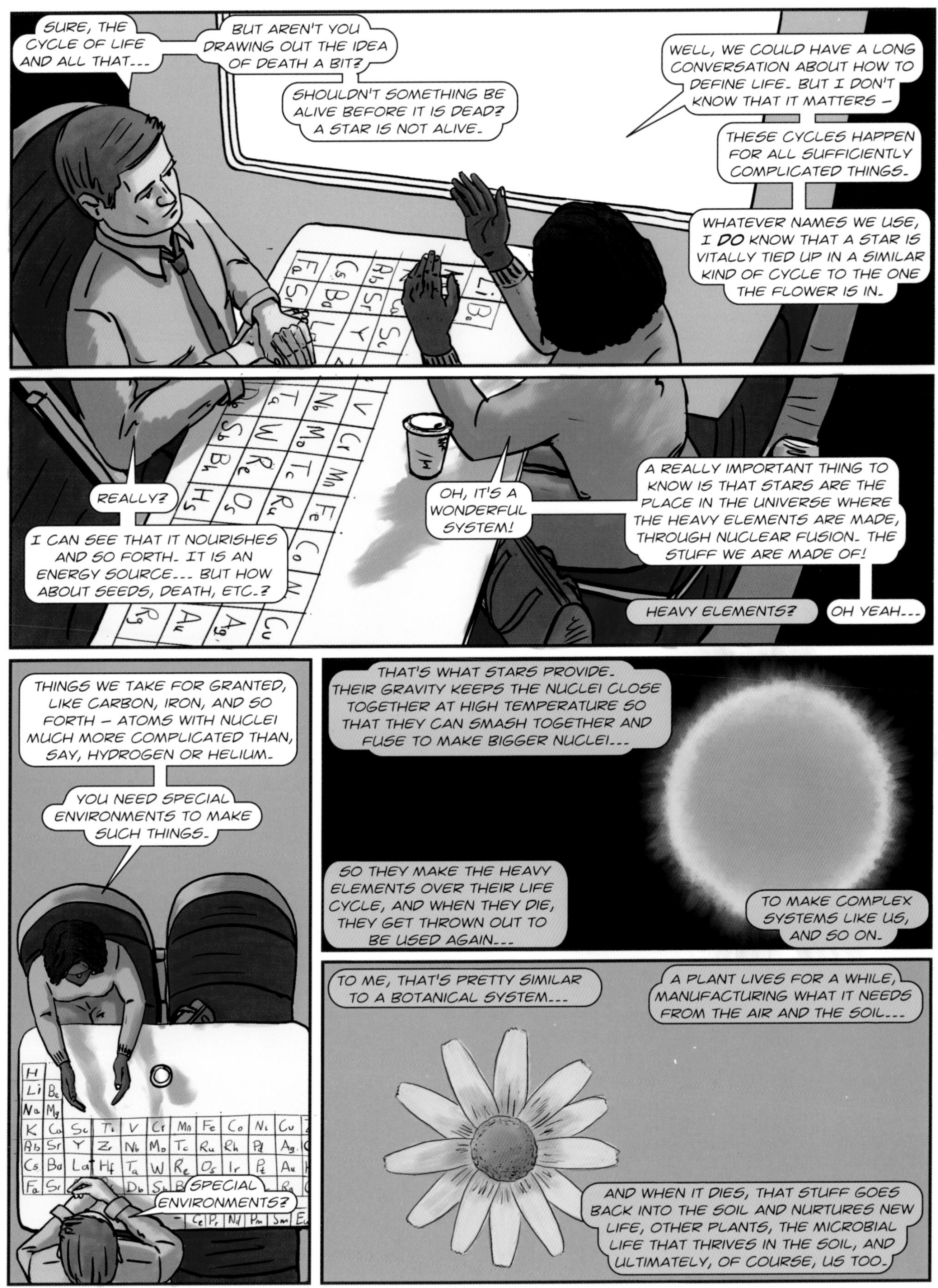
SURE, THE CYCLE OF LIFE AND ALL THAT...
BUT AREN'T YOU DRAWING OUT THE IDEA OF DEATH A BIT?
SHOULDN'T SOMETHING BE ALIVE BEFORE IT IS DEAD? A STAR IS NOT ALIVE.
WELL, WE COULD HAVE A LONG CONVERSATION ABOUT HOW TO DEFINE LIFE. BUT I DON'T KNOW THAT IT MATTERS –
THESE CYCLES HAPPEN FOR ALL SUFFICIENTLY COMPLICATED THINGS.
WHATEVER NAMES WE USE, I **DO** KNOW THAT A STAR IS VITALLY TIED UP IN A SIMILAR KIND OF CYCLE TO THE ONE THE FLOWER IS IN.
REALLY?
I CAN SEE THAT IT NOURISHES AND SO FORTH. IT IS AN ENERGY SOURCE... BUT HOW ABOUT SEEDS, DEATH, ETC.?
OH, IT'S A WONDERFUL SYSTEM!
A REALLY IMPORTANT THING TO KNOW IS THAT STARS ARE THE PLACE IN THE UNIVERSE WHERE THE HEAVY ELEMENTS ARE MADE, THROUGH NUCLEAR FUSION. THE STUFF WE ARE MADE OF!
HEAVY ELEMENTS?
OH YEAH...
THINGS WE TAKE FOR GRANTED, LIKE CARBON, IRON, AND SO FORTH – ATOMS WITH NUCLEI MUCH MORE COMPLICATED THAN, SAY, HYDROGEN OR HELIUM.
YOU NEED SPECIAL ENVIRONMENTS TO MAKE SUCH THINGS.
SPECIAL ENVIRONMENTS?
THAT'S WHAT STARS PROVIDE. THEIR GRAVITY KEEPS THE NUCLEI CLOSE TOGETHER AT HIGH TEMPERATURE SO THAT THEY CAN SMASH TOGETHER AND FUSE TO MAKE BIGGER NUCLEI...
SO THEY MAKE THE HEAVY ELEMENTS OVER THEIR LIFE CYCLE, AND WHEN THEY DIE, THEY GET THROWN OUT TO BE USED AGAIN...
TO MAKE COMPLEX SYSTEMS LIKE US, AND SO ON.
TO ME, THAT'S PRETTY SIMILAR TO A BOTANICAL SYSTEM...
A PLANT LIVES FOR A WHILE, MANUFACTURING WHAT IT NEEDS FROM THE AIR AND THE SOIL...
AND WHEN IT DIES, THAT STUFF GOES BACK INTO THE SOIL AND NURTURES NEW LIFE, OTHER PLANTS, THE MICROBIAL LIFE THAT THRIVES IN THE SOIL, AND ULTIMATELY, OF COURSE, US TOO.

WELL, THAT'S ALL FINE, BUT YOU PLAYED DOWN THE LIVING VS. NONLIVING ISSUE, AND I DON'T THINK YOU CAN...
ALSO, DID WE NOT START OUT WITH LOVE, AND WHETHER THAT CAN BE FOREVER?
LOVE?
WELL, LIVING THINGS FEEL LOVE, RIGHT?
YES.
SO THEY DIE EVENTUALLY, AND SO THE LOVE DIES AT THAT POINT...
AND PROBABLY SOONER.
SOONER?
WELL, WHAT HAPPENS WHEN PEOPLE SAY THEY HAVE FALLEN OUT OF LOVE?
ARE THEY LYING?
THEY MAYBE WERE MISTAKEN ABOUT BEING IN LOVE IN THE FIRST PLACE. IT WAS MAYBE A DIFFERENT L-WORD, LIKE LUST, OR LIKE.
REALLY?
YEAH.
TRUE LOVE STAYS FOREVER.
NEVER DIES.
EVEN IF THEY DIE?
WELL, IF THEIR ESSENCE LIVES ON...
ESSENCE?
YOU MEAN LIKE A SOUL? A SPIRIT?
YES.
YOU REALLY BELIEVE ALL THIS?
WELL, YES!
IT FEELS RIGHT TO ME. THE ALTERNATIVE SORT OF MAKES ME...
FRIGHTENED?
THE CLAIM THAT WITHOUT AN AFTERLIFE IT IS ALL MEANINGLESS?
PERHAPS YOU'RE LOOKING AT IT THE WRONG WAY.
WHO ARE YOU TO SAY WHICH IS THE RIGHT WAY TO LOOK AT THINGS?
I'M NOT, AND I SAID "PERHAPS."
I'M SUGGESTING A GLIMPSE OF ANOTHER WAY OF LOOKING AT IT.
YOU DON'T HAVE TO DO IT MY WAY, BUT AT LEAST HEAR WHAT I HAVE TO SAY.
OK, GO ON...

HEAR THAT MUSIC?
YES.
KNOW WHAT IT IS?
SURE, IT'S HER CLASSIC...
WHAT'S YOUR POINT?
WELL, WHAT IF SHE'D NEVER LIVED? WOULD THAT MUSIC HAVE BEEN WRITTEN?
ISN'T THE UNIVERSE PROFOUNDLY DIFFERENT AS A RESULT OF IT HAVING BEEN WRITTEN?
PROFOUNDLY?
WELL, PROFOUNDLY ENOUGH, FOR THE PEOPLE IT HAS TOUCHED, AND FOR THE PEOPLE WHO'VE BEEN AFFECTED BY THOSE PEOPLE WHO WERE TOUCHED BY IT, AND SO ON...
YES, I SUPPOSE SO.
WHAT LASTS AND WHAT REALLY MATTERS IS THAT WE TOUCH EACH OTHER SOMEHOW.
WE CONTRIBUTE. WE **BE** IN THE UNIVERSE.... THAT'S HOW WE LIVE ON,
WHETHER AS A PERSON IN THE WORLD OR AS A SPECIES IN THE GALAXY, IN THE UNIVERSE...

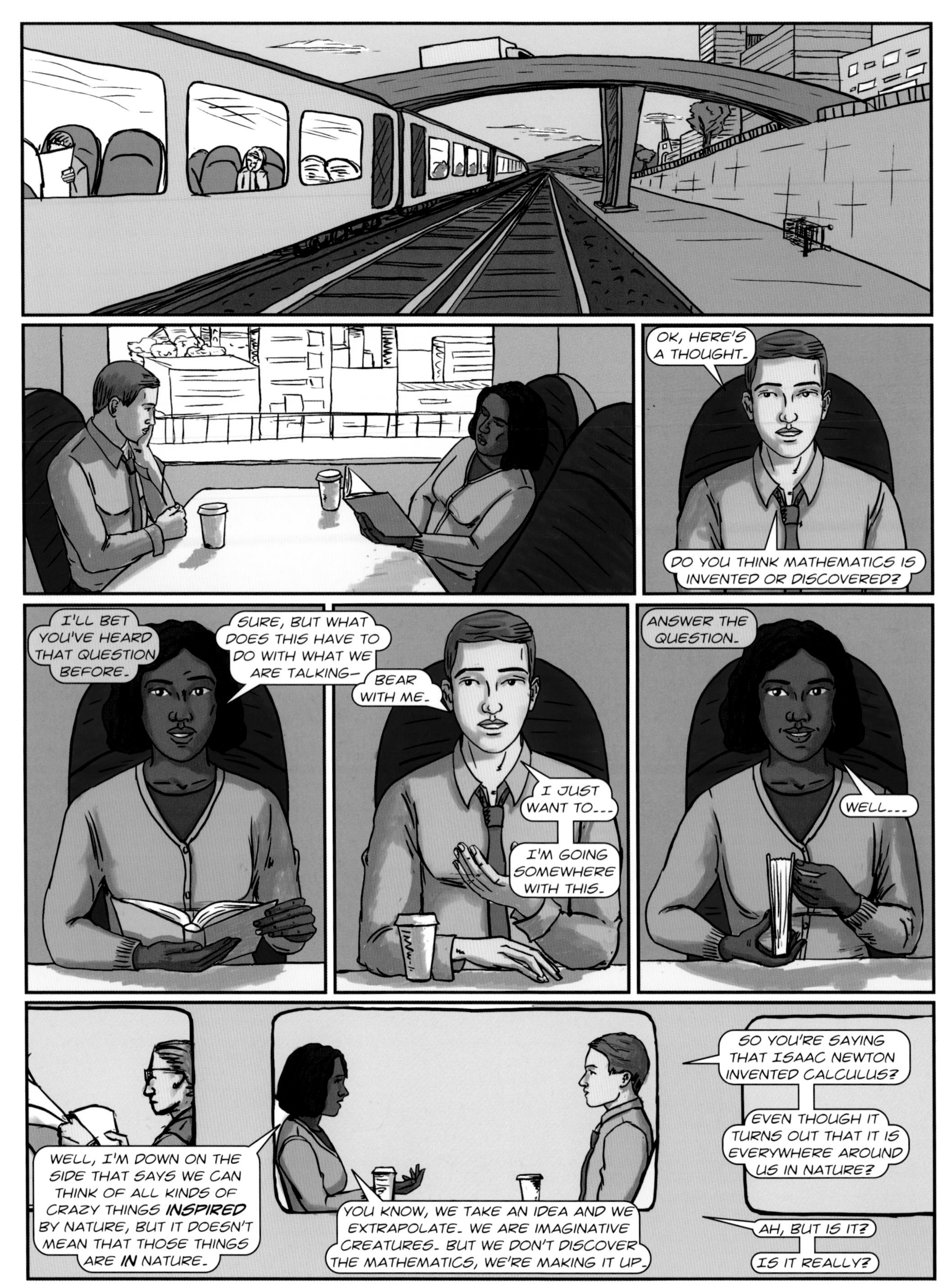

OK, HERE'S A THOUGHT.
DO YOU THINK MATHEMATICS IS INVENTED OR DISCOVERED?
I'LL BET YOU'VE HEARD THAT QUESTION BEFORE.
SURE, BUT WHAT DOES THIS HAVE TO DO WITH WHAT WE ARE TALKING—
BEAR WITH ME.
I JUST WANT TO...
I'M GOING SOMEWHERE WITH THIS.
ANSWER THE QUESTION.
WELL...
WELL, I'M DOWN ON THE SIDE THAT SAYS WE CAN THINK OF ALL KINDS OF CRAZY THINGS **INSPIRED** BY NATURE, BUT IT DOESN'T MEAN THAT THOSE THINGS ARE **IN** NATURE.
YOU KNOW, WE TAKE AN IDEA AND WE EXTRAPOLATE. WE ARE IMAGINATIVE CREATURES. BUT WE DON'T DISCOVER THE MATHEMATICS, WE'RE MAKING IT UP.
SO YOU'RE SAYING THAT ISAAC NEWTON INVENTED CALCULUS?
EVEN THOUGH IT TURNS OUT THAT IT IS EVERYWHERE AROUND US IN NATURE?
AH, BUT IS IT?
IS IT REALLY?

BY THE WAY, IT IS NEWTON AND LEIBNIZ...
HMMM?
INVENTORS OR DISCOVERERS OF CALCULUS...
OK, BUT TO MY POINT...
SURE...ANYWAY, I THINK THAT IT IS IMPORTANT TO NOT MIX UP OUR IDEALIZED DESCRIPTIONS OF NATURE WITH THE REAL THING...
BUT THE CONCEPT OF THE DERIVATIVE – DIFFERENTIATION – IS OUT THERE IN PHYSICAL THINGS LIKE SPEED, ACCELERATION...
WELL, YES AND NO. I THINK THAT THEY ARE GOOD APPROXIMATIONS TO THE MATHEMATICAL IDEAL...
ALSO, MATHEMATICALLY YOU CAN BUILD ALL KINDS OF EXTRA STUFF BEYOND BASIC CALCULUS THAT PROBABLY ISN'T OUT THERE...
SO THAT TELLS ME THAT NATURE IS A GOOD SOURCE OF INSPIRATION FOR MATHEMATICAL IDEALIZATIONS, AND SOMETIMES THE THINGS IN IT ARE APPROXIMATIONS TO THE IDEALS...
BUT THE IDEALS ARE IN OUR MINDS, NOT IN THE WORLD EXACTLY.
I DON'T THINK I AGREE WITH YOU.
I THINK TH–
OK, SHOW ME A PERFECT CIRCLE.
WHAT?
A PERFECT CIRCLE.
MY POINT IS THAT THE MATHEMATICAL IDEA OF A PERFECT CIRCLE EXISTS...
SURE...
BUT THERE ARE NO PERFECT CIRLCES OUT THERE IN THE UNIVERSE.
HOW DO YOU KNOW?
WELL, DO YOU HAVE EVIDENCE THEY DO EXIST?
WELL, NO...

WELL, YOU SEE, I TAKE THE OPPOSITE VIEW.
I THINK THAT FUNDAMENTALLY WE ARE MADE OF NATURE'S STUFF. WE ARE PART OF NATURE, SO HOW CAN WE THINK OF THINGS THAT ARE OUTSIDE OF NATURE?
BY DEFINITION WE'RE ALMOST BOUND TO BE COMING UP WITH THINGS THAT ARE IN NATURE.
SO THAT'S WHAT I THINK, AND SO—
WELL LET'S AGREE TO DISAGREE. BUT, WHAT DOES THIS HAVE TO DO WITH WHAT WE WERE TALKING ABOUT?
I THINK IT HAS EVERYTHING TO DO WITH WHAT WE'RE TALKING ABOUT.
BECAUSE OF INFINITY.
INFINITY?
YES, INFINITY.
THAT IS A WELL-DEFINED CONCEPT IN MATHEMATICS, RIGHT?
SURE.
WELL, IF THE CONCEPT EXISTS IN MATHEMATICS, AND IF WE DISCOVER MATHEMATICAL THINGS THAT ARE ACTUALLY OUT THERE IN NATURE...
THEN INFINITY IS REALLY OUT THERE.
YOU MIGHT SAY WE MIGHT FIND THINGS THAT REMIND US OF INFINITY OR INSPIRE US TO INVENT INFINITY... BUT I WOULD SAY INFINITY'S REALLY OUT THERE!
UM, I DON'T KNOW WHAT TO DO WITH THAT.
AND I STILL DON'T SEE WHAT IT HAS TO DO WITH WHAT WE ARE TALKING ABOUT.

WELL, ACTUALLY I THINK IT HAS EVERYTHING TO DO WITH IT BECAUSE WE CAN APPLY INFINITY TO TIME.
WHY ISN'T IT TRUE THAT SOMEWHERE OUT THERE, INFINITY –
SOMETHING THAT NATURE KNOWS ABOUT –
AND TIME ARE COMBINED?
AND SO ALL WE NEED IS TO COMBINE INFINITY AND TIME AND LIFE AND THEN WE HAVE THE REALITY OF SOMETHING LIVING FOREVER.
WAIT, WAIT...
SO YOU'RE USING MATHEMATICS TO PROVE THE EXISTENCE OF GOD?
WELL...
YOU'RE TAKING MATHEMATICS AND SOMEHOW TWISTING IT TO SAY "THEREFORE GOD"?
NOW YOU'RE MAKING FUN OF ME.
I DIDN'T SAY ANYTHING ABOUT GOD.
I SAID THE IDEA OF IMMORTALITY MAYBE ISN'T SO UNNATURAL IF INFINITY IS PART OF NATURE.
WELL, I THINK IT'S ALL JUST...
IT'S JUST TOO MUCH. YOU'RE... NO...**NO**.
I DON'T KNOW... I WANT A BETTER ARGUMENT THAT JUST "NO." IS THAT ALL YOU HAVE TO REFUTE MY ARGUMENT?
WELL IT ISN'T REALLY AN ARGUMENT!
YOU'RE STRINGING TOGETHER THINGS THAT MAY OR MAY NOT HAVE **ANYTHING** TO DO WITH EACH OTHER.
SEEMS TO ME THAT THESE THINGS **COULD** HAVE SOMETHING TO DO WITH EACH OTHER...
YOU SCIENTISTS ARE ALWAYS SAYING THAT IT'S A BIG UNIVERSE.
WELL, IT JUST SEEMS... LIKE WE'RE NOT TALKING SCIENCE ANYMORE.

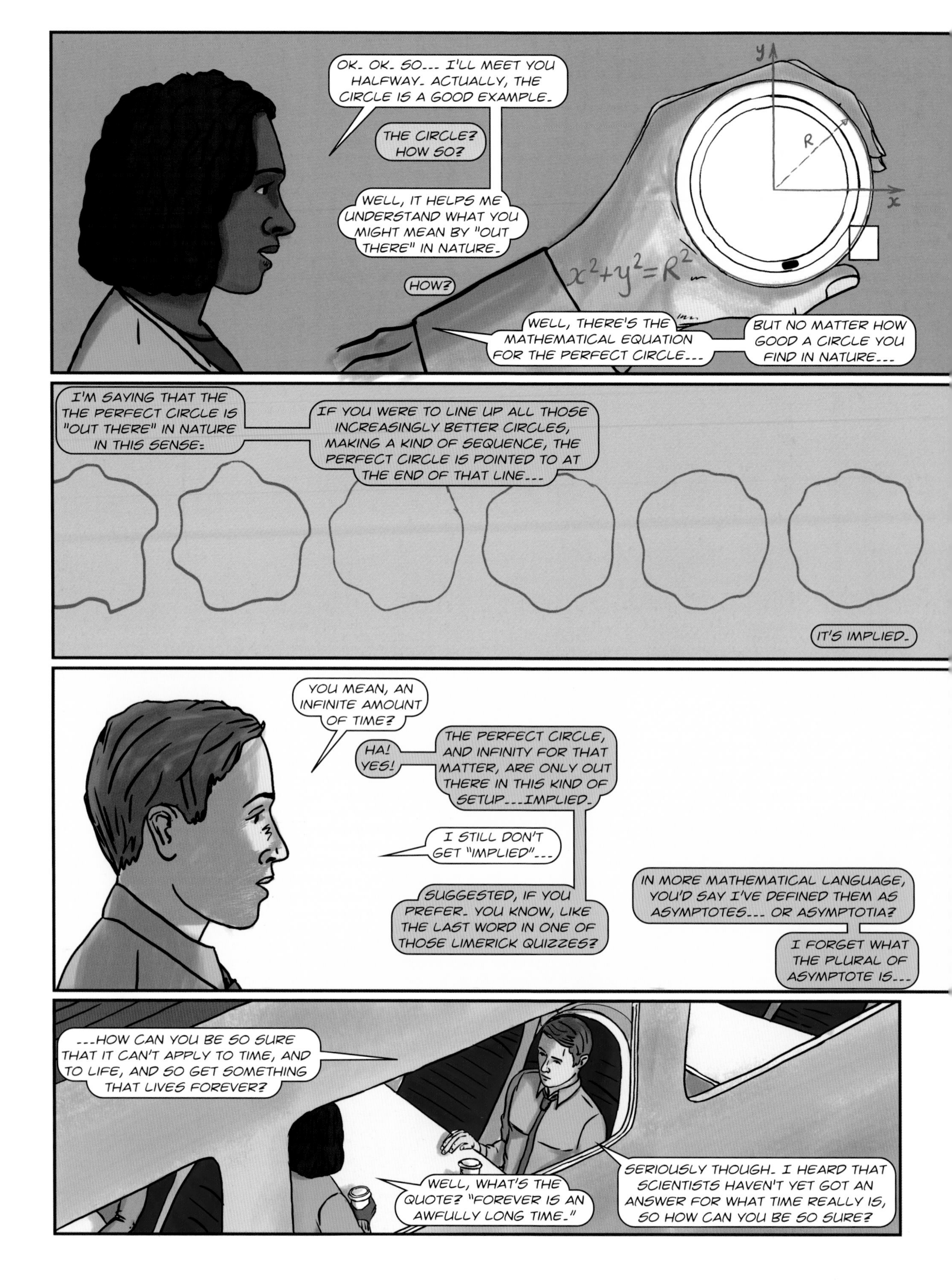
OK. OK. SO... I'LL MEET YOU HALFWAY. ACTUALLY, THE CIRCLE IS A GOOD EXAMPLE.
THE CIRCLE? HOW SO?
WELL, IT HELPS ME UNDERSTAND WHAT YOU MIGHT MEAN BY "OUT THERE" IN NATURE.
HOW?
WELL, THERE'S THE MATHEMATICAL EQUATION FOR THE PERFECT CIRCLE...
BUT NO MATTER HOW GOOD A CIRCLE YOU FIND IN NATURE...
y
x
R
$x^2+y^2=R^2$
I'M SAYING THAT THE THE PERFECT CIRCLE IS "OUT THERE" IN NATURE IN THIS SENSE:
IF YOU WERE TO LINE UP ALL THOSE INCREASINGLY BETTER CIRCLES, MAKING A KIND OF SEQUENCE, THE PERFECT CIRCLE IS POINTED TO AT THE END OF THAT LINE...
IT'S IMPLIED.
YOU MEAN, AN INFINITE AMOUNT OF TIME?
HA! YES!
THE PERFECT CIRCLE, AND INFINITY FOR THAT MATTER, ARE ONLY OUT THERE IN THIS KIND OF SETUP...IMPLIED.
I STILL DON'T GET "IMPLIED"...
SUGGESTED, IF YOU PREFER. YOU KNOW, LIKE THE LAST WORD IN ONE OF THOSE LIMERICK QUIZZES?
IN MORE MATHEMATICAL LANGUAGE, YOU'D SAY I'VE DEFINED THEM AS ASYMPTOTES... OR ASYMPTOTIA?
I FORGET WHAT THE PLURAL OF ASYMPTOTE IS...
...HOW CAN YOU BE SO SURE THAT IT CAN'T APPLY TO TIME, AND TO LIFE, AND SO GET SOMETHING THAT LIVES FOREVER?
WELL, WHAT'S THE QUOTE? "FOREVER IS AN AWFULLY LONG TIME."
SERIOUSLY THOUGH. I HEARD THAT SCIENTISTS HAVEN'T YET GOT AN ANSWER FOR WHAT TIME REALLY IS, SO HOW CAN YOU BE SO SURE?

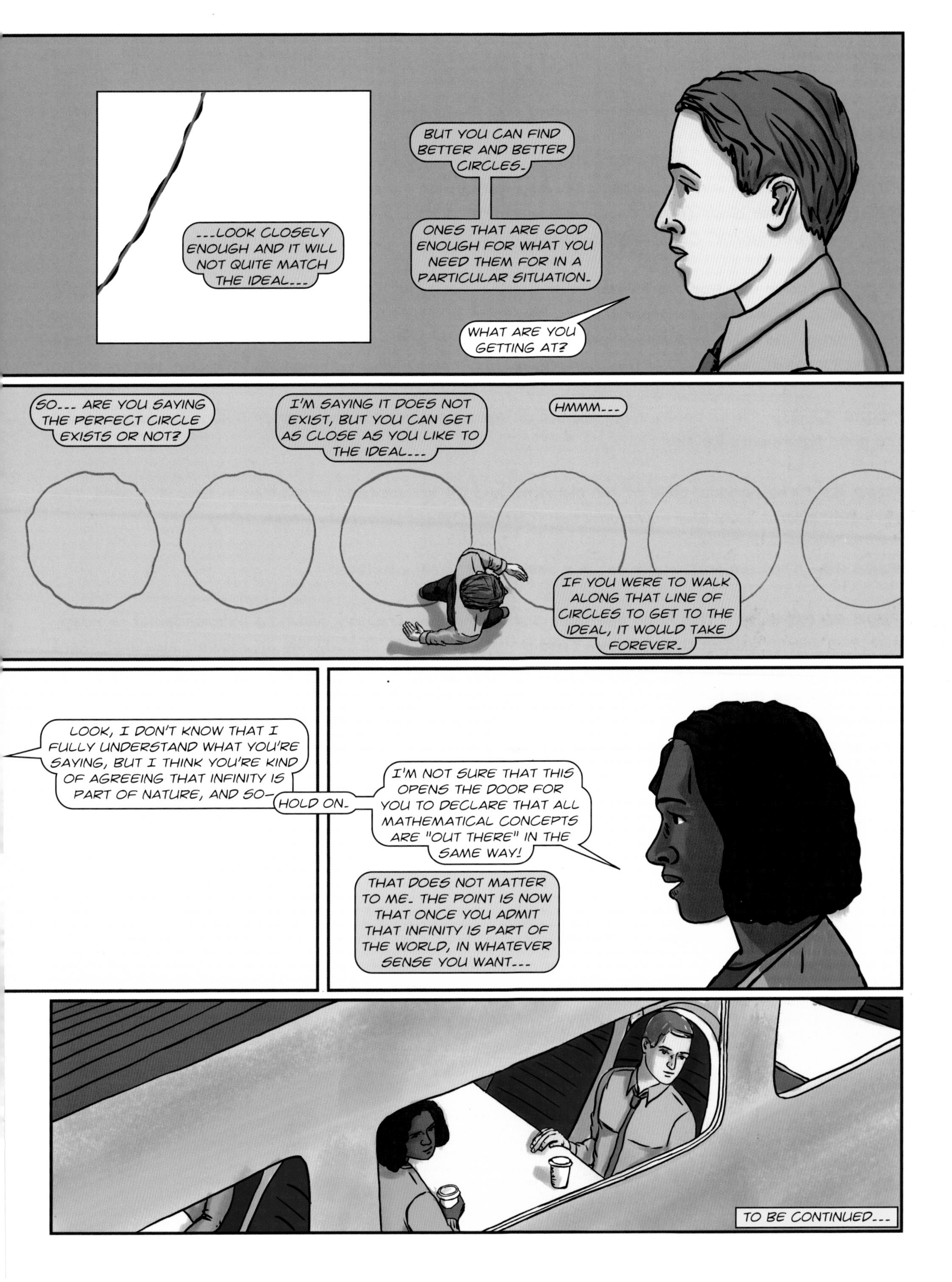
...LOOK CLOSELY ENOUGH AND IT WILL NOT QUITE MATCH THE IDEAL...
BUT YOU CAN FIND BETTER AND BETTER CIRCLES.
ONES THAT ARE GOOD ENOUGH FOR WHAT YOU NEED THEM FOR IN A PARTICULAR SITUATION.
WHAT ARE YOU GETTING AT?
SO... ARE YOU SAYING THE PERFECT CIRCLE EXISTS OR NOT?
I'M SAYING IT DOES NOT EXIST, BUT YOU CAN GET AS CLOSE AS YOU LIKE TO THE IDEAL...
HMMM...
IF YOU WERE TO WALK ALONG THAT LINE OF CIRCLES TO GET TO THE IDEAL, IT WOULD TAKE FOREVER.
LOOK, I DON'T KNOW THAT I FULLY UNDERSTAND WHAT YOU'RE SAYING, BUT I THINK YOU'RE KIND OF AGREEING THAT INFINITY IS PART OF NATURE, AND SO–
HOLD ON.
I'M NOT SURE THAT THIS OPENS THE DOOR FOR YOU TO DECLARE THAT ALL MATHEMATICAL CONCEPTS ARE "OUT THERE" IN THE SAME WAY!
THAT DOES NOT MATTER TO ME. THE POINT IS NOW THAT ONCE YOU ADMIT THAT INFINITY IS PART OF THE WORLD, IN WHATEVER SENSE YOU WANT...
TO BE CONTINUED...

HEY THERE!
OH, HI! HOW ARE YOU?
I'M PRETTY GOOD! WHAT'S UP WITH YOU?
I'M GOOD, THANKS.
GETTING A WALK AND MAYBE SOME COFFEE.
ANGEL'S FLIGHT
FLIGHT
I'M GOING TO UNION STATION. WHERE ARE YOU GOING FOR COFFEE?

GROUNDWIRE, IT'S—
THAT CAFÉ ON MAIN?
RIGHT!
IT'S ON MY WAY, KIND OF.
CAN I WALK WITH YOU?
I'VE GOT A QUESTION TO ASK YOU.
SURE.
ANGELS FLIGHT

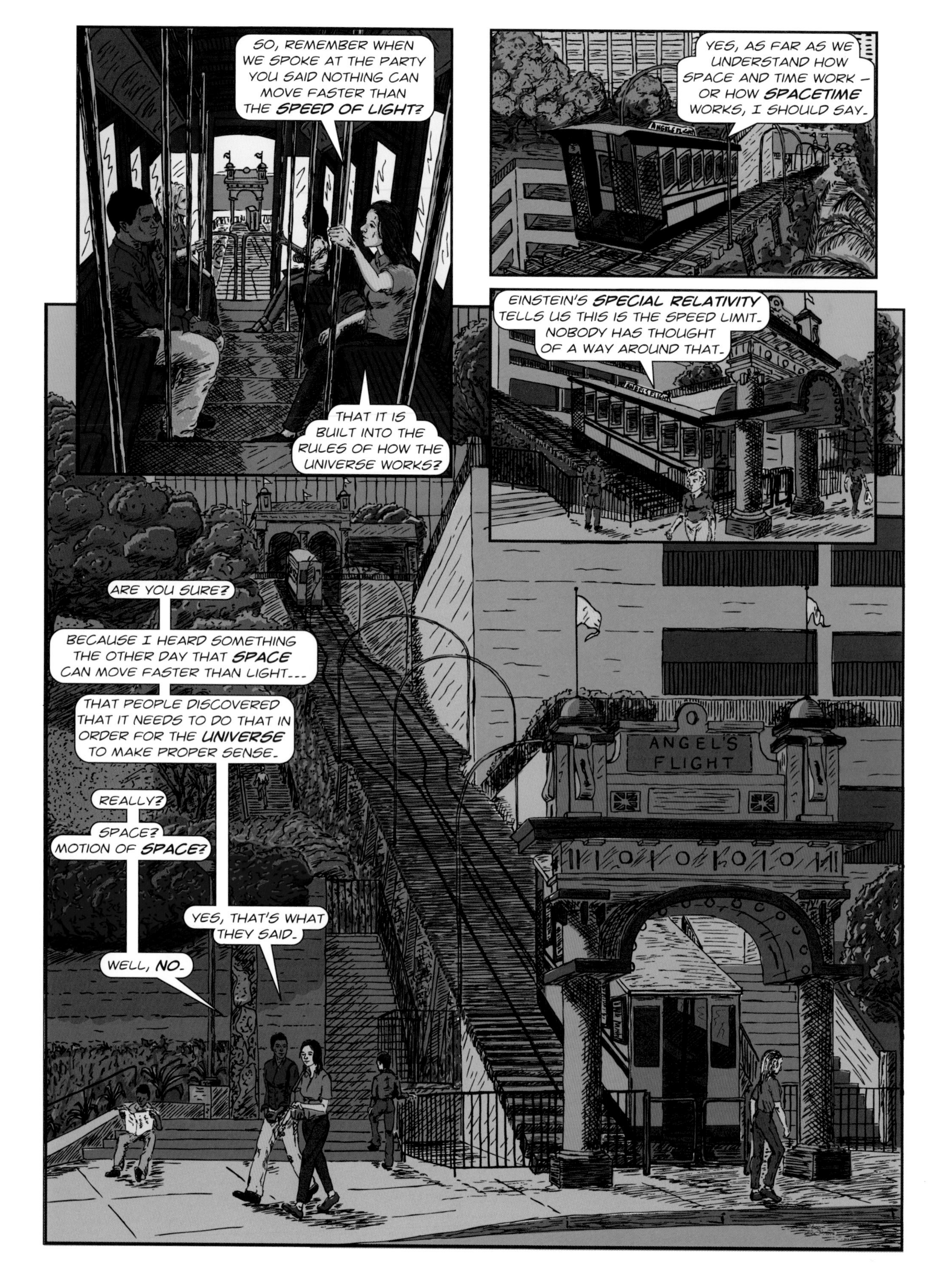
SO, REMEMBER WHEN WE SPOKE AT THE PARTY YOU SAID NOTHING CAN MOVE FASTER THAN THE SPEED OF LIGHT?
THAT IT IS BUILT INTO THE RULES OF HOW THE UNIVERSE WORKS?
YES, AS FAR AS WE UNDERSTAND HOW SPACE AND TIME WORK – OR HOW SPACETIME WORKS, I SHOULD SAY.
ANGEL'S FLIGHT
EINSTEIN'S SPECIAL RELATIVITY TELLS US THIS IS THE SPEED LIMIT. NOBODY HAS THOUGHT OF A WAY AROUND THAT.
ANGEL'S FLIGHT
ARE YOU SURE?
BECAUSE I HEARD SOMETHING THE OTHER DAY THAT SPACE CAN MOVE FASTER THAN LIGHT...
THAT PEOPLE DISCOVERED THAT IT NEEDS TO DO THAT IN ORDER FOR THE UNIVERSE TO MAKE PROPER SENSE.
REALLY?
SPACE? MOTION OF SPACE?
YES, THAT'S WHAT THEY SAID.
WELL, NO.
ANGEL'S FLIGHT

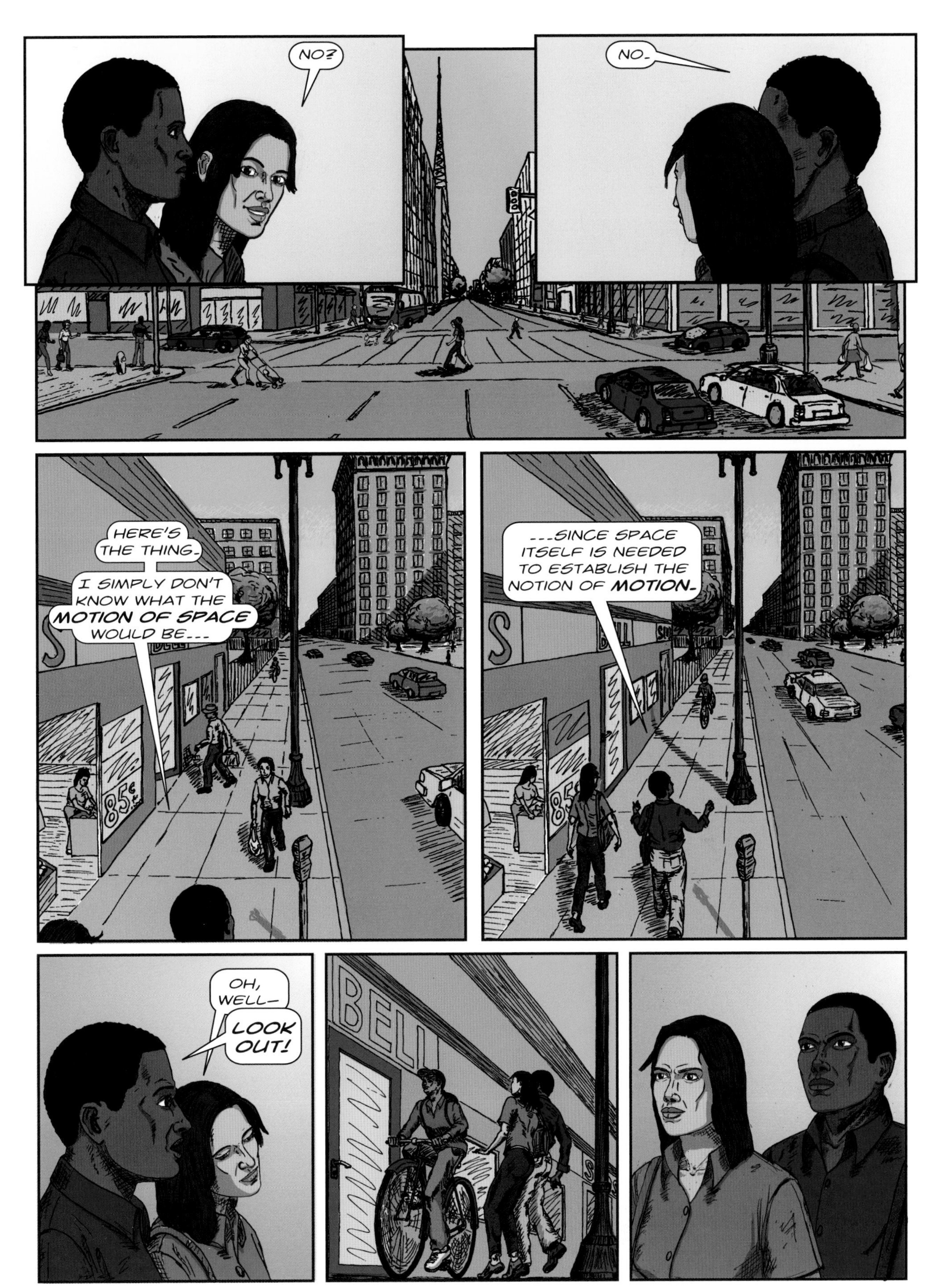
NO?
NO.
HERE'S THE THING.
I SIMPLY DON'T KNOW WHAT THE **MOTION OF SPACE** WOULD BE...
...SINCE SPACE ITSELF IS NEEDED TO ESTABLISH THE NOTION OF **MOTION**.
OH, WELL—
**LOOK OUT!**

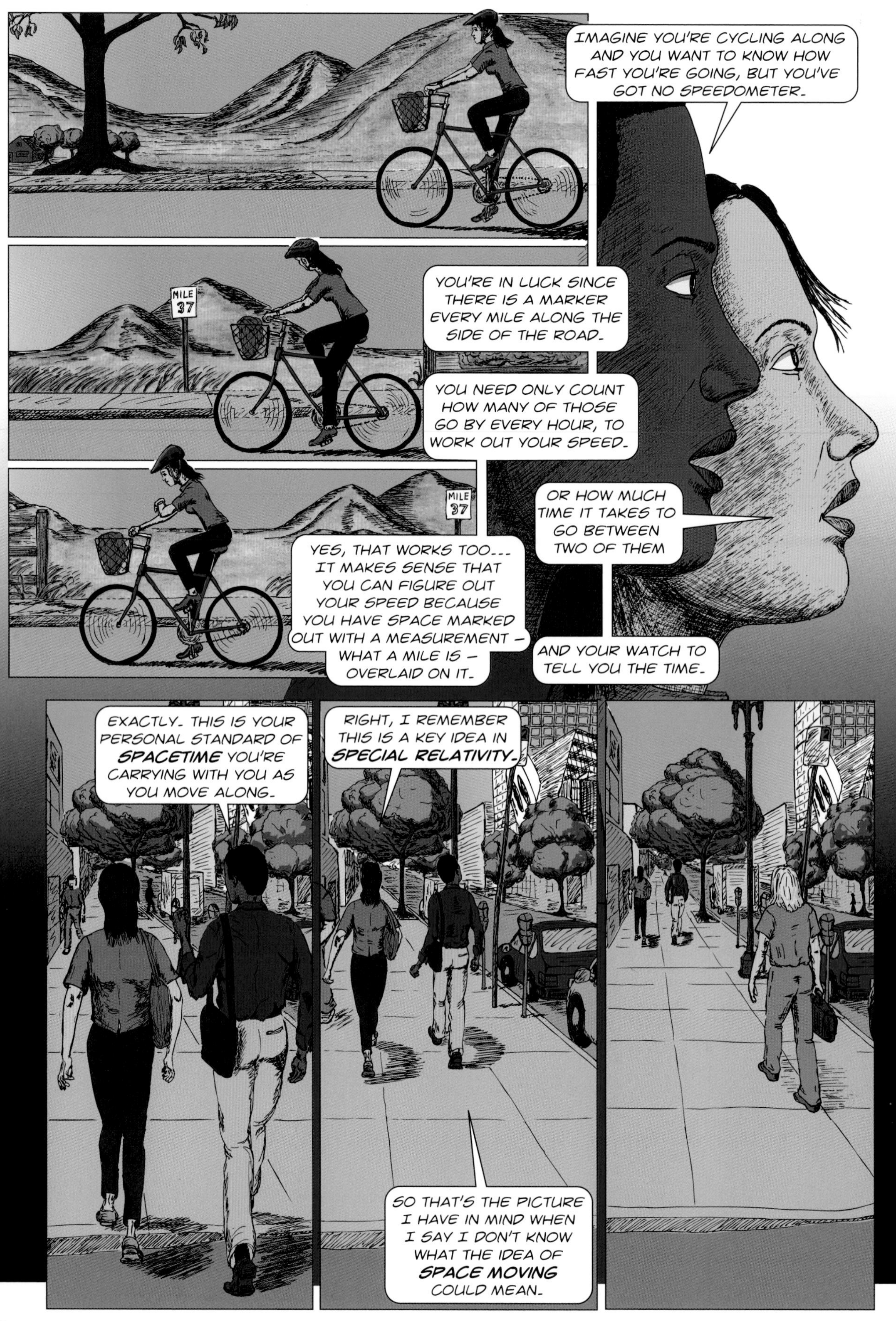

IMAGINE YOU'RE CYCLING ALONG AND YOU WANT TO KNOW HOW FAST YOU'RE GOING, BUT YOU'VE GOT NO SPEEDOMETER.
MILE 37
YOU'RE IN LUCK SINCE THERE IS A MARKER EVERY MILE ALONG THE SIDE OF THE ROAD.
YOU NEED ONLY COUNT HOW MANY OF THOSE GO BY EVERY HOUR, TO WORK OUT YOUR SPEED.
MILE 37
OR HOW MUCH TIME IT TAKES TO GO BETWEEN TWO OF THEM
YES, THAT WORKS TOO... IT MAKES SENSE THAT YOU CAN FIGURE OUT YOUR SPEED BECAUSE YOU HAVE SPACE MARKED OUT WITH A MEASUREMENT — WHAT A MILE IS — OVERLAID ON IT.
AND YOUR WATCH TO TELL YOU THE TIME.
EXACTLY. THIS IS YOUR PERSONAL STANDARD OF **SPACETIME** YOU'RE CARRYING WITH YOU AS YOU MOVE ALONG.
RIGHT, I REMEMBER THIS IS A KEY IDEA IN **SPECIAL RELATIVITY**.
SO THAT'S THE PICTURE I HAVE IN MIND WHEN I SAY I DON'T KNOW WHAT THE IDEA OF **SPACE MOVING** COULD MEAN.

OH!
I SEE...
THEY'RE PROBABLY TALKING ABOUT **INFLATION**.
IN INFLATION, THE SIZE OF THE UNIVERSE INCREASES VERY RAPIDLY AS TIME GOES BY.
ASTRONOMICAL OBSERVATIONS SHOW THAT THE UNIVERSE DID THIS IN THE FIRST TINY FRACTIONS OF A SECOND OF ITS EXISTENCE.
THIS IS NOT THE REGULAR EXPANSION GOING ON **NOW**, BUT INSTEAD A SUPER-FAST EXPANSION RIGHT NEAR THE BEGINNING...
YES!
THAT'S WHAT I WAS HEARING ABOUT. YOU NEED SOMETHING LIKE COSMIC INFLATION TO EXPLAIN WHY TOTALLY DIFFERENT PARTS OF THE UNIVERSE THAT WE CAN SEE HAVE SIMILAR PROPERTIES...
...BUT I'M NOT SURE WHY THAT'S A PROBLEM THAT NEEDED TO BE SOLVED. I GOT CONFUSED.
GROUNDWIRE
coffee

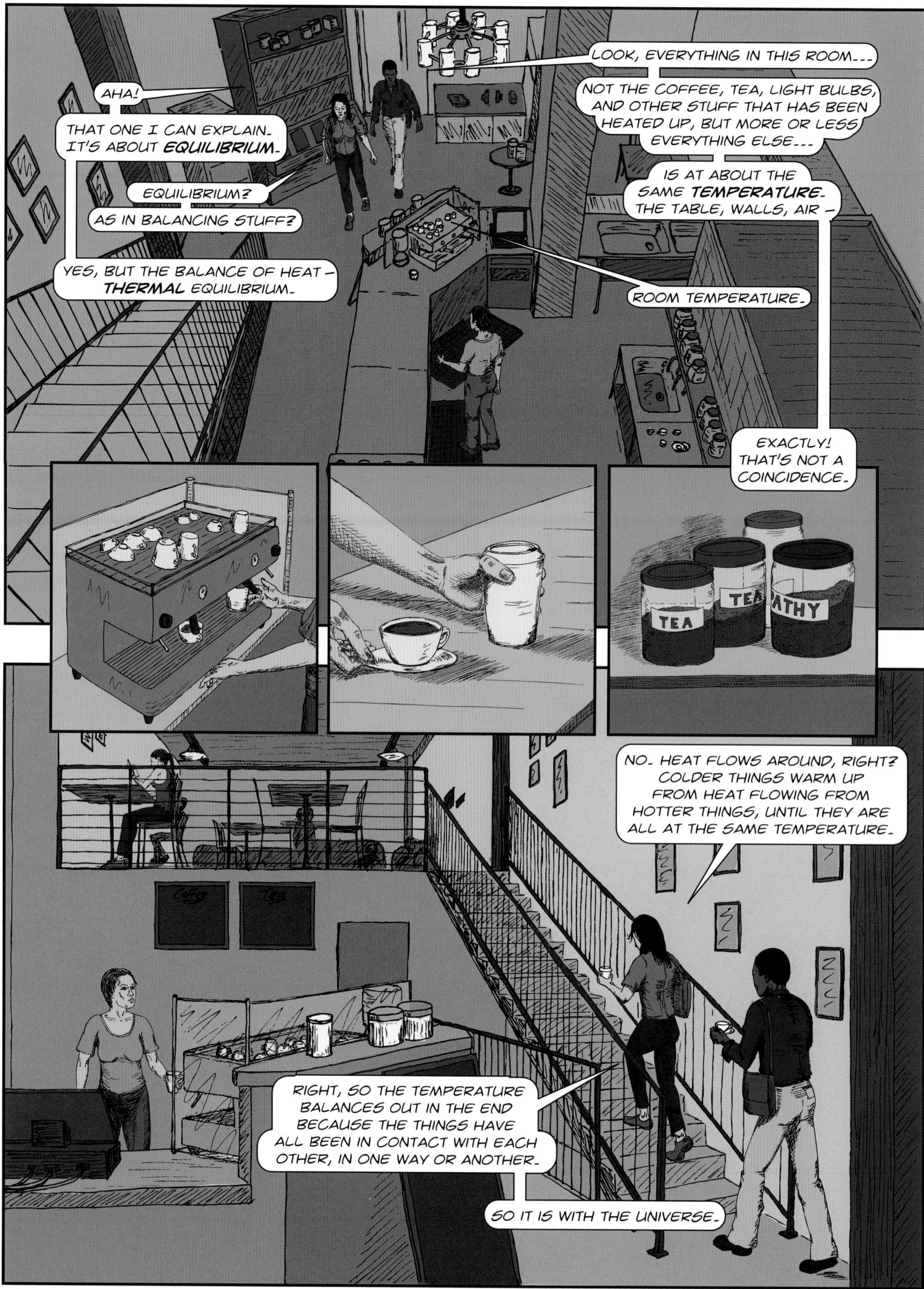
AHA!
THAT ONE I CAN EXPLAIN. IT'S ABOUT EQUILIBRIUM.
EQUILIBRIUM?
AS IN BALANCING STUFF?
YES, BUT THE BALANCE OF HEAT – THERMAL EQUILIBRIUM.
LOOK, EVERYTHING IN THIS ROOM...
NOT THE COFFEE, TEA, LIGHT BULBS, AND OTHER STUFF THAT HAS BEEN HEATED UP, BUT MORE OR LESS EVERYTHING ELSE...
IS AT ABOUT THE SAME TEMPERATURE. THE TABLE, WALLS, AIR –
ROOM TEMPERATURE.
EXACTLY! THAT'S NOT A COINCIDENCE.
TEA
TEA
ATHY
NO. HEAT FLOWS AROUND, RIGHT? COLDER THINGS WARM UP FROM HEAT FLOWING FROM HOTTER THINGS, UNTIL THEY ARE ALL AT THE SAME TEMPERATURE.
RIGHT, SO THE TEMPERATURE BALANCES OUT IN THE END BECAUSE THE THINGS HAVE ALL BEEN IN CONTACT WITH EACH OTHER, IN ONE WAY OR ANOTHER.
SO IT IS WITH THE UNIVERSE.

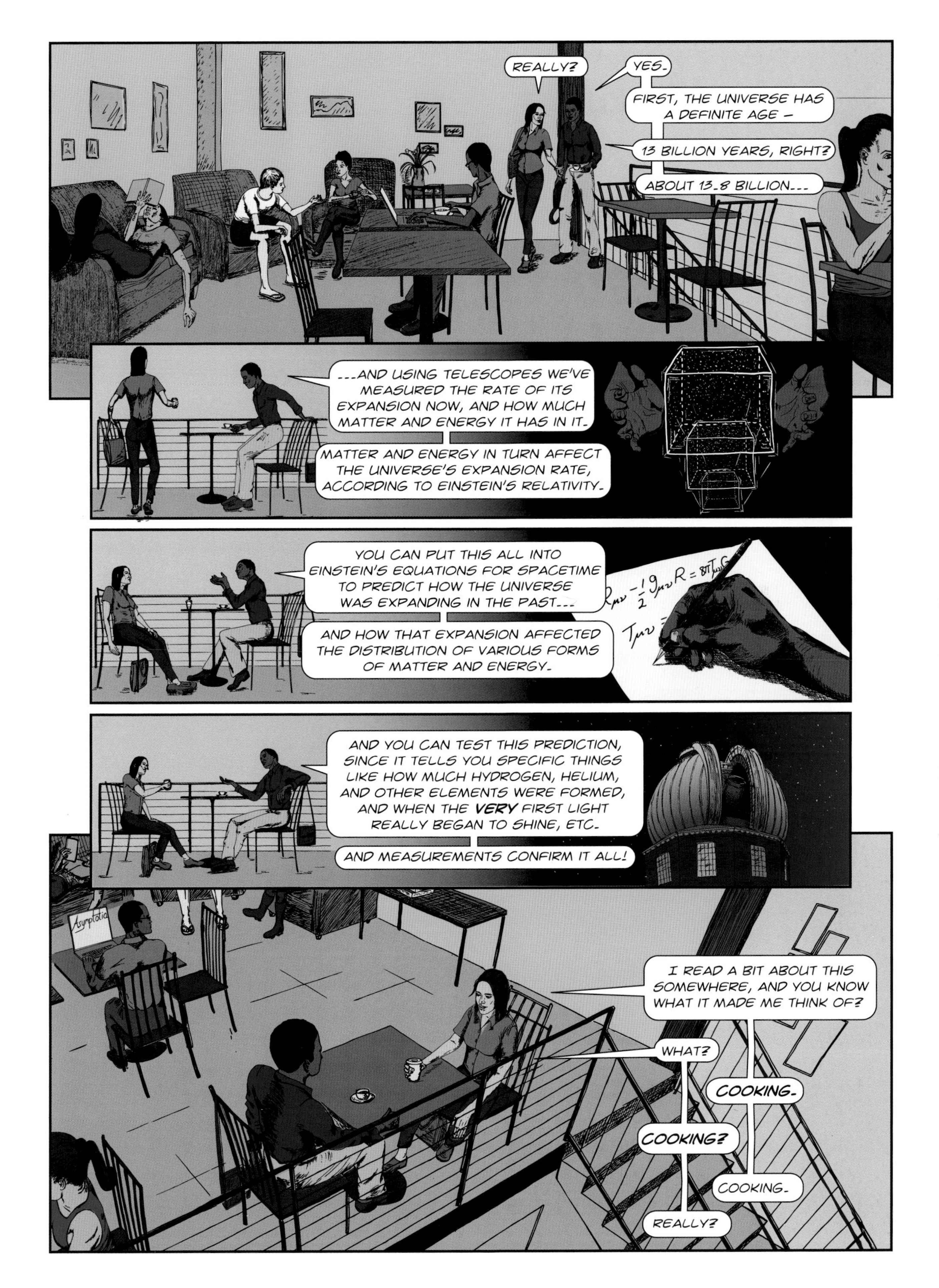
REALLY?
YES.
FIRST, THE UNIVERSE HAS A DEFINITE AGE –
13 BILLION YEARS, RIGHT?
ABOUT 13.8 BILLION...
...AND USING TELESCOPES WE'VE MEASURED THE RATE OF ITS EXPANSION NOW, AND HOW MUCH MATTER AND ENERGY IT HAS IN IT.
MATTER AND ENERGY IN TURN AFFECT THE UNIVERSE'S EXPANSION RATE, ACCORDING TO EINSTEIN'S RELATIVITY.
YOU CAN PUT THIS ALL INTO EINSTEIN'S EQUATIONS FOR SPACETIME TO PREDICT HOW THE UNIVERSE WAS EXPANDING IN THE PAST...
AND HOW THAT EXPANSION AFFECTED THE DISTRIBUTION OF VARIOUS FORMS OF MATTER AND ENERGY.
AND YOU CAN TEST THIS PREDICTION, SINCE IT TELLS YOU SPECIFIC THINGS LIKE HOW MUCH HYDROGEN, HELIUM, AND OTHER ELEMENTS WERE FORMED, AND WHEN THE **VERY** FIRST LIGHT REALLY BEGAN TO SHINE, ETC.
AND MEASUREMENTS CONFIRM IT ALL!
Asymptotia
I READ A BIT ABOUT THIS SOMEWHERE, AND YOU KNOW WHAT IT MADE ME THINK OF?
WHAT?
**COOKING.**
**COOKING?**
COOKING.
REALLY?

WELL, YES.
YOU'RE COMBINING VARIOUS INGREDIENTS TO MAKE – I DON'T KNOW – LET'S SAY A STEW.
YOU'VE GOT TO GET THEM IN AT THE RIGHT STAGES OF COOKING.
...AND SOMETIMES YOU HAVE TO HAVE THE TEMPERATURE UP HIGH, MAYBE AT THE BEGINNING TO SAUTÉ THE ONIONS AND GARLIC
WHILE LATER YOU MIGHT TURN IT DOWN, TO SIMMER.
OK, SO–
AND SOMETIMES YOU JUST **TASTE** THE STEW, OR EVEN JUST **SMELL** IT – THAT'S LIKE YOUR TELESCOPE MEASUREMENTS – TO TELL WHICH WAY THE INGREDIENTS HAVE BEEN COMBINED...
IS THERE TOO MUCH SALT?
SO WHAT IS THE ANALOGUE OF EINSTEIN'S EQUATIONS HERE? IS IT THE **RECIPE** FOR THE STEW?

NO, I DON'T THINK SO.
THE EQUATIONS SEEM TO ME MORE LIKE AN UNDERLYING LOGIC THAT UNDERPINS THE WHOLE THING. NOT JUST A BUNCH OF RULES, LIKE A RECIPE.
RECIPES COME FROM EXPERIENCE, CULTURE, TRADITION, AND SO ON, AS WELL AS THE SCIENCE OF HOW INGREDIENTS WORK TOGETHER.
ALL OF THAT IS IN THE CHEF'S TRAINING AND EXPERIENCE. IT'S ALL THAT STUFF THAT'S LIKE THE EQUATIONS.
REALLY? WELL TO MAKE THE ANALOGY COMPLETE, THE CHEF OUGHT TO BE ABLE TO TASTE THE STEW AND FIGURE OUT WHAT INGREDIENTS ARE IN IT, AND ALL THAT WAS DONE TO MAKE IT.
THAT SEEMS A BIT OF A TALL ORDER...
ARE YOU KIDDING!? THAT'S EXACTLY WHAT A GOOD CHEF CAN DO!
THEY'VE EVEN MADE TV SHOWS ABOUT IT!
YOU PUT A DISH IN FRONT OF THEM AND THEY CAN TASTE IT AND WRITE DOWN ALL THE INGREDIENTS.
THEY'D KNOW FROM THE FLAVORS AND TEXTURES HOW TO GET VARIOUS INGREDIENTS TO COMBINE TO GIVE THAT, WHAT COOKING TEMPERATURES AND SIMMERING TIMES ARE NEEDED, AND SO ON.
IT SEEMS TO ME THAT IT'S EXACTLY LIKE PHYSICISTS DOING ASTRONOMY MEASUREMENTS NOW AND, KNOWING THE RIGHT PHYSICS EQUATIONS, WORKING OUT WHAT THE EARLY UNIVERSE WAS LIKE.

WELL, OK.
BUT I DOUBT THEY CAN GET A PRECISE ANSWER DOWN TO THE LAST DETAIL.
WELL, CAN THE PHYSICISTS? IF YOU WERE PERFECT AT IT, YOU'D BE DONE BY NOW...
RIGHT?
WE DO PRETTY WELL.
REALLY WELL.
SO DO THE GOOD CHEFS.
ESPECIALLY IF THEY'RE TASTING A DISH FROM A COOKING TRADITION THEY KNOW WELL.
AHA!
IT'S LIKE IT BEING IMPORTANT FOR US TO WORK WITH THE RIGHT SETS OF EQUATIONS FOR THE RIGHT PROBLEMS.
YEAH, A MASTER FRENCH CHEF IS NOT GOING TO FIGURE EVERYTHING OUT ABOUT A POLYNESIAN DISH.
SO LOOKING FOR BETTER UNDERSTANDING OF ASPECTS OF OUR UNIVERSE MIGHT ONE DAY MEAN LOOKING AT DIFFERENT TRADITIONS FOR IDEAS...
TOTALLY DIFFERENT PHYSICS APPROACHES FROM OTHER FIELDS.

OK...I BUY IT.

SO...
WHERE WERE WE?
INGREDIENTS OF THE UNIVERSE.
OH RIGHT!
WELL THE POINT IS THAT IT DIDN'T ALL FIT NEATLY TOGETHER.
THE STORY OF WHAT COULD BE EXPLAINED WAS INCOMPLETE.
HOW SO?
WELL, THERE ARE PARTS OF THE UNIVERSE THAT ARE HUGELY SEPARATED FROM EACH OTHER, BUT THEY'RE AT MORE OR LESS THE SAME TEMPERATURE.
EVEN THOUGH THEY ARE SO FAR APART THAT THEY COULD NOT HAVE BEEN IN CONTACT WITH EACH OTHER TO BE IN EQUILIBRIUM.
THEY'D HAVE HAD TO APPARENTLY MOVE FASTER THAN THE SPEED OF LIGHT TO GET SO FAR APART DURING THE 13.8 BILLION YEARS OF THE LIFE OF THE UNIVERSE.
I DON'T THINK I UNDERSTAND.
LET ME PUT IT THIS WAY...
IN ORDER FOR THE UNIVERSE TO BE AS BIG AS IT IS TODAY, WITH ALL ITS PARTS AT ROUGHLY THE SAME TEMPERATURE, BUT BE ONLY 13.8 BILLION YEARS OLD...
THOSE PARTS MUST HAVE BEEN IN CONTACT, BUT THEN LATER SEPARATED FASTER THAN THE SPEED OF LIGHT, BUT THAT'S NOT IN THE RULES.
IT'S LIKE WE WERE SAYING ABOUT THE STUFF IN THIS ROOM ALL BEING AT THE SAME TEMPERATURE BECAUSE THEY'VE ALL BEEN IN CONTACT IN ONE WAY OR ANOTHER?
YES.

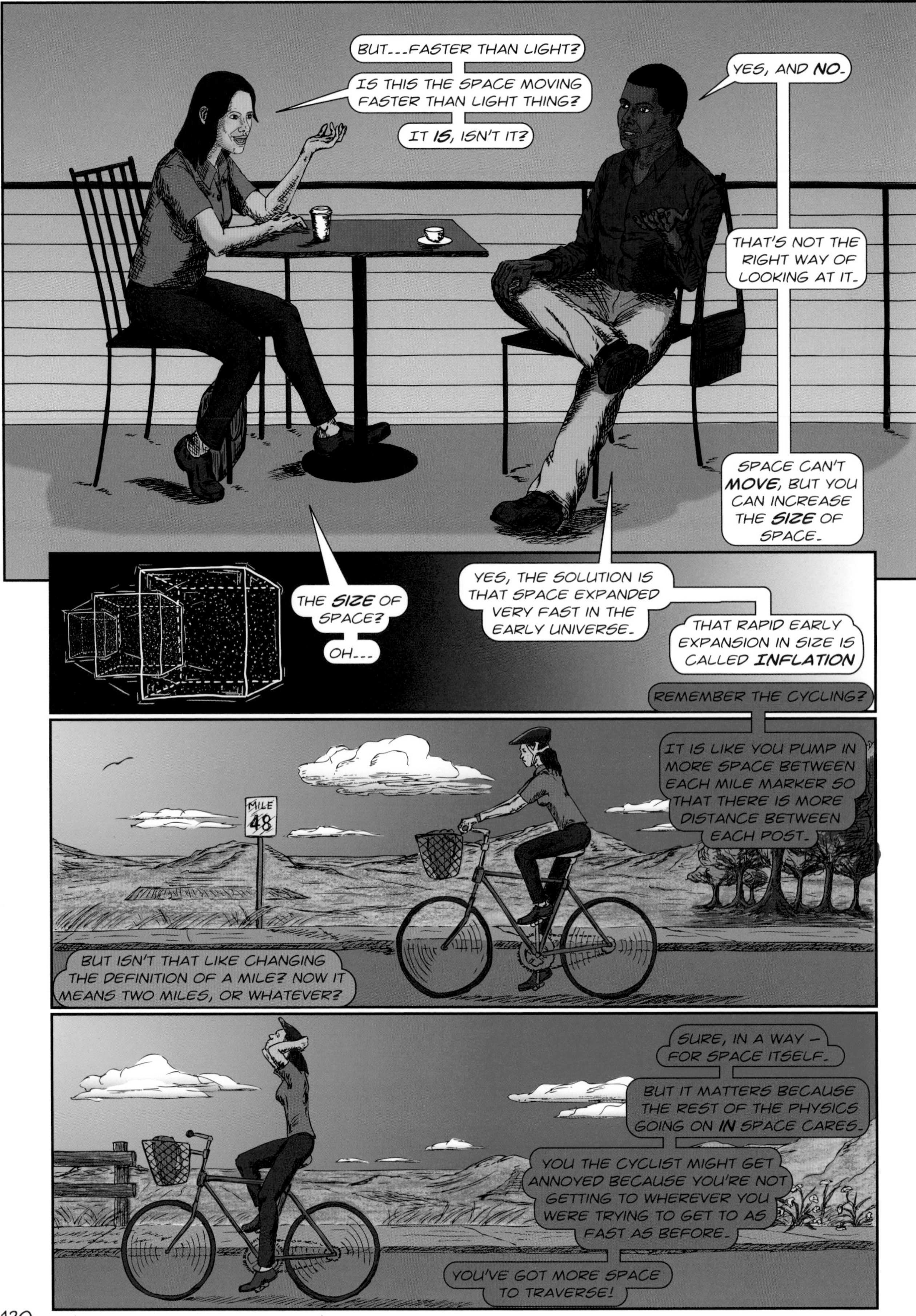
BUT...FASTER THAN LIGHT?
IS THIS THE SPACE MOVING FASTER THAN LIGHT THING?
IT IS, ISN'T IT?
YES, AND NO.
THAT'S NOT THE RIGHT WAY OF LOOKING AT IT.
SPACE CAN'T MOVE, BUT YOU CAN INCREASE THE SIZE OF SPACE.
THE SIZE OF SPACE?
OH...
YES, THE SOLUTION IS THAT SPACE EXPANDED VERY FAST IN THE EARLY UNIVERSE.
THAT RAPID EARLY EXPANSION IN SIZE IS CALLED INFLATION
REMEMBER THE CYCLING?
IT IS LIKE YOU PUMP IN MORE SPACE BETWEEN EACH MILE MARKER SO THAT THERE IS MORE DISTANCE BETWEEN EACH POST.
MILE
48
BUT ISN'T THAT LIKE CHANGING THE DEFINITION OF A MILE? NOW IT MEANS TWO MILES, OR WHATEVER?
SURE, IN A WAY — FOR SPACE ITSELF.
BUT IT MATTERS BECAUSE THE REST OF THE PHYSICS GOING ON IN SPACE CARES.
YOU THE CYCLIST MIGHT GET ANNOYED BECAUSE YOU'RE NOT GETTING TO WHEREVER YOU WERE TRYING TO GET TO AS FAST AS BEFORE.
YOU'VE GOT MORE SPACE TO TRAVERSE!

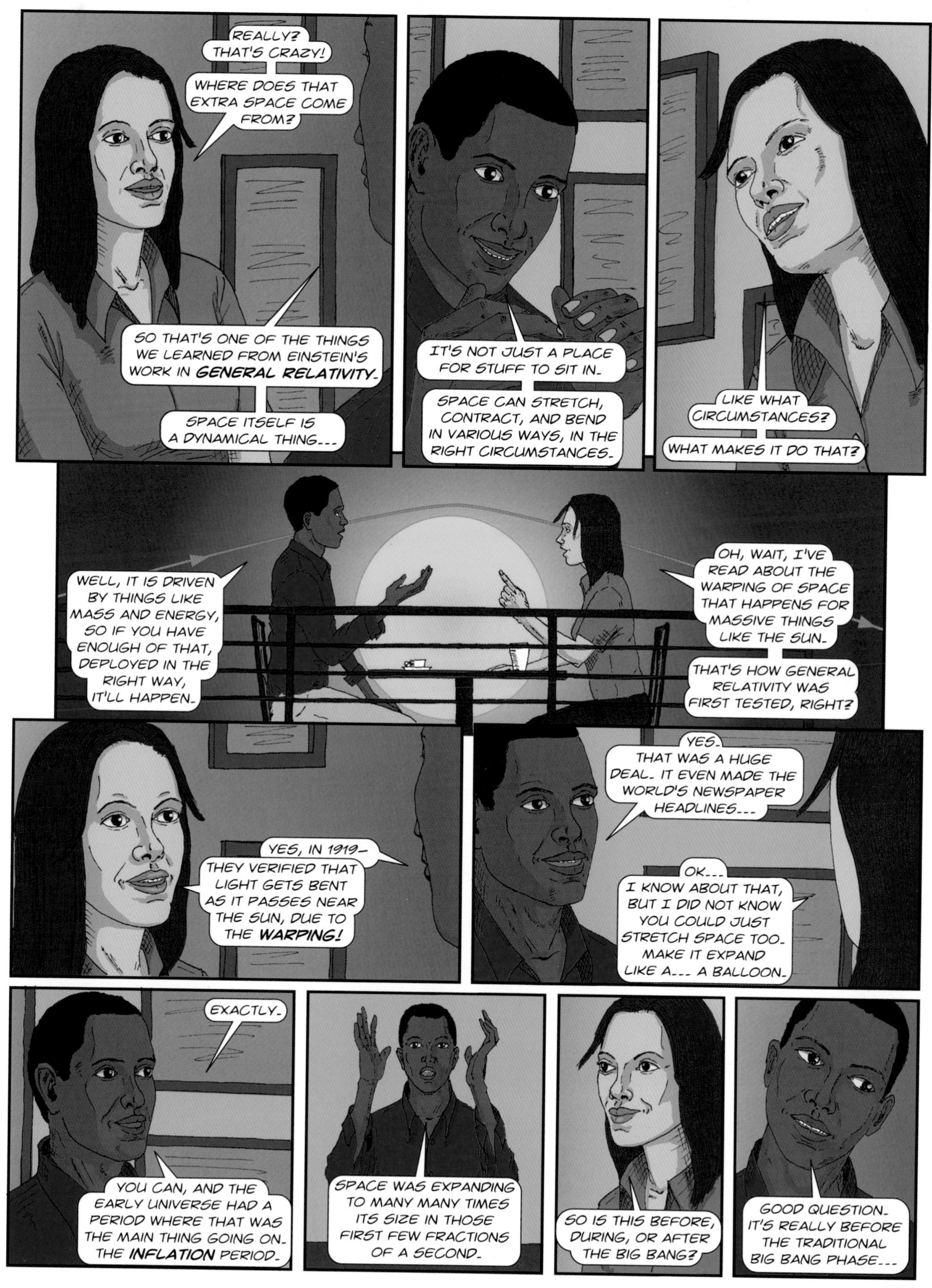
REALLY? THAT'S CRAZY!
WHERE DOES THAT EXTRA SPACE COME FROM?
SO THAT'S ONE OF THE THINGS WE LEARNED FROM EINSTEIN'S WORK IN GENERAL RELATIVITY.
SPACE ITSELF IS A DYNAMICAL THING...
IT'S NOT JUST A PLACE FOR STUFF TO SIT IN.
SPACE CAN STRETCH, CONTRACT, AND BEND IN VARIOUS WAYS, IN THE RIGHT CIRCUMSTANCES.
LIKE WHAT CIRCUMSTANCES?
WHAT MAKES IT DO THAT?
WELL, IT IS DRIVEN BY THINGS LIKE MASS AND ENERGY, SO IF YOU HAVE ENOUGH OF THAT, DEPLOYED IN THE RIGHT WAY, IT'LL HAPPEN.
OH, WAIT, I'VE READ ABOUT THE WARPING OF SPACE THAT HAPPENS FOR MASSIVE THINGS LIKE THE SUN.
THAT'S HOW GENERAL RELATIVITY WAS FIRST TESTED, RIGHT?
YES. THAT WAS A HUGE DEAL. IT EVEN MADE THE WORLD'S NEWSPAPER HEADLINES...
YES, IN 1919—
THEY VERIFIED THAT LIGHT GETS BENT AS IT PASSES NEAR THE SUN, DUE TO THE WARPING!
OK... I KNOW ABOUT THAT, BUT I DID NOT KNOW YOU COULD JUST STRETCH SPACE TOO. MAKE IT EXPAND LIKE A... A BALLOON.
EXACTLY.
YOU CAN, AND THE EARLY UNIVERSE HAD A PERIOD WHERE THAT WAS THE MAIN THING GOING ON. THE INFLATION PERIOD.
SPACE WAS EXPANDING TO MANY MANY TIMES ITS SIZE IN THOSE FIRST FEW FRACTIONS OF A SECOND.
SO IS THIS BEFORE, DURING, OR AFTER THE BIG BANG?
GOOD QUESTION. IT'S REALLY BEFORE THE TRADITIONAL BIG BANG PHASE...

BUT HOW ARE WE SO SURE THAT REALLY HAPPENED? IT IS SO LONG AGO.
WELL, ALL THAT STRETCHING HAS REAL CONSEQUENCES. EFFECTS WE CAN MEASURE. THAT EQUILIBRIUM PROBLEM WAS JUST THE FIRST CLUE...
...AND INFLATION SHOWS HOW ALL THOSE PARTS OF THE LARGE UNIVERSE WE SEE NOW WERE IN CONTACT EARLIER ON...
BUT IT AFFECTED MORE SUBTLE THINGS TOO.
YOU STRETCH SPACE THAT MUCH AND YOU'LL MAKE IT VERY FLAT –
SOFTENING ANY LUMPS IN IT AND THE STUFF IN IT IN VERY PARTICULAR WAYS THAT THE EQUATIONS PREDICT.
AMAZING INSTRUMENTS LIKE THE **WMAP** AND **PLANCK** OBSERVATORIES IN SPACE HAVE CONFIRMED THOSE AND OTHER FEATURES WITH **INCREDIBLE** ACCURACY.
I SEE.
IT'S THAT TASTING OF THE STEW NOW AND FIGURING OUT HOW IT WAS COOKED.
HA! YES...
YOU'RE TAKING A LITTLE BIT OF THE SAUCE, YOU'RE LOOKING AT ITS COLOR, INHALING ITS AROMA, AND PUTTING IT ON YOUR TONGUE...
YOU CAN PICK UP HINTS – STRONG HINTS – OF THE BEGINNING OF THE WHOLE STEW...
THAT HOT SAUTÉEING OF THE ONIONS IN BUTTER, THAT SPLASH OF WINE, SPRINKLING OF SALT...
YEAH. SOUNDS TASTY...!
AND MAYBE A TOTALLY DIFFERENT TASTE WOULD HAVE MEANT A TOTALLY DIFFERENT START TO THE STEW...
EXACTLY...
IN THIS CASE, FOR OUR UNIVERSE-STEW, THAT INITIAL COOKING PERIOD HAD A RAPID INFLATION OF SPACE –
PLANCK AND WMAP CAN TASTE IT.
RIGHT. I SEE.

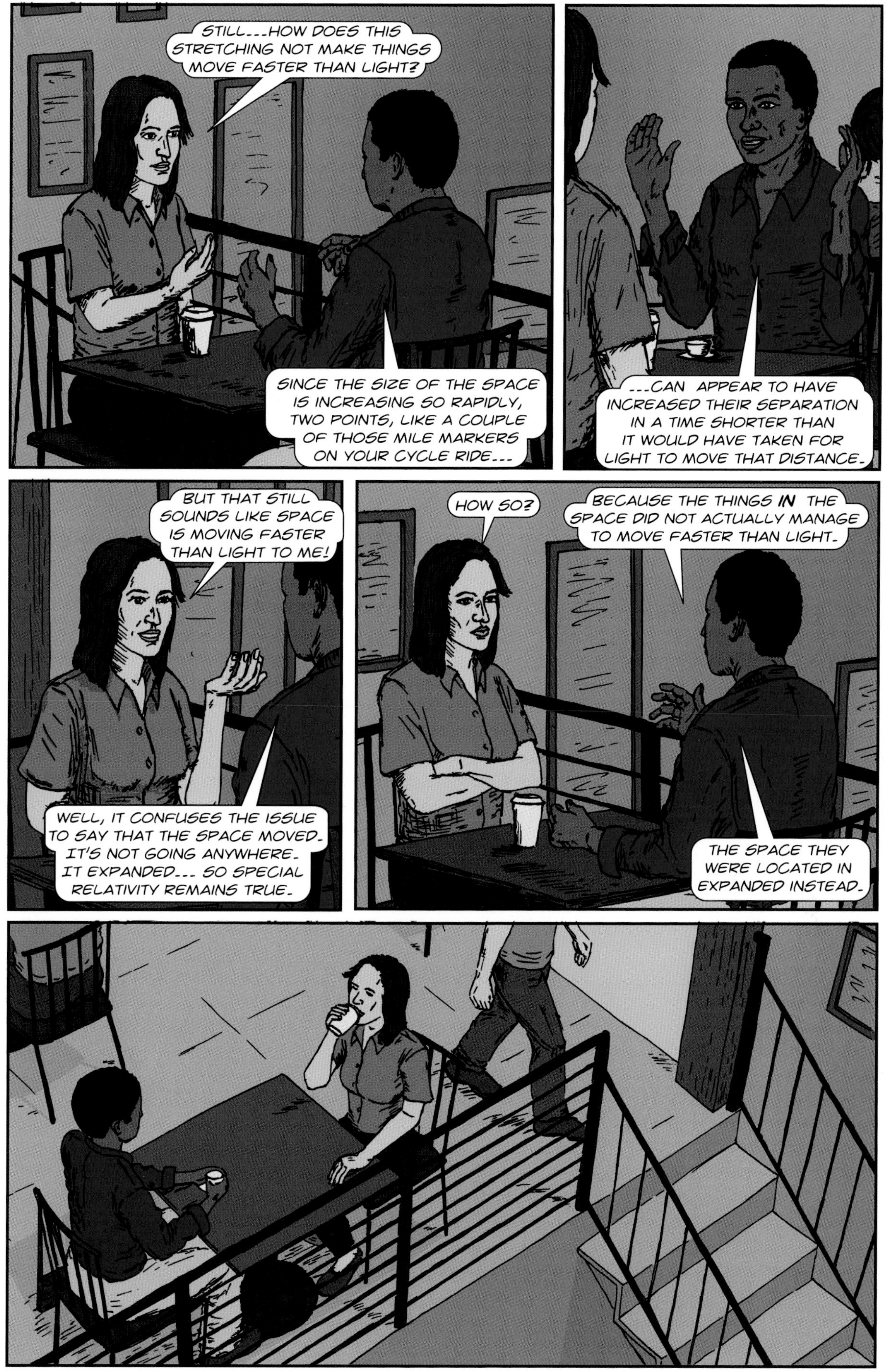
STILL... HOW DOES THIS STRETCHING NOT MAKE THINGS MOVE FASTER THAN LIGHT?
SINCE THE SIZE OF THE SPACE IS INCREASING SO RAPIDLY, TWO POINTS, LIKE A COUPLE OF THOSE MILE MARKERS ON YOUR CYCLE RIDE...
...CAN APPEAR TO HAVE INCREASED THEIR SEPARATION IN A TIME SHORTER THAN IT WOULD HAVE TAKEN FOR LIGHT TO MOVE THAT DISTANCE.
BUT THAT STILL SOUNDS LIKE SPACE IS MOVING FASTER THAN LIGHT TO ME!
WELL, IT CONFUSES THE ISSUE TO SAY THAT THE SPACE MOVED. IT'S NOT GOING ANYWHERE. IT EXPANDED... SO SPECIAL RELATIVITY REMAINS TRUE.
HOW SO?
BECAUSE THE THINGS **IN** THE SPACE DID NOT ACTUALLY MANAGE TO MOVE FASTER THAN LIGHT.
THE SPACE THEY WERE LOCATED IN EXPANDED INSTEAD.

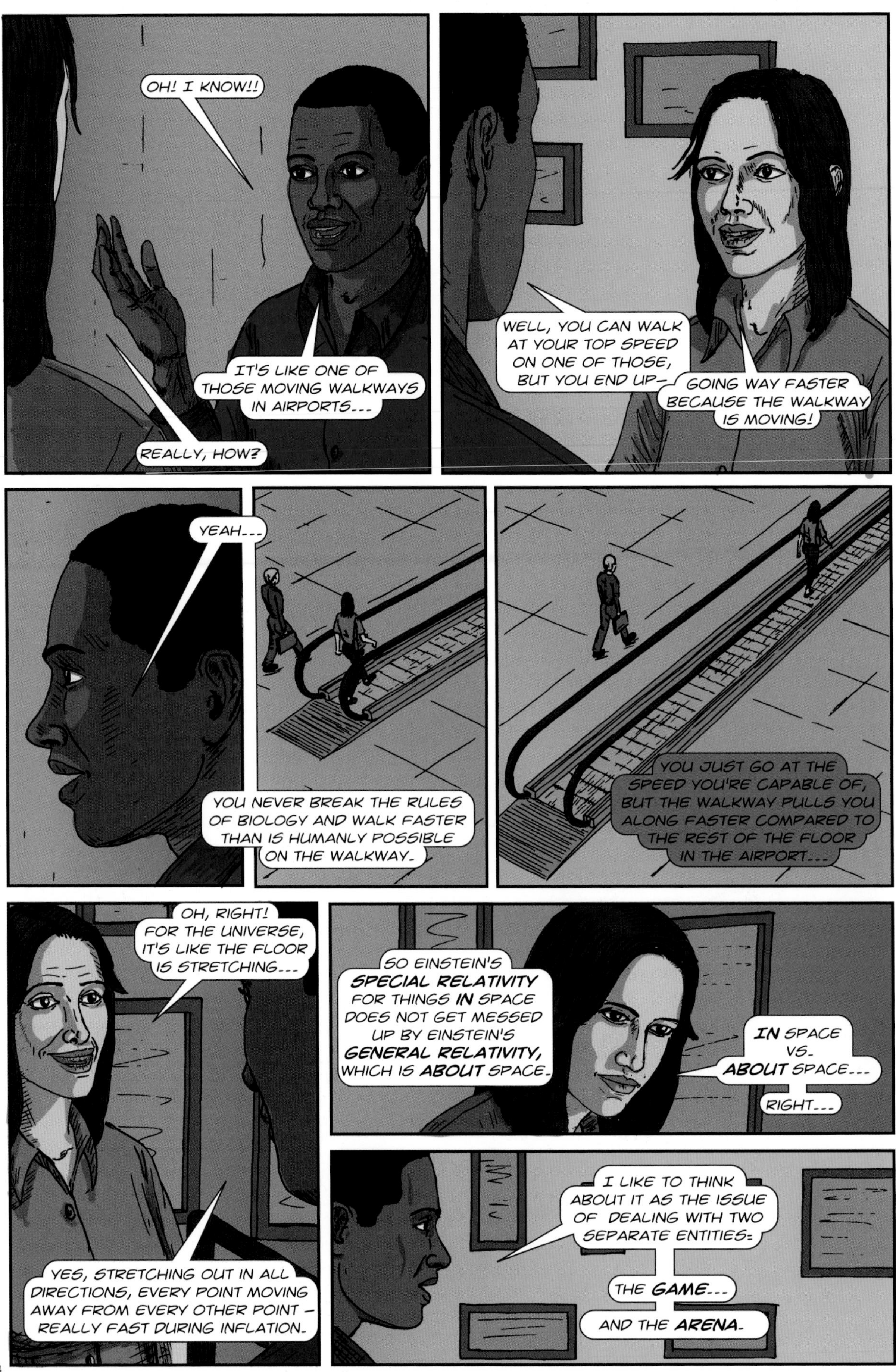
OH! I KNOW!!
IT'S LIKE ONE OF THOSE MOVING WALKWAYS IN AIRPORTS...
REALLY, HOW?
WELL, YOU CAN WALK AT YOUR TOP SPEED ON ONE OF THOSE, BUT YOU END UP—
GOING WAY FASTER BECAUSE THE WALKWAY IS MOVING!
YEAH...
YOU NEVER BREAK THE RULES OF BIOLOGY AND WALK FASTER THAN IS HUMANLY POSSIBLE ON THE WALKWAY.
YOU JUST GO AT THE SPEED YOU'RE CAPABLE OF, BUT THE WALKWAY PULLS YOU ALONG FASTER COMPARED TO THE REST OF THE FLOOR IN THE AIRPORT...
OH, RIGHT! FOR THE UNIVERSE, IT'S LIKE THE FLOOR IS STRETCHING...
YES, STRETCHING OUT IN ALL DIRECTIONS, EVERY POINT MOVING AWAY FROM EVERY OTHER POINT — REALLY FAST DURING INFLATION.
SO EINSTEIN'S **SPECIAL RELATIVITY** FOR THINGS **IN** SPACE DOES NOT GET MESSED UP BY EINSTEIN'S **GENERAL RELATIVITY,** WHICH IS **ABOUT** SPACE.
**IN** SPACE VS. **ABOUT** SPACE...
RIGHT...
I LIKE TO THINK ABOUT IT AS THE ISSUE OF DEALING WITH TWO SEPARATE ENTITIES:
THE **GAME**...
AND THE **ARENA**.

THE GAME HAS CERTAIN RULES YOU DON'T BREAK.
THE GAME IS GOVERNED BY THE LAWS OF PHYSICS ABOUT OBJECTS – THE STUFF WE HAVE AROUND US. LIKE THIS CUP, OR YOUR PHONE...
ALL THE ELEMENTARY PARTICLES THAT MAKE UP ALL THESE THINGS – AND OURSELVES TOO...
AND THE DIFFERENT KINDS OF FORCES THAT HOLD THEM TOGETHER...
ONE OF THE RULES OF THE GAME THAT WE'VE LEARNED IS THAT THE SPEED OF LIGHT IS THE BEST YOU CAN DO.
THE ARENA IS SPACE AND TIME – OR SPACETIME – WHERE EVERYTHING SITS.
WHERE THE GAME TAKES PLACE.
SPACE TELLS US WHERE THINGS ARE, WHILE TIME HELPS US KEEP TRACK OF HOW THOSE POSITIONS CHANGE, AND SO ON...
SURE, BUT WHAT'S THE MAIN POINT HERE?
GETTING THERE...
A LOT OF SCIENCE DEALS WITH THE GAME, FIGURING OUT THE RULES AND THE PLAYERS. LIKE CHEMISTRY, NUCLEAR PHYSICS, BIOLOGY... FOUNDED ON THINGS LIKE QUANTUM MECHANICS, GENETICS, THERMODYNAMICS, AND SO ON.
BUT AHA! SOMETIMES THE ISSUES YOU ARE DEALING WITH ARE ABOUT THE ARENA ITSELF. SINCE EINSTEIN, WE'VE BEEN APPRECIATING MORE THAT THE ARENA IS NOT A FIXED THING. IT CAN CHANGE, AND THERE ARE LAWS OF PHYSICS FOR THAT TOO. ACTUALLY, THAT IS WHAT GENERAL RELATIVITY IS ALL ABOUT.

BUT WAIT, ISN'T THAT ALREADY TRUE IN SPECIAL RELATIVITY?
WHAT'S ALREADY TRUE?
THAT SPACE AND TIME CAN CHANGE. YOU DON'T NEED GENERAL RELATIVITY TO SEE THAT, RIGHT?
ER...
IF I MOVE AT SOME SPEED RELATIVE TO YOU, I MEASURE TIME DIFFERENTLY FROM YOU. I ALSO MEASURE SPACE DIFFERENTLY FROM YOU...
...SO MY MOVING CHANGES BOTH SPACE AND TIME. SPECIAL RELATIVITY.
WE TALKED ABOUT THIS AT THE PARTY!
OH, YES. I SEE NOW.
BUT THAT'S DIFFERENT.
HOW?
FOR ME...
SPECIAL RELATIVITY JUST SAYS THAT DIFFERENT OBSERVERS SLICE SPACETIME DIFFERENTLY. A SLICE DETERMINES WHAT'S SPACE AND WHAT'S TIME...
...BUT SPACETIME ITSELF IS KIND OF A GIVEN THERE.
OH, OK.
GENERAL RELATIVITY SHOWS HOW SPACETIME ITSELF CHANGES.
IT'S THE BIRTH OF MODERN COSMOLOGY.
THE WHOLE BUSINESS OF THE UNIVERSE CHANGING?
YES.
AND NOT JUST THE STUFF IN IT CHANGING, BUT ALSO SPACE AND TIME — OR SPACETIME.
I SEE.
I LIKE THAT.
WELL, SPEAKING OF TIME, I'D BETTER HEAD FOR MY TRAIN.

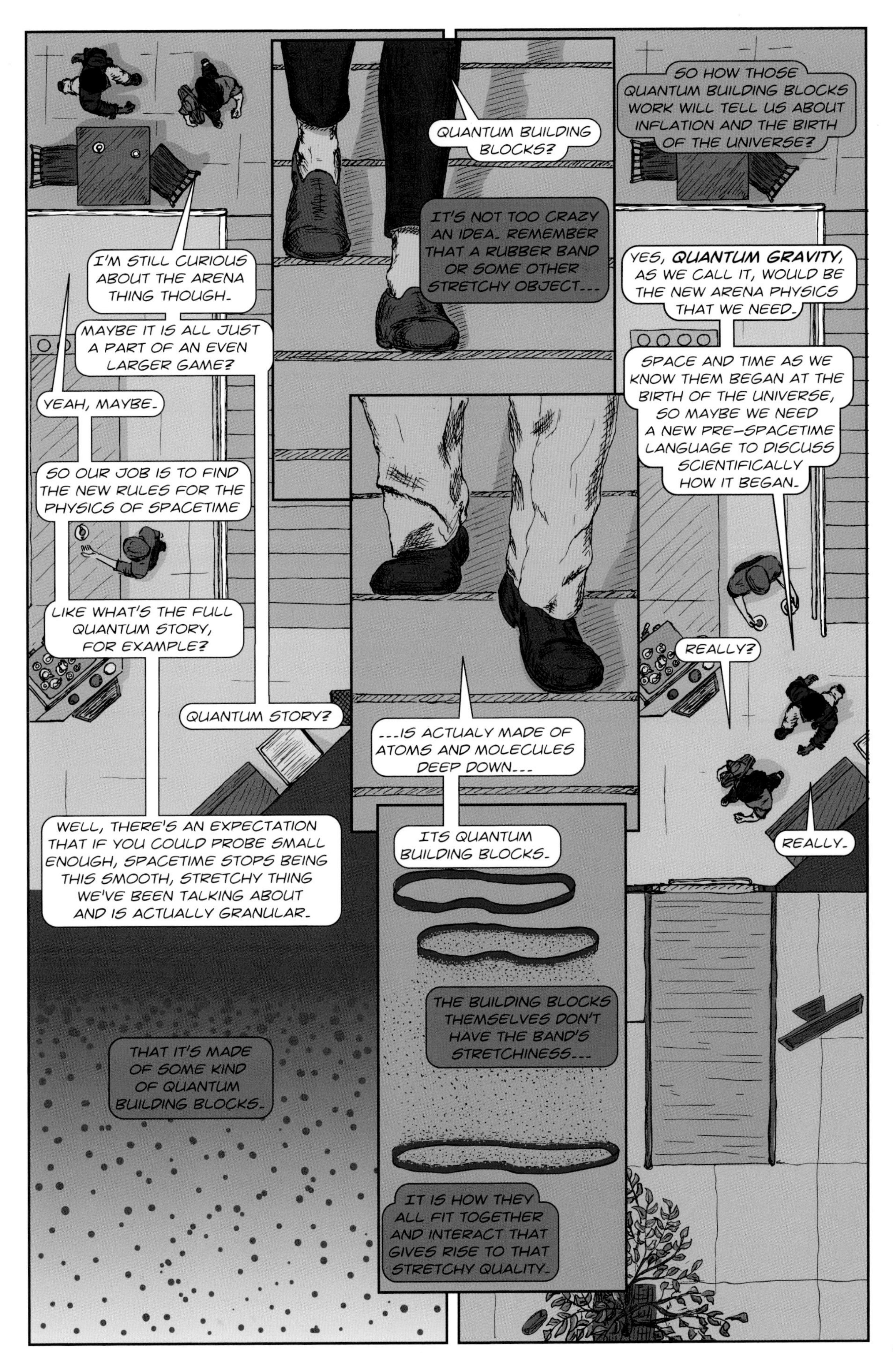
I'M STILL CURIOUS ABOUT THE ARENA THING THOUGH.
MAYBE IT IS ALL JUST A PART OF AN EVEN LARGER GAME?
YEAH, MAYBE.
SO OUR JOB IS TO FIND THE NEW RULES FOR THE PHYSICS OF SPACETIME
LIKE WHAT'S THE FULL QUANTUM STORY, FOR EXAMPLE?
QUANTUM STORY?
WELL, THERE'S AN EXPECTATION THAT IF YOU COULD PROBE SMALL ENOUGH, SPACETIME STOPS BEING THIS SMOOTH, STRETCHY THING WE'VE BEEN TALKING ABOUT AND IS ACTUALLY GRANULAR.
THAT IT'S MADE OF SOME KIND OF QUANTUM BUILDING BLOCKS.
QUANTUM BUILDING BLOCKS?
IT'S NOT TOO CRAZY AN IDEA. REMEMBER THAT A RUBBER BAND OR SOME OTHER STRETCHY OBJECT...
...IS ACTUALY MADE OF ATOMS AND MOLECULES DEEP DOWN...
ITS QUANTUM BUILDING BLOCKS.
THE BUILDING BLOCKS THEMSELVES DON'T HAVE THE BAND'S STRETCHINESS...
IT IS HOW THEY ALL FIT TOGETHER AND INTERACT THAT GIVES RISE TO THAT STRETCHY QUALITY.
SO HOW THOSE QUANTUM BUILDING BLOCKS WORK WILL TELL US ABOUT INFLATION AND THE BIRTH OF THE UNIVERSE?
YES, **QUANTUM GRAVITY**, AS WE CALL IT, WOULD BE THE NEW ARENA PHYSICS THAT WE NEED.
SPACE AND TIME AS WE KNOW THEM BEGAN AT THE BIRTH OF THE UNIVERSE, SO MAYBE WE NEED A NEW PRE-SPACETIME LANGUAGE TO DISCUSS SCIENTIFICALLY HOW IT BEGAN.
REALLY?
REALLY.

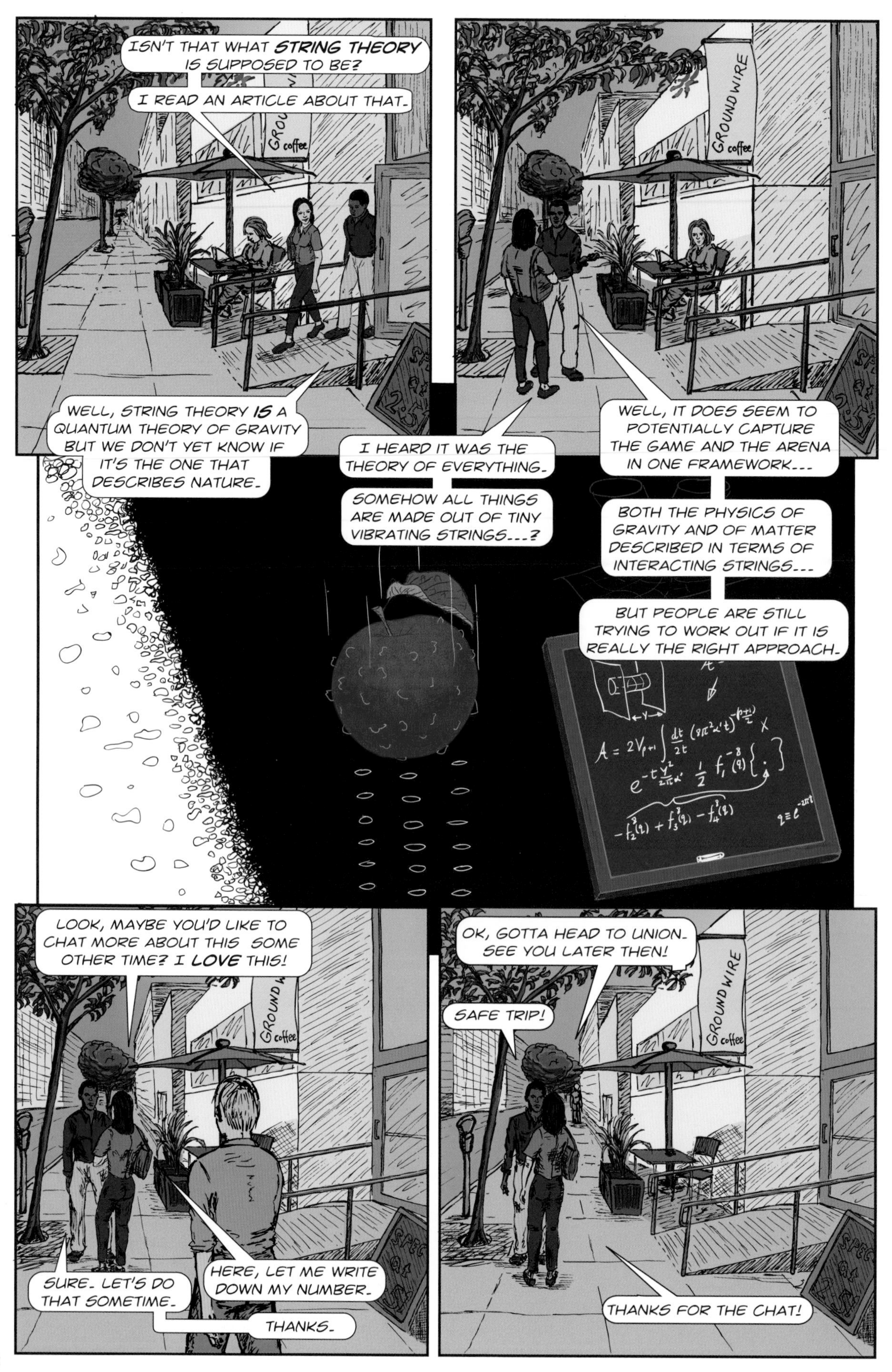
ISN'T THAT WHAT STRING THEORY IS SUPPOSED TO BE?
I READ AN ARTICLE ABOUT THAT.
WELL, STRING THEORY IS A QUANTUM THEORY OF GRAVITY BUT WE DON'T YET KNOW IF IT'S THE ONE THAT DESCRIBES NATURE.
I HEARD IT WAS THE THEORY OF EVERYTHING.
SOMEHOW ALL THINGS ARE MADE OUT OF TINY VIBRATING STRINGS...?
WELL, IT DOES SEEM TO POTENTIALLY CAPTURE THE GAME AND THE ARENA IN ONE FRAMEWORK...
BOTH THE PHYSICS OF GRAVITY AND OF MATTER DESCRIBED IN TERMS OF INTERACTING STRINGS...
BUT PEOPLE ARE STILL TRYING TO WORK OUT IF IT IS REALLY THE RIGHT APPROACH.
LOOK, MAYBE YOU'D LIKE TO CHAT MORE ABOUT THIS SOME OTHER TIME? I LOVE THIS!
SURE. LET'S DO THAT SOMETIME.
HERE, LET ME WRITE DOWN MY NUMBER.
THANKS.
OK, GOTTA HEAD TO UNION. SEE YOU LATER THEN!
SAFE TRIP!
THANKS FOR THE CHAT!

GROUNDWIRE
coffee

YOU KNOW, HERE'S THE THING.
...
HMMM...?
WE'VE LEARNED A LOT ABOUT THE PHYSICS OF TIME, EVEN IF WE DO NOT KNOW WHAT IT IS.
LIKE WHAT?
WELL, TIME CAN APPEAR DIFFERENT TO DIFFERENT OBSERVERS...
THAT'S ONE OF THE THINGS WE LEARNED FROM EINSTEIN'S RELATIVITY.
DIFFERENT OBSERVERS?
AND WHAT DO YOU MEAN BY "APPEAR DIFFERENT"?
TIME RUNNING FASTER OR SLOWER.
OH.
YOU SEE, IT'S TO DO WITH BEING IN DIFFERENT STATES OF MOTION...
REALLY?
OH!
WHAT?
I KNOW HOW TO GET WHAT YOU WANT!
IT'S REALLY ABOUT TIME **AND** SPACE.
WHETHER IT'S RELATIVE MOTION, OR GRAVITY – THEY'RE BOTH ABOUT DIFFERENT WAYS OF EXPERIENCING SPACE...
OR DIFFERENT GRAVITATIONAL FIELDS.
I... SEE...
SO I CAN CREATE YOUR IMMORTAL BEING FOR YOU...
HUH?
OR CLOSE ENOUGH TO IMMORTAL!
WE JUST TAKE A LIVING BEING AND PUT IT INTO THE RIGHT SITUATION SO THAT OTHER OBSERVERS SEE IT AS LIVING FOREVER.
OR AS CLOSE TO FOREVER AS YOU LIKE.
WHAT KIND OF SITUATION IS THAT?

A BLACK HOLE WOULD BE HELPFUL.
REMEMBER I MENTIONED STARS THAT COLLAPSE AND EXPLODE?
PLANTING THE SEEDS FOR NEW LIFE TO BEGIN?
HEAVY ELEMENTS AND ALL THAT?
YES
WELL, ANOTHER OUTPUT OF THAT PROCESS CAN OFTEN BE A BLACK HOLE.
WHAT'S LEFT OF THE STAR THAT DOES NOT GET THROWN OFF, COLLAPSES FURTHER UNDER GRAVITY, AND MAKES THIS NEW, REMARKABLE OBJECT.
SURE, I'VE HEARD OF BLACK HOLES.
THE GRAVITY GETS SO STRONG THAT EVEN LIGHT CAN'T ESCAPE FROM IT.
YES. THAT'S THE CLASSICAL STORY.
THERE'S A HORIZON...
IT'S A POINT OF CLOSEST APPROACH...
PAST WHICH YOU CANNOT RETURN.
IT'S JUST WHAT WE NEED TO MAKE YOUR NEAR-ETERNAL BEING...
THE GOD OF THIS STAR SYSTEM I'M COOKING UP FOR YOU...
YOU'RE GOING TO KEEP TEASING ME ABOUT GOD AREN'T YOU?
HA!
I CAN USE A DIFFERENT TERM IF YOU PREFER...
BUT YOU'LL SEE THAT THIS SITUATION WILL MAKE A GODLIKE BEING OF THE KIND YOU SEEM TO WANT.
WELL, I THINK THAT GOD IS MORE THAN JUST...
WELL, GO ON...
OK! HERE GOES!
SO IMAGINE ANOTHER CIVILIZATION...

YOU'VE GOT THIS CIVILIZATION ON A PLANET IN ORBIT AROUND A POWERFUL SOURCE OF GRAVITY, LIKE A HUGE BLACK HOLE...
HOW DID THAT COME ABOUT?
AND WHERE DO THEY GET THEIR LIGHT FROM?
WELL, THE DETAILS AREN'T SUPER-IMPORTANT, BUT...
OK, LET'S SAY THEY ORBIT A REGULAR STAR...
BUT THAT STAR IS ITSELF IN ORBIT AROUND THE BLACK HOLE.
OK...

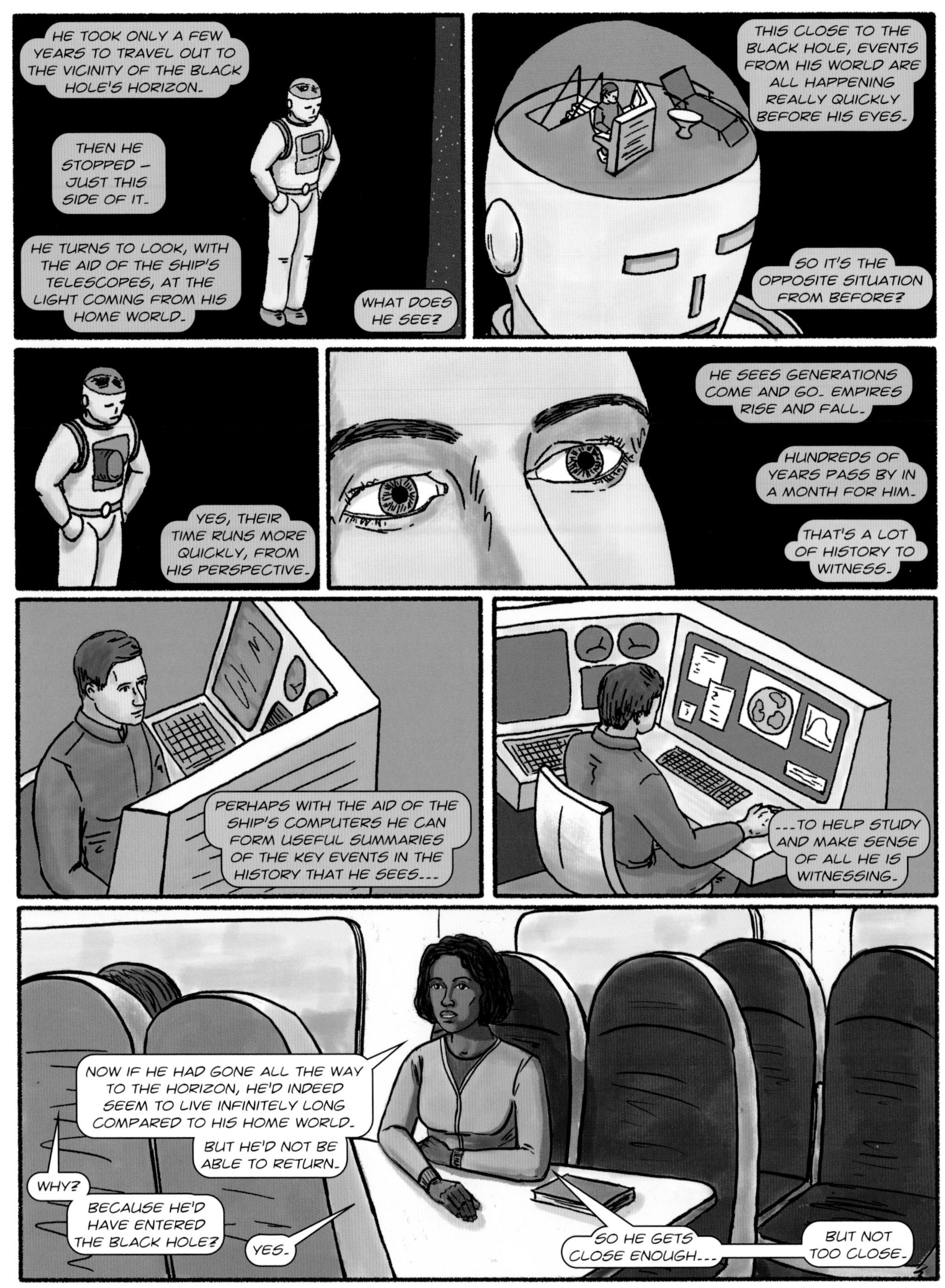
HE TOOK ONLY A FEW YEARS TO TRAVEL OUT TO THE VICINITY OF THE BLACK HOLE'S HORIZON.
THEN HE STOPPED – JUST THIS SIDE OF IT.
HE TURNS TO LOOK, WITH THE AID OF THE SHIP'S TELESCOPES, AT THE LIGHT COMING FROM HIS HOME WORLD.
WHAT DOES HE SEE?
THIS CLOSE TO THE BLACK HOLE, EVENTS FROM HIS WORLD ARE ALL HAPPENING REALLY QUICKLY BEFORE HIS EYES.
SO IT'S THE OPPOSITE SITUATION FROM BEFORE?
YES, THEIR TIME RUNS MORE QUICKLY, FROM HIS PERSPECTIVE.
HE SEES GENERATIONS COME AND GO. EMPIRES RISE AND FALL.
HUNDREDS OF YEARS PASS BY IN A MONTH FOR HIM.
THAT'S A LOT OF HISTORY TO WITNESS.
PERHAPS WITH THE AID OF THE SHIP'S COMPUTERS HE CAN FORM USEFUL SUMMARIES OF THE KEY EVENTS IN THE HISTORY THAT HE SEES...
...TO HELP STUDY AND MAKE SENSE OF ALL HE IS WITNESSING.
NOW IF HE HAD GONE ALL THE WAY TO THE HORIZON, HE'D INDEED SEEM TO LIVE INFINITELY LONG COMPARED TO HIS HOME WORLD.
BUT HE'D NOT BE ABLE TO RETURN.
WHY?
BECAUSE HE'D HAVE ENTERED THE BLACK HOLE?
YES.
SO HE GETS CLOSE ENOUGH...
BUT NOT TOO CLOSE.

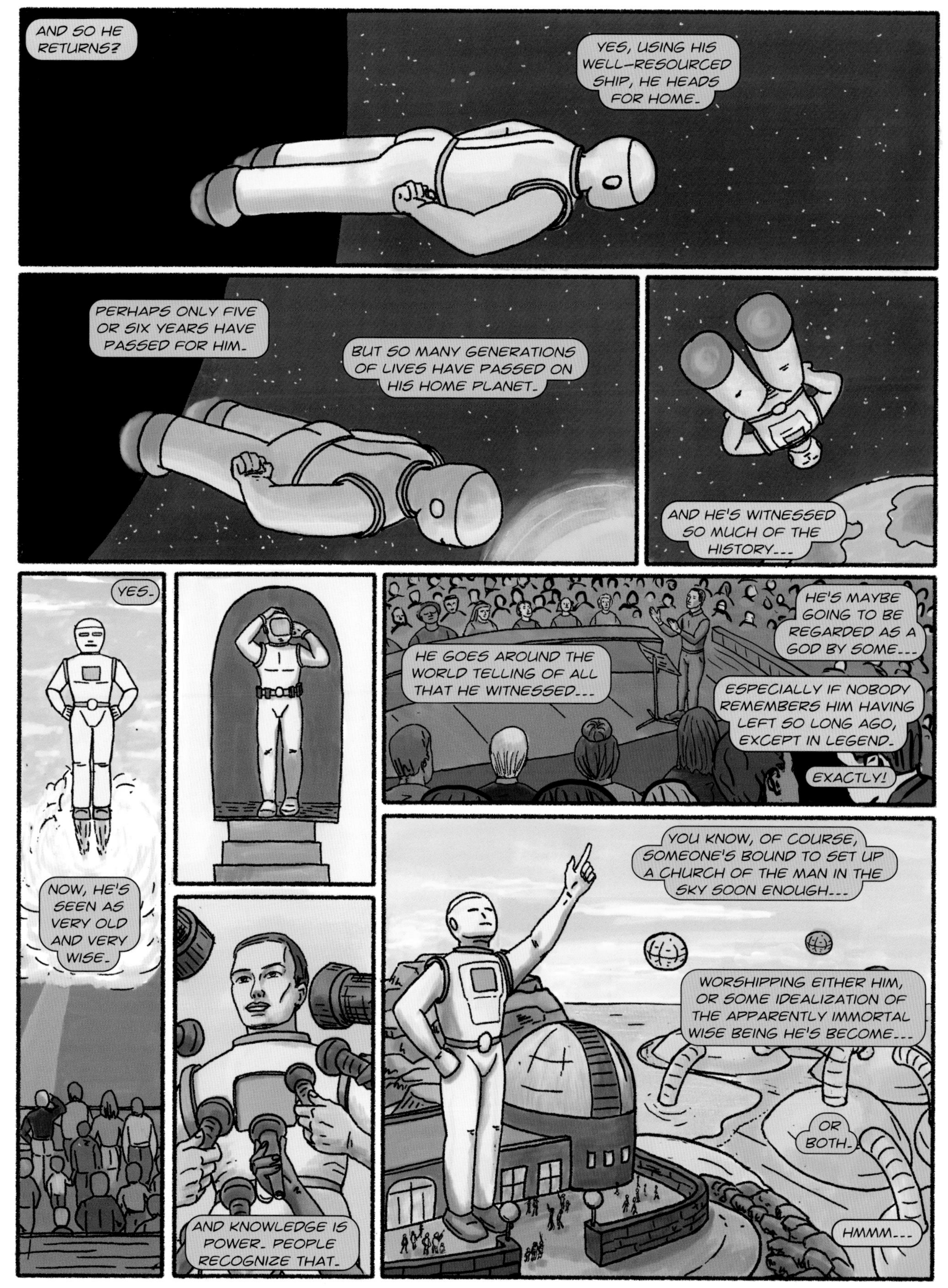
AND SO HE RETURNS?
YES, USING HIS WELL-RESOURCED SHIP, HE HEADS FOR HOME.
PERHAPS ONLY FIVE OR SIX YEARS HAVE PASSED FOR HIM.
BUT SO MANY GENERATIONS OF LIVES HAVE PASSED ON HIS HOME PLANET.
AND HE'S WITNESSED SO MUCH OF THE HISTORY...
YES.
NOW, HE'S SEEN AS VERY OLD AND VERY WISE.
HE GOES AROUND THE WORLD TELLING OF ALL THAT HE WITNESSED...
HE'S MAYBE GOING TO BE REGARDED AS A GOD BY SOME...
ESPECIALLY IF NOBODY REMEMBERS HIM HAVING LEFT SO LONG AGO, EXCEPT IN LEGEND.
EXACTLY!
AND KNOWLEDGE IS POWER. PEOPLE RECOGNIZE THAT.
YOU KNOW, OF COURSE, SOMEONE'S BOUND TO SET UP A CHURCH OF THE MAN IN THE SKY SOON ENOUGH...
WORSHIPPING EITHER HIM, OR SOME IDEALIZATION OF THE APPARENTLY IMMORTAL WISE BEING HE'S BECOME...
OR BOTH.
HMMM...

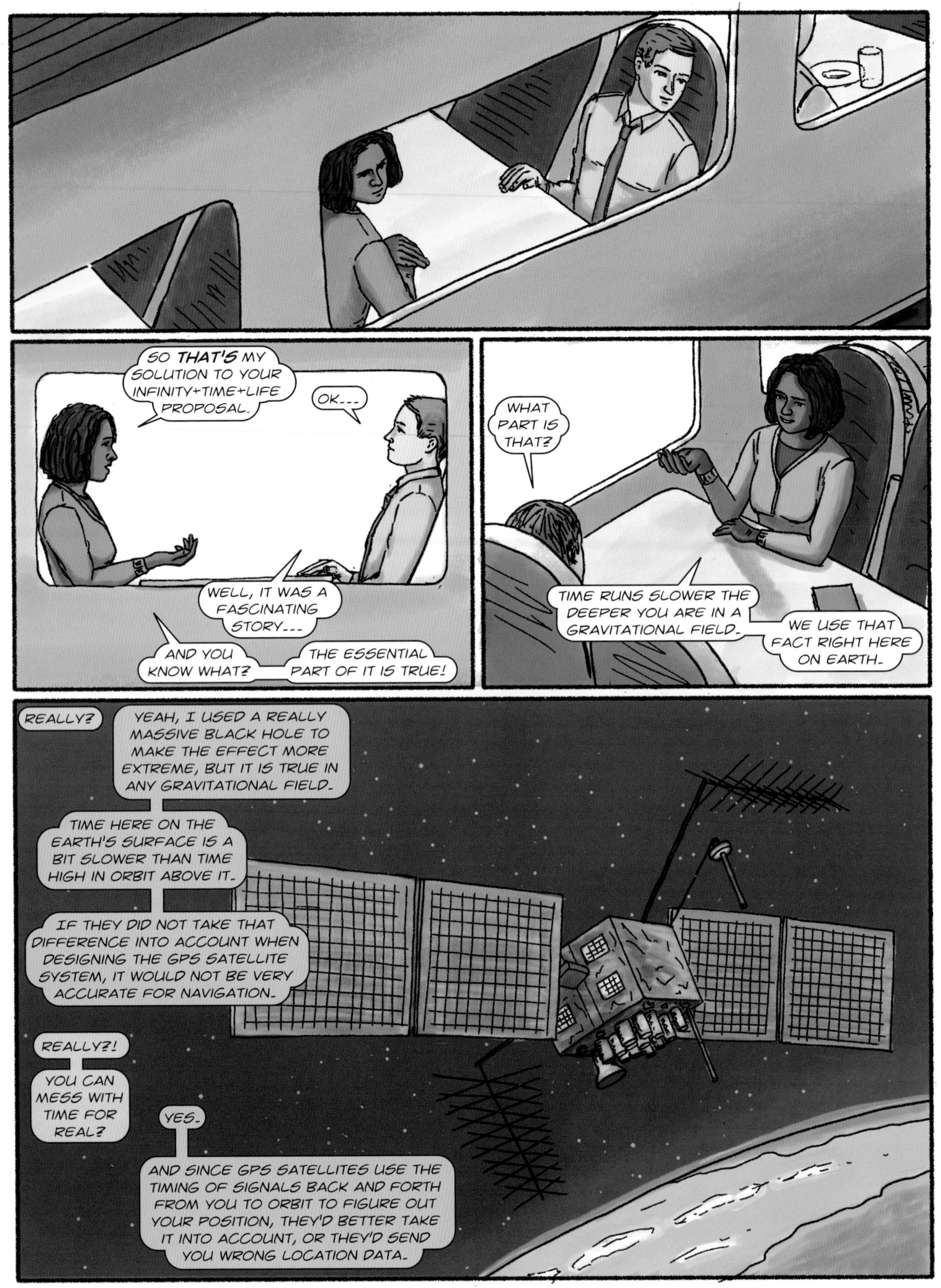
SO THAT'S MY SOLUTION TO YOUR INFINITY+TIME+LIFE PROPOSAL.
OK...
WELL, IT WAS A FASCINATING STORY...
AND YOU KNOW WHAT?
THE ESSENTIAL PART OF IT IS TRUE!
WHAT PART IS THAT?
TIME RUNS SLOWER THE DEEPER YOU ARE IN A GRAVITATIONAL FIELD.
WE USE THAT FACT RIGHT HERE ON EARTH.
REALLY?
YEAH, I USED A REALLY MASSIVE BLACK HOLE TO MAKE THE EFFECT MORE EXTREME, BUT IT IS TRUE IN ANY GRAVITATIONAL FIELD.
TIME HERE ON THE EARTH'S SURFACE IS A BIT SLOWER THAN TIME HIGH IN ORBIT ABOVE IT.
IF THEY DID NOT TAKE THAT DIFFERENCE INTO ACCOUNT WHEN DESIGNING THE GPS SATELLITE SYSTEM, IT WOULD NOT BE VERY ACCURATE FOR NAVIGATION.
REALLY?!
YOU CAN MESS WITH TIME FOR REAL?
YES.
AND SINCE GPS SATELLITES USE THE TIMING OF SIGNALS BACK AND FORTH FROM YOU TO ORBIT TO FIGURE OUT YOUR POSITION, THEY'D BETTER TAKE IT INTO ACCOUNT, OR THEY'D SEND YOU WRONG LOCATION DATA.

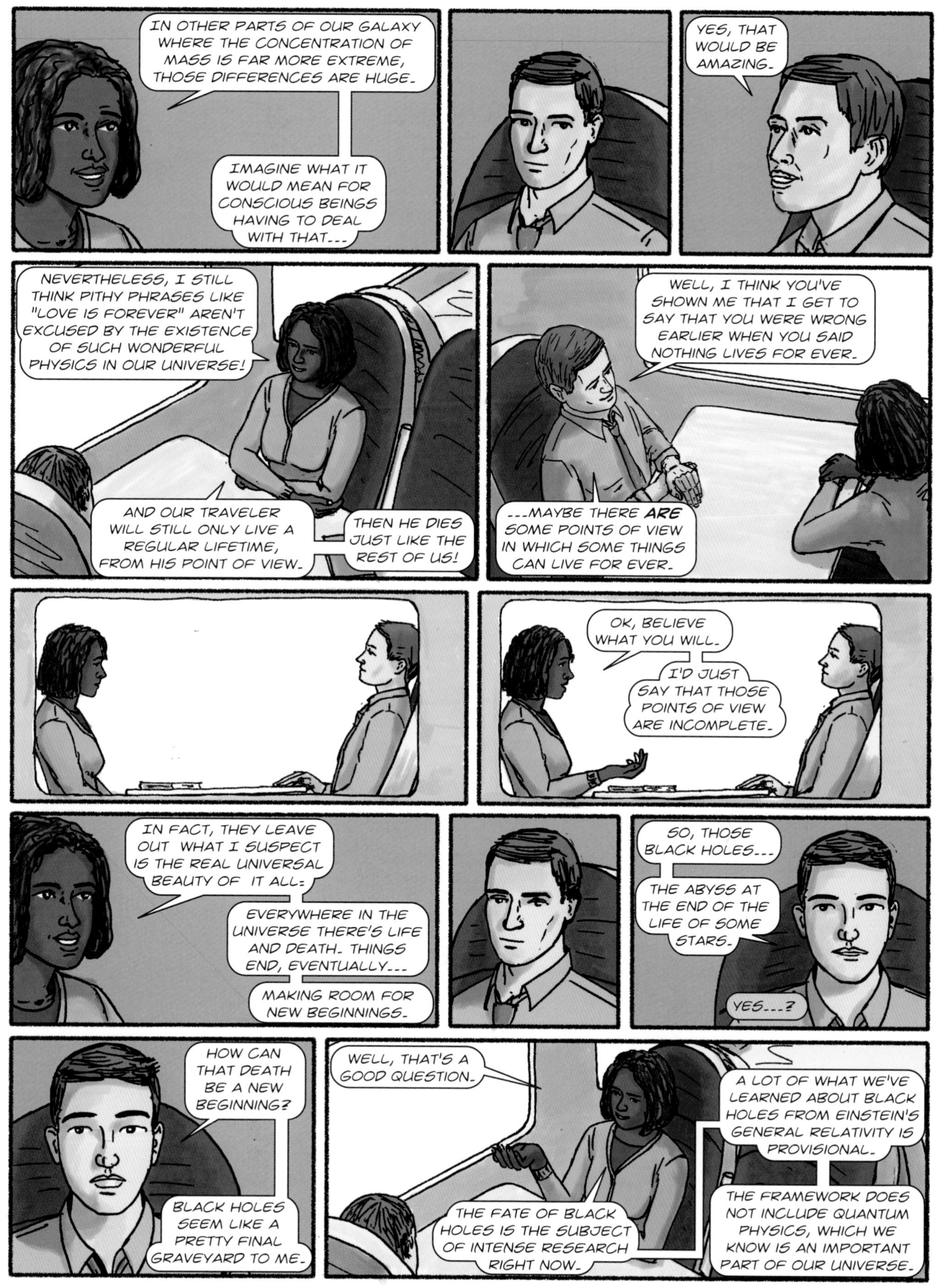
IN OTHER PARTS OF OUR GALAXY WHERE THE CONCENTRATION OF MASS IS FAR MORE EXTREME, THOSE DIFFERENCES ARE HUGE.
IMAGINE WHAT IT WOULD MEAN FOR CONSCIOUS BEINGS HAVING TO DEAL WITH THAT...
YES, THAT WOULD BE AMAZING.
NEVERTHELESS, I STILL THINK PITHY PHRASES LIKE "LOVE IS FOREVER" AREN'T EXCUSED BY THE EXISTENCE OF SUCH WONDERFUL PHYSICS IN OUR UNIVERSE!
AND OUR TRAVELER WILL STILL ONLY LIVE A REGULAR LIFETIME, FROM HIS POINT OF VIEW.
THEN HE DIES JUST LIKE THE REST OF US!
WELL, I THINK YOU'VE SHOWN ME THAT I GET TO SAY THAT YOU WERE WRONG EARLIER WHEN YOU SAID NOTHING LIVES FOR EVER.
...MAYBE THERE **ARE** SOME POINTS OF VIEW IN WHICH SOME THINGS CAN LIVE FOR EVER.
OK, BELIEVE WHAT YOU WILL.
I'D JUST SAY THAT THOSE POINTS OF VIEW ARE INCOMPLETE.
IN FACT, THEY LEAVE OUT WHAT I SUSPECT IS THE REAL UNIVERSAL BEAUTY OF IT ALL:
EVERYWHERE IN THE UNIVERSE THERE'S LIFE AND DEATH. THINGS END, EVENTUALLY...
MAKING ROOM FOR NEW BEGINNINGS.
SO, THOSE BLACK HOLES...
THE ABYSS AT THE END OF THE LIFE OF SOME STARS.
YES...?
HOW CAN THAT DEATH BE A NEW BEGINNING?
BLACK HOLES SEEM LIKE A PRETTY FINAL GRAVEYARD TO ME.
WELL, THAT'S A GOOD QUESTION.
A LOT OF WHAT WE'VE LEARNED ABOUT BLACK HOLES FROM EINSTEIN'S GENERAL RELATIVITY IS PROVISIONAL.
THE FRAMEWORK DOES NOT INCLUDE QUANTUM PHYSICS, WHICH WE KNOW IS AN IMPORTANT PART OF OUR UNIVERSE.
THE FATE OF BLACK HOLES IS THE SUBJECT OF INTENSE RESEARCH RIGHT NOW.

YOU MEAN EINSTEIN'S THEORY IS WRONG?
A MORE ACCURATE STATEMENT IS THAT IT IS INCOMPLETE.
HOW? WAS ALL THAT STUFF ABOUT TIME IN YOUR STORY WRONG THEN?
NO.
ACTUALLY IT IS JUST ABOUT AT THE HORIZON WHERE QUANTUM PHYSICS BEGINS TO CHANGE THE PICTURE.
REMEMBER THAT OUR TRAVELER CAME UP TO THE HORIZON, BUT WAS CAREFUL TO NOT GO ALL THE WAY, SO WE'RE OK.
SO WHAT'S SO SPECIAL ABOUT THE HORIZON, THEN?
WELL, PERFECT ONE-WAY BARRIERS DON'T PLAY WELL WITH QUANTUM MECHANICS.
TURNS OUT THAT THERE'S A PROBABILITY THAT A PARTICLE CAN QUANTUM TUNNEL PAST IT AND GET OUT...
JUST LIKE HAPPENS IN OTHER PHYSICS SITUATIONS, LIKE NUCLEAR PHYSICS, OR ATOMIC PHYSICS.
SO...
BLACK HOLES AREN'T BLACK?
STUFF CAN GET OUT?
RIGHT.
NOW, FOR LARGE BLACK HOLES THAT I STUDY AS AN ASTRONOMER, IT IS A TINY EFFECT, AND SO YOU CAN IGNORE IT.
WAIT...CAN A PERSON GET BACK OUT?
AHA! I KNEW THAT'S WHERE YOU'D GO NEXT.
REMEMBER WE'RE TALKING ABOUT TINY INDIVIDUAL PARTICLES HERE...
BUT A PERSON IS MADE OF LOTS OF TINY INDIVIDUAL PARTICLES...
YES, THAT'S TRUE, BUT...
IT IS NO MORE LIKELY FOR A COMPLEX STRUCTURE LIKE A WHOLE PERSON BEING ABLE TO QUANTUM-TUNNEL BACK OUT OF A BLACK HOLE THAN IT IS FOR YOU TO QUANTUM-TUNNEL TO THE OTHER SIDE OF THAT WALL!
OH...
SO THAT'S IT?
PARTICLES CAN LEAK OUT A LITTLE BIT?
OH, NO, THAT'S JUST THE BEGINNING OF A LONG STORY.
AND THINGS GET PRETTY INTERESTING INSIDE GENERAL RELATIVITY'S BLACK HOLE TOO.
EVEN BEFORE YOU GET TO QUANTUM PHYSICS.

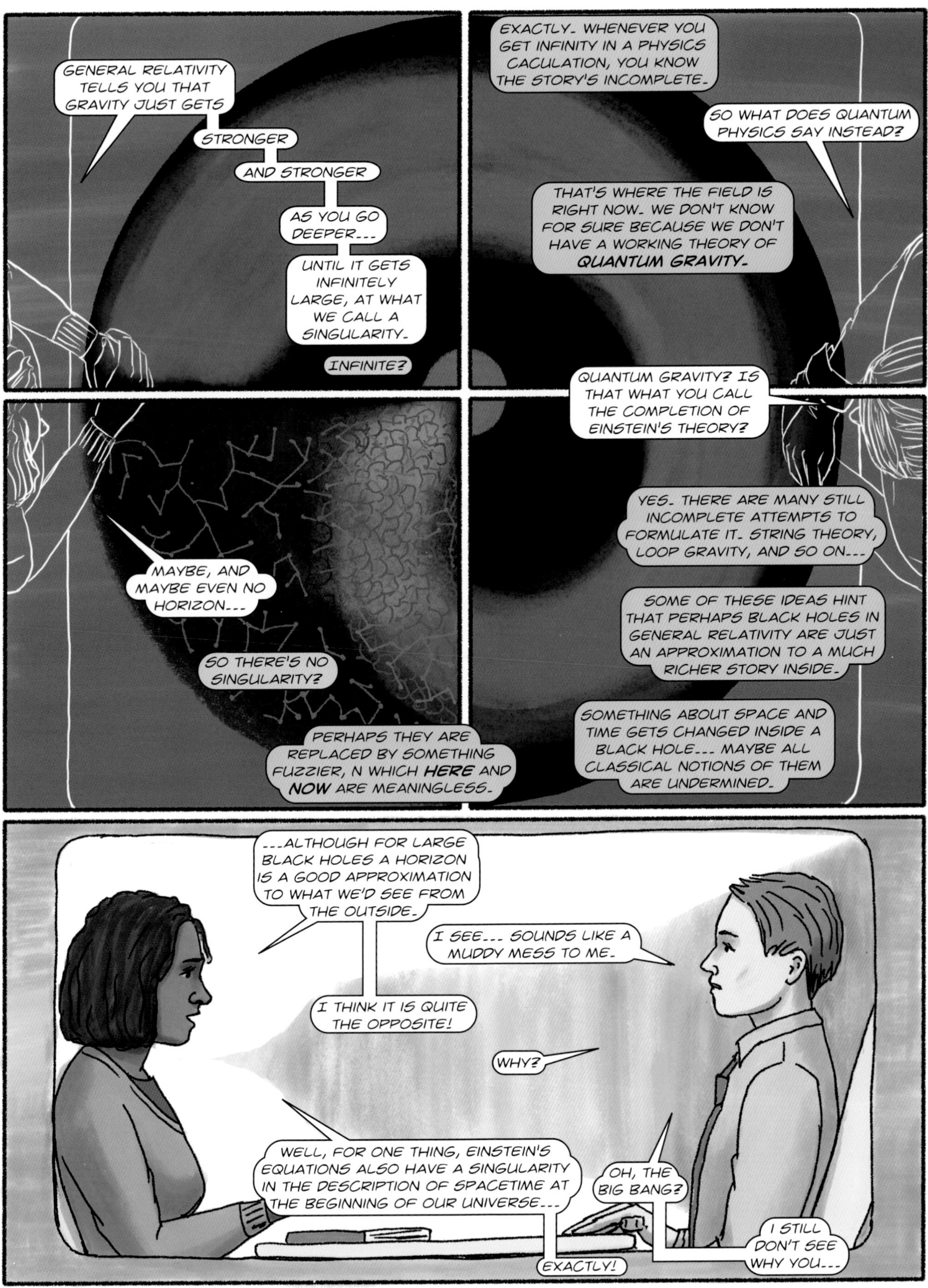

GENERAL RELATIVITY TELLS YOU THAT GRAVITY JUST GETS
STRONGER
AND STRONGER
AS YOU GO DEEPER...
UNTIL IT GETS INFINITELY LARGE, AT WHAT WE CALL A SINGULARITY.
INFINITE?
EXACTLY. WHENEVER YOU GET INFINITY IN A PHYSICS CACULATION, YOU KNOW THE STORY'S INCOMPLETE.
SO WHAT DOES QUANTUM PHYSICS SAY INSTEAD?
THAT'S WHERE THE FIELD IS RIGHT NOW. WE DON'T KNOW FOR SURE BECAUSE WE DON'T HAVE A WORKING THEORY OF **QUANTUM GRAVITY.**
QUANTUM GRAVITY? IS THAT WHAT YOU CALL THE COMPLETION OF EINSTEIN'S THEORY?
YES. THERE ARE MANY STILL INCOMPLETE ATTEMPTS TO FORMULATE IT. STRING THEORY, LOOP GRAVITY, AND SO ON...
SOME OF THESE IDEAS HINT THAT PERHAPS BLACK HOLES IN GENERAL RELATIVITY ARE JUST AN APPROXIMATION TO A MUCH RICHER STORY INSIDE.
SOMETHING ABOUT SPACE AND TIME GETS CHANGED INSIDE A BLACK HOLE... MAYBE ALL CLASSICAL NOTIONS OF THEM ARE UNDERMINED.
SO THERE'S NO SINGULARITY?
MAYBE, AND MAYBE EVEN NO HORIZON...
PERHAPS THEY ARE REPLACED BY SOMETHING FUZZIER, N WHICH **HERE** AND **NOW** ARE MEANINGLESS.
...ALTHOUGH FOR LARGE BLACK HOLES A HORIZON IS A GOOD APPROXIMATION TO WHAT WE'D SEE FROM THE OUTSIDE.
I SEE... SOUNDS LIKE A MUDDY MESS TO ME.
I THINK IT IS QUITE THE OPPOSITE!
WHY?
WELL, FOR ONE THING, EINSTEIN'S EQUATIONS ALSO HAVE A SINGULARITY IN THE DESCRIPTION OF SPACETIME AT THE BEGINNING OF OUR UNIVERSE...
OH, THE BIG BANG?
EXACTLY!
I STILL DON'T SEE WHY YOU...

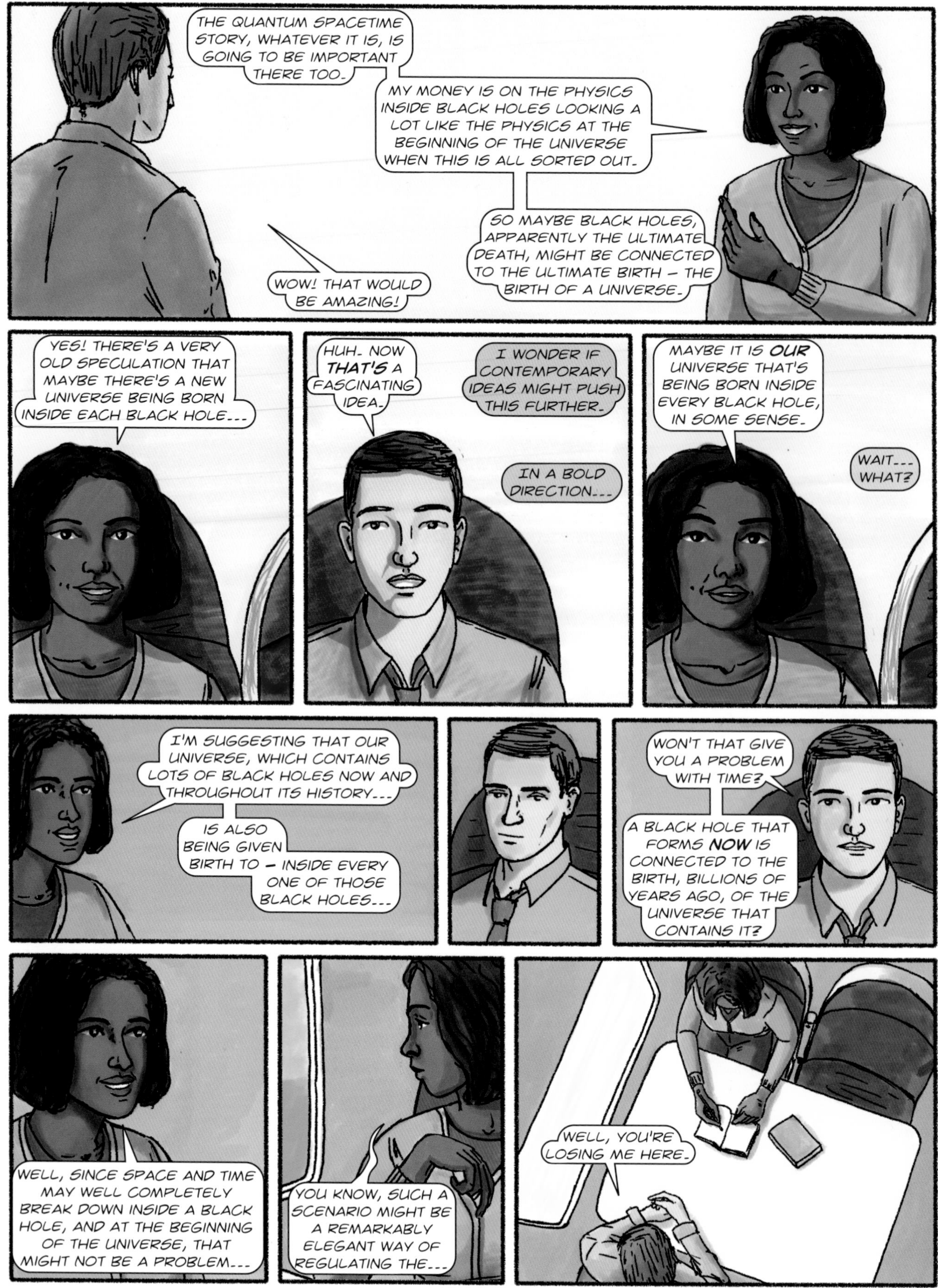
THE QUANTUM SPACETIME STORY, WHATEVER IT IS, IS GOING TO BE IMPORTANT THERE TOO.
MY MONEY IS ON THE PHYSICS INSIDE BLACK HOLES LOOKING A LOT LIKE THE PHYSICS AT THE BEGINNING OF THE UNIVERSE WHEN THIS IS ALL SORTED OUT.
SO MAYBE BLACK HOLES, APPARENTLY THE ULTIMATE DEATH, MIGHT BE CONNECTED TO THE ULTIMATE BIRTH – THE BIRTH OF A UNIVERSE.
WOW! THAT WOULD BE AMAZING!
YES! THERE'S A VERY OLD SPECULATION THAT MAYBE THERE'S A NEW UNIVERSE BEING BORN INSIDE EACH BLACK HOLE...
HUH. NOW **THAT'S** A FASCINATING IDEA.
I WONDER IF CONTEMPORARY IDEAS MIGHT PUSH THIS FURTHER.
IN A BOLD DIRECTION...
MAYBE IT IS **OUR** UNIVERSE THAT'S BEING BORN INSIDE EVERY BLACK HOLE, IN SOME SENSE.
WAIT... WHAT?
I'M SUGGESTING THAT OUR UNIVERSE, WHICH CONTAINS LOTS OF BLACK HOLES NOW AND THROUGHOUT ITS HISTORY...
IS ALSO BEING GIVEN BIRTH TO – INSIDE EVERY ONE OF THOSE BLACK HOLES...
WON'T THAT GIVE YOU A PROBLEM WITH TIME?
A BLACK HOLE THAT FORMS **NOW** IS CONNECTED TO THE BIRTH, BILLIONS OF YEARS AGO, OF THE UNIVERSE THAT CONTAINS IT?
WELL, SINCE SPACE AND TIME MAY WELL COMPLETELY BREAK DOWN INSIDE A BLACK HOLE, AND AT THE BEGINNING OF THE UNIVERSE, THAT MIGHT NOT BE A PROBLEM...
YOU KNOW, SUCH A SCENARIO MIGHT BE A REMARKABLY ELEGANT WAY OF REGULATING THE...
WELL, YOU'RE LOSING ME HERE.

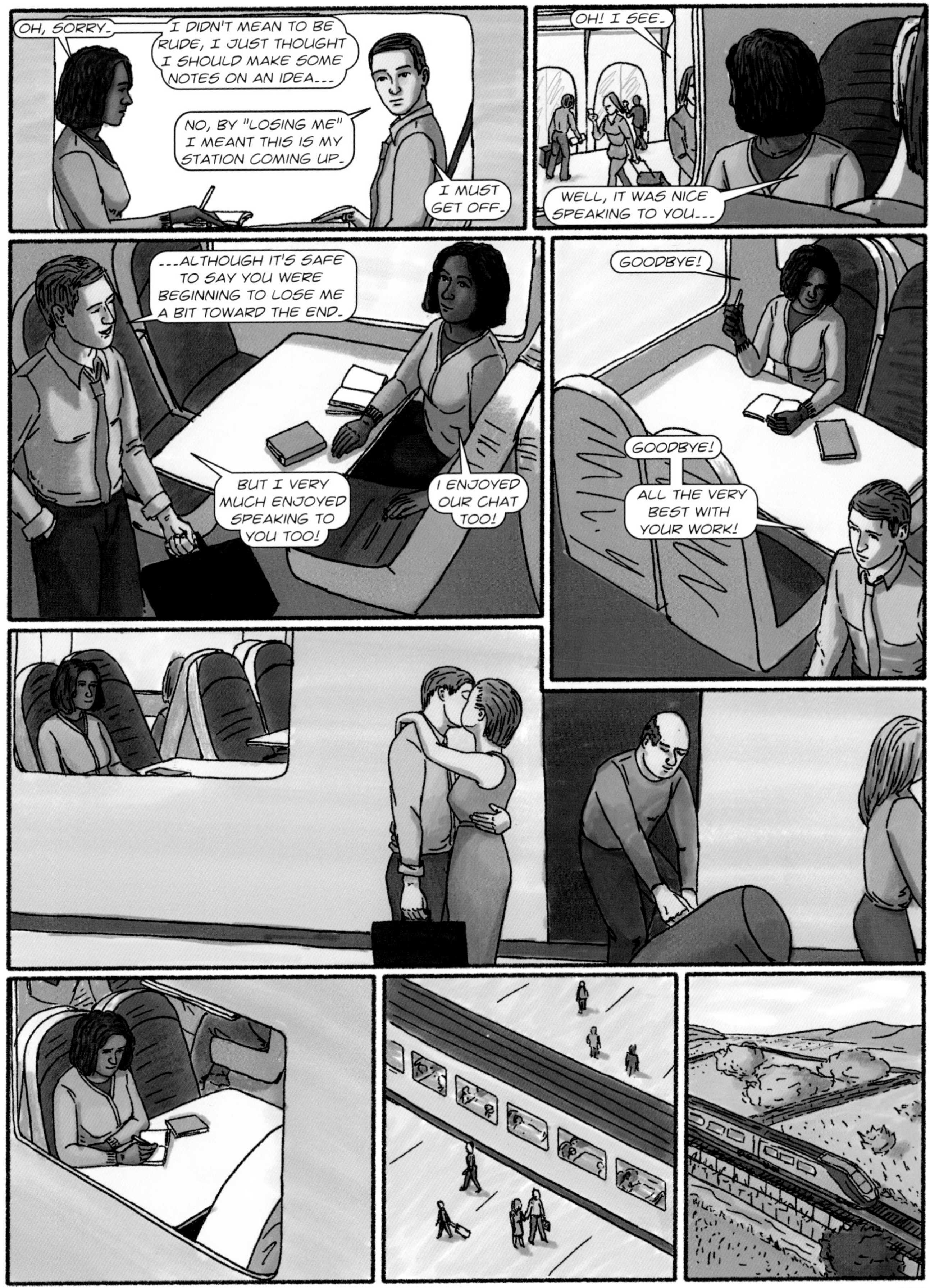
OH, SORRY.
I DIDN'T MEAN TO BE RUDE, I JUST THOUGHT I SHOULD MAKE SOME NOTES ON AN IDEA...
NO, BY "LOSING ME" I MEANT THIS IS MY STATION COMING UP.
I MUST GET OFF.
OH! I SEE.
WELL, IT WAS NICE SPEAKING TO YOU...
...ALTHOUGH IT'S SAFE TO SAY YOU WERE BEGINNING TO LOSE ME A BIT TOWARD THE END.
BUT I VERY MUCH ENJOYED SPEAKING TO YOU TOO!
I ENJOYED OUR CHAT TOO!
GOODBYE!
GOODBYE!
ALL THE VERY BEST WITH YOUR WORK!

## Notes

***Page 136 and beyond*** – There are numerous accessible treatments of black holes. Some of the books already mentioned in notes on earlier chapters have good discussions, such as Garfinkle and Garfinkle (chapter 3 note to pages 46–47), Tyson, Strauss, and Gott (chapter 1 note to page 12), and especially Thorne (chapter 4 note to pages 79–82). For a swift survey with an emphasis on real astrophysical black holes (including an account of supermassive ones at the center our own galaxy) here's another: Katherine M. Blundell, *Black Holes: A Very Short Introduction* (Oxford: Oxford University Press, 2015).

***Page 139*** – The key piece of physics here is called gravitational time dilation, and it is a real effect, from general relativity as mentioned on page 142 (see also the reading material on relativity in chapter 4). You might have seen it in action (also in fiction) in the 2014 film *Interstellar* (Paramount, Warner Bros.). The physics of that scenario is discussed extensively in Kip Thorne, *The Science of Interstellar* (New York: W. W. Norton & Co., 2014).

***Page 142*** – The essence of the calculations needed to see the effect in the context of GPS is discussed in an excellent undergraduate text: James B. Hartle, *Gravity: An Introduction to Einstein's General Relativity* (San Francisco: Addison-Wesley, 2003). The book for nonexperts by Randall mentioned in the chapter 3 notes for page 45 discusses GPS and time dilation too.

***Page 144*** (panel 4) – Quantum tunneling, being able to pass through a barrier that classically would be impenetrable, is considered one of the most surprising aspects of quantum physics when people first learn about it. The next step is to wonder whether what's true for a subatomic particle could be true for a person. A tunneling process is highly statistical, however, so to get all the person's particles to do that same thing (tunnel) at the same time has almost vanishing probability. This does not mean that tunneling is one of those minor bits of oddness that does not affect our everyday lives—our sun would not work without it! Tunneling is not mentioned explicitly in a lot of the books already suggested, including the suggestions for quantum physics given in the chapter 8 notes. So for another quick introduction to quantum physics, see John Polkinghorne, *Quantum Theory: A Very Short Introduction* (Oxford: Oxford University Press, 2002).

***Pages 144–145*** – The quantum nature of black holes is currently a subject of intense research. It is hard to find a good book summarizing the state of the art. The core ideas such as Hawking radiation are covered in the early parts of, for example, Leonard Susskind, *The Black Hole War: My Battle with Stephen Hawking to Make the World Safe for Quantum Mechanics* (New York: Little, Brown, 2008). Much more has happened in the field since that book appeared, and not only because of the leaps in our understanding of the kind of quantum gravity found in the string theory approach. Here are a few magazine articles that help give a flavor of what's going on: Juan Maldacena, "Black Holes and Wormholes and the Secrets of Quantum Spacetime," *Scientific American* (November 2016): 26–31; Jennifer Ouellette, "Alice and Bob Meet the Wall of Fire," *Quanta Magazine* (December 2012), https://www.quantamagazine.org/20121221-alice-and-bob-meet-the-wall-of-fire/; Jennifer Ouellette, "The Fuzzball Fix for a Black Hole Paradox," *Quanta Magazine*, (June 2015), https://www.quantamagazine.org/20150623-fuzzballs-black-hole-firewalls/; K. C. Cole, "Wormholes Untangle a Black Hole Paradox," *Quanta Magazine* (April 2015), https://www.quantamagazine.org/20150424-wormholes-entanglement-firewalls-er-epr/. Note that the various approaches described within are revisiting (sometimes radically) long-held ideas about the entire black hole, including the neighborhood of the horizon, not just the interior.

***Page 145*** - "Holographic" approaches such as AdS/CFT (which will come up in chapter 9) are one way of realizing the old idea that spacetime is somehow emergent (not necessarily fundamental). There are several ways of constructing/modeling black hole physics in string theory, and spacetime's emergent nature is evident in lots of these different string theory models. All this provides new insights and understanding of black holes, at least in these approaches. (Some of these ideas may well be applied directly to nature one day, but we are far from that stage at present.) Ideas that underlie emergence that you can look for when exploring various books on string theory (like those of Greene and of Gubser, mentioned in the chapter 3 notes) are T-duality, S-duality, world-sheet vs. spacetime, gauge/gravity duality, and matrix theory. It is worth mentioning again that some of the other approaches to quantum gravity such as loop quantum gravity (see the summary in the book by Smolin mentioned in the chapter 6 note to pages 126–127) also contain hints that spacetime breaks down into some other description at short distances. There'll be much more on all this in a later conversation in chapter 9 and its accompanying notes.

***Page 145*** (panels 1–4) - A little note on the storytelling medium we're using. Panels and other elements in sequential art form a kind of spacetime (see my remarks in the preface): For example, the panels invoke space and their relative positioning (crucially, including the gaps between them), gives a notion of time passing.* You've likely been reading them in the conventional order (or you wouldn't have ventured this far into the book). It was entirely intentional to disrupt the conventional flow of the book's spacetime in these four panels given what was being discussed.

*See the following books for a discussion of aspects of the workings of comics: Will Eisner, *Comics and Sequential Art* (New York: W. W. Norton & Co., 2008); Will Eisner, *Graphic Storytelling and Visual Narrative* (New York: W. W. Norton & Co., 2008); Scott McCloud, *Understanding Comics: The Invisible Art* (New York: Harper, 1994).

***Page 146*** (panel 2) - This goes back at least to V. P. Frolov, M. A. Markov, and V. F. Mukhanov, "Through a Black Hole into a New Universe?," *Phys. Lett.* *B216* (1989): 272–276, where, motivated by the idea of quantum modifications to spacetime, the authors do the *classical* computation that adjoins the spacetime geometry of the interior of a black hole to that of an expanding universe. Such computations are not proofs of anything, but can be suggestive.

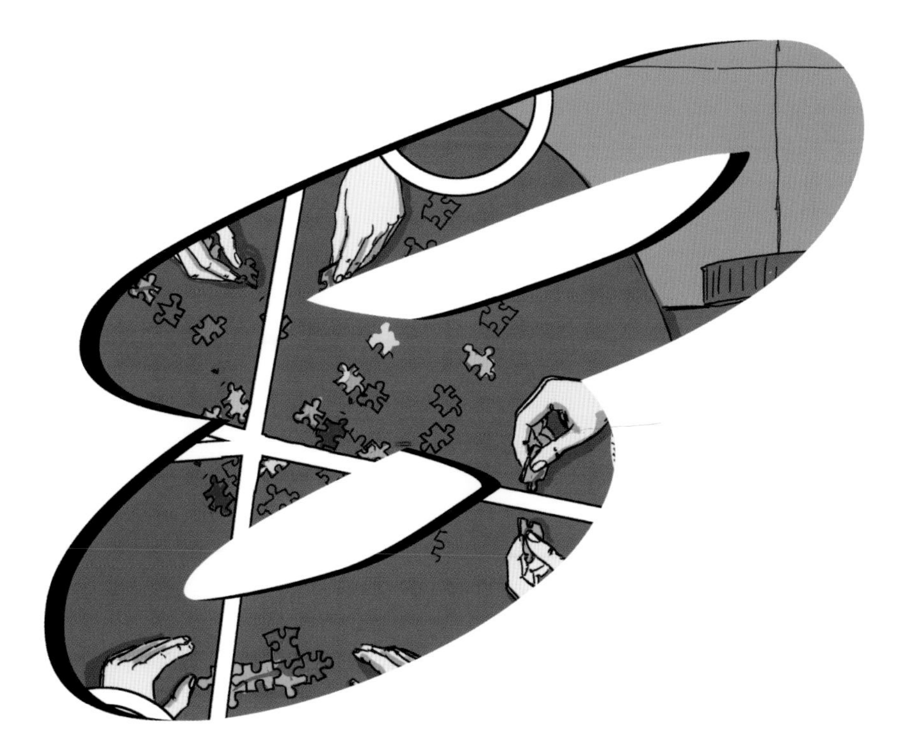

And God said
$\nabla \cdot B = 0$

HI!
HI!
JUST WANTED TO SAY:
AWESOME T-SHIRT!
OH! THANK YOU!
YOU MIGHT BE THE FIRST PERSON WHO'S...
...EVER APPRECIATED IT SINCE I'VE HAD IT!
REALLY? IT'S SO GREAT...
And God said...
$\nabla \cdot E = \frac{\rho}{\epsilon_0}$
$\nabla \cdot B = 0$
$\nabla \times E = -\frac{\partial B}{\partial t}$
$\nabla \times B = \mu_0 J + \mu_0 \epsilon_0 \frac{\partial E}{\partial t}$
... and there was Light
THEY'RE ON MY SHORTLIST OF ALL-TIME GREATEST EQUATIONS, AND THE JOKE IS PERFECT.
I'VE HEARD OF THESE T-SHIRTS, BUT I'VE NEVER ACTUALLY SEEN ONE OUT IN THE WILD.
LET ME GUESS, THIS PROBABLY MEANS YOU'RE A SCIENTIST?
OR MAYBE A BIG SCIENCE FAN LIKE ME?
SCIENTIST.
I'M A THEORETICAL PHYSICIST.
YOU'RE A SCIENCE FAN?
REALLY HARDCORE, WITH A MAXWELL'S EQUATIONS T-SHIRT AND EVERYTHING. I LIKE IT!
WELL, WHY NOT?
TOTALLY WITH YOU!
MY FAVORITE GYM T-SHIRT IS THE "SCIENCE, IT WORKS BITCHES" ONE...
THE ONE WITH THE WMAP BLACK BODY CURVE ON IT...?
OH, RIGHT! THAT'S GREAT.
WHAT KIND OF PHYSICS DO YOU WORK ON?
DO YOU WANT TO JOIN ME?
SURE, FOR A LITTLE WHILE.
I'M MEETING A FRIEND OVER AT THE LIBRARY LATER...
BUT I CAME EARLY TO SIT IN THIS SPACE AND WORK.
SO I'VE GOT SOME TIME.
IF YOU WERE BUSY, DON'T LET ME...
NO, NO.
I NEED A BREAK ANYWAY.
I GOT STUCK ON A COMPUTATION.

I HOPE YOUR COMPUTATION WORKS OUT...
WOULD I EVEN UNDERSTAND WHAT IT WAS ABOUT, OR IS IT A SUPER-TECHNICAL ISSUE?
WELL, I DON'T KNOW HOW MUCH PHYSICS YOU KNOW...
IT'S A NEW APPROACH TO UNDERSTANDING THE COSMOLOGICAL CONSTANT...
OH! COSMOLOGY!
I'VE BEEN READING ABOUT THAT RECENTLY...
OH?
SO AM I ALLOWED TO ASK IF YOU'RE PRO- OR ANTI- THE MULTIVERSE IDEA?
SEEMS TO BE THE BIG DIVIDER THESE DAYS...
PEOPLE ASK ME THAT A LOT!
I THINK I'M AN AGNOSTIC RIGHT NOW.
I DON'T THINK WE'RE ASKING THE RIGHT QUESTIONS YET.
REALLY? LIKE WHAT QUESTIONS?
I THINK WE NEED A BETTER UNDERSTANDING OF WHAT SPACETIME REALLY IS...
WHERE IT COMES FROM...
HOW IT COMES TO BE.
HUH...
YOU DON'T THINK THAT WE KNOW ENOUGH FROM EINSTEIN'S WORK ON RELATIVITY?
NO, WE DON'T.
IT'S LIKE TRYING TO ANSWER THE QUESTION OF WHY THE SKY IS BLUE, BEFORE KNOWING ABOUT THE MOLECULAR COMPOSITION OF THE ATMOSPHERE...
BEFORE EVEN KNOWING WHAT MOLECULES ARE...
SO...
YOU'RE SAYING WE'RE NOT EVEN CLOSE TO GETTING AT THE RIGHT QUESTIONS YET THEN...
I DON'T KNOW FOR SURE, BUT THAT'S MY GUT FEELING.
YOU KNOW...
I GET THE SENSE THAT THERE'S ALL THESE PEOPLE AROUND THE WORLD WORKING ON ONE APPROACH OR THE OTHER...
SO MANY THEORIES, MODELS...
I SOMETIMES WONDER...
YOU WONDER WHAT?

NOW WE'RE IN THIS INTERCONNECTED WORLD, ISN'T THERE A NEW APPROACH WE CAN TRY?
LIKE WHAT?
WELL, NOW MORE THAN EVER, THERE MUST BE A WAY TO GET EVERYONE WORKING TOGETHER TO SOLVE SOME OF THESE PROBLEMS WITH OUR UNDERSTANDING OF SPACE AND TIME.
HOW?
WELL, GET EVERYONE CONNECTED USING SOCIAL MEDIA OR VIRTUAL REALITY OR SOMETHING...
YOU KNOW, ORGANIZE THEM PROPERLY...
BY PUTTING PEOPLE WITH THE RIGHT EXPERTISE ON THE RIGHT PARTS OF THE PROBLEM, AND FIGURE IT OUT ONCE AND FOR ALL?
WELL, IT'S A NICE THOUGHT, BUT I DON'T SEE HOW IT WILL WORK.
NOBODY KNOWS WHICH IS THE RIGHT APPROACH, SO HOW IS THIS PROGRAM GOING TO BE RUN?
WELL, MAYBE A PANEL OF SENIOR PEOPLE IN THE FIELD CAN GET TOGETHER TO DECIDE...
BASED ON THEIR EXPERTISE, KNOWLEDGE OF THE HISTORY OF THE FIELD...
WELL, I DON'T THINK THAT I'M GOING TO TRUST THAT ANY GROUP OF PEOPLE HAS A HANDLE ON WHAT THE RIGHT APPROACH IS...
IN FACT, THERE PROBABLY **IS** NO RIGHT APPROACH.
REALLY?
DOESN'T THAT MEAN THERE'S NO ANSWER TO ALL THOSE QUESTIONS PEOPLE ARE WORKING ON?
NO, NO. IT JUST MEANS IT IS DIFFICULT TO FIND ANSWERS DIRECTLY...
IT'S LIKE...
IT'S LIKE SOLVING A **JIGSAW PUZZLE.**
ACTUALLY A **LOT** LIKE THAT...
REALLY?

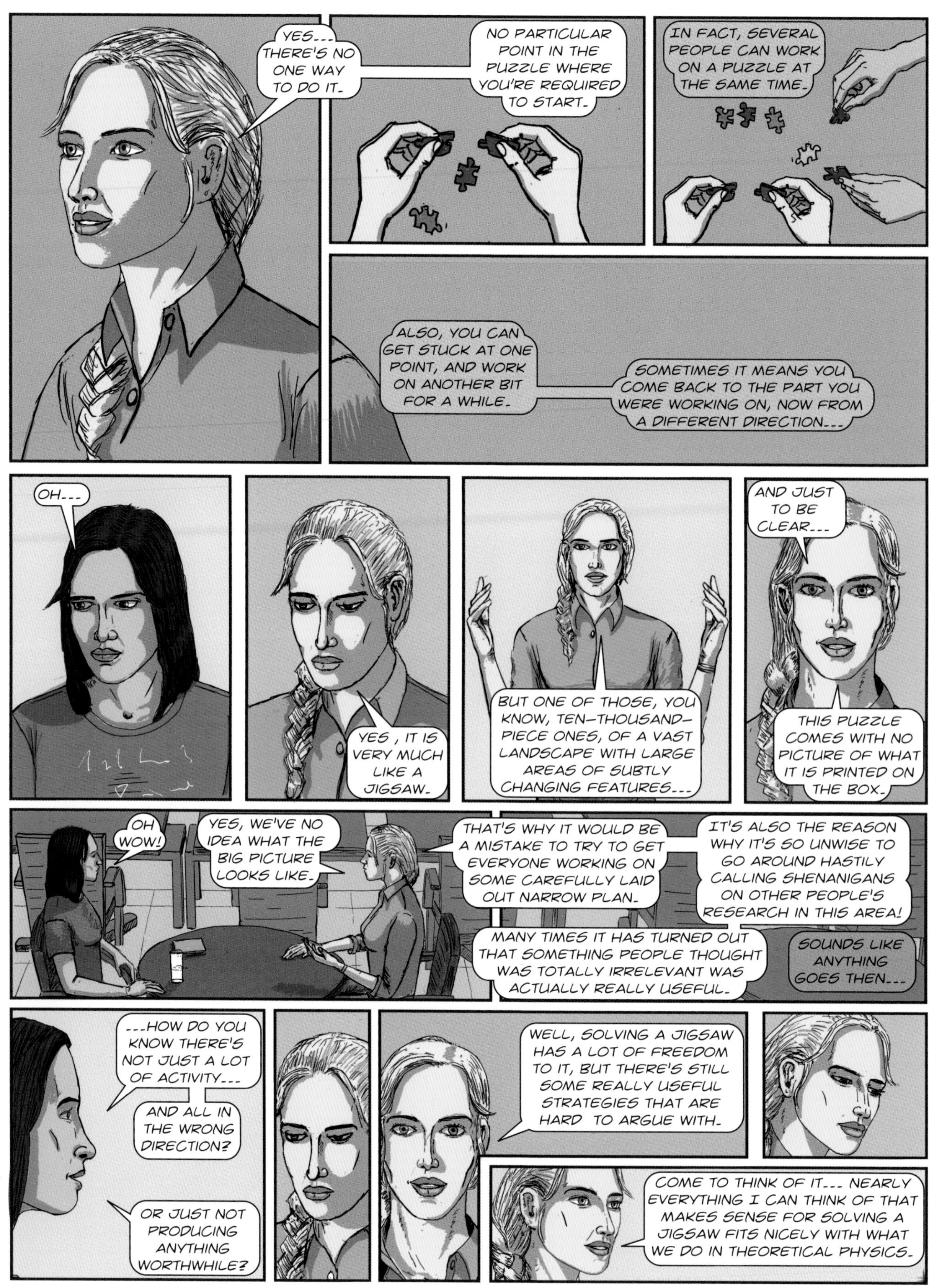
YES... THERE'S NO ONE WAY TO DO IT.
NO PARTICULAR POINT IN THE PUZZLE WHERE YOU'RE REQUIRED TO START.
IN FACT, SEVERAL PEOPLE CAN WORK ON A PUZZLE AT THE SAME TIME.
ALSO, YOU CAN GET STUCK AT ONE POINT, AND WORK ON ANOTHER BIT FOR A WHILE.
SOMETIMES IT MEANS YOU COME BACK TO THE PART YOU WERE WORKING ON, NOW FROM A DIFFERENT DIRECTION...
OH...
YES , IT IS VERY MUCH LIKE A JIGSAW.
BUT ONE OF THOSE, YOU KNOW, TEN-THOUSAND-PIECE ONES, OF A VAST LANDSCAPE WITH LARGE AREAS OF SUBTLY CHANGING FEATURES...
AND JUST TO BE CLEAR...
THIS PUZZLE COMES WITH NO PICTURE OF WHAT IT IS PRINTED ON THE BOX.
OH WOW!
YES, WE'VE NO IDEA WHAT THE BIG PICTURE LOOKS LIKE.
THAT'S WHY IT WOULD BE A MISTAKE TO TRY TO GET EVERYONE WORKING ON SOME CAREFULLY LAID OUT NARROW PLAN.
IT'S ALSO THE REASON WHY IT'S SO UNWISE TO GO AROUND HASTILY CALLING SHENANIGANS ON OTHER PEOPLE'S RESEARCH IN THIS AREA!
MANY TIMES IT HAS TURNED OUT THAT SOMETHING PEOPLE THOUGHT WAS TOTALLY IRRELEVANT WAS ACTUALLY REALLY USEFUL.
SOUNDS LIKE ANYTHING GOES THEN...
...HOW DO YOU KNOW THERE'S NOT JUST A LOT OF ACTIVITY...
AND ALL IN THE WRONG DIRECTION?
OR JUST NOT PRODUCING ANYTHING WORTHWHILE?
WELL, SOLVING A JIGSAW HAS A LOT OF FREEDOM TO IT, BUT THERE'S STILL SOME REALLY USEFUL STRATEGIES THAT ARE HARD TO ARGUE WITH.
COME TO THINK OF IT... NEARLY EVERYTHING I CAN THINK OF THAT MAKES SENSE FOR SOLVING A JIGSAW FITS NICELY WITH WHAT WE DO IN THEORETICAL PHYSICS.

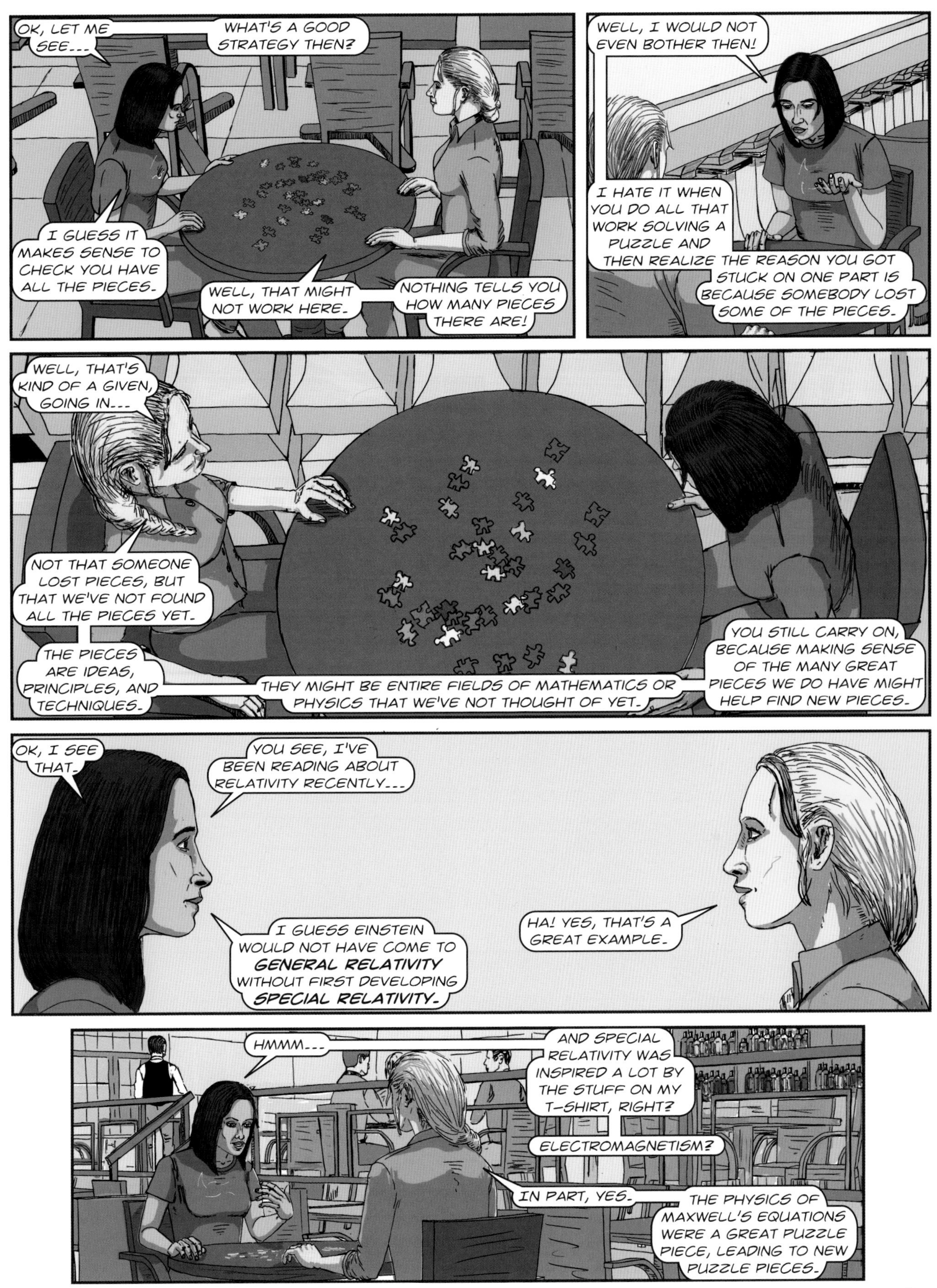
OK, LET ME SEE...
WHAT'S A GOOD STRATEGY THEN?
I GUESS IT MAKES SENSE TO CHECK YOU HAVE ALL THE PIECES.
WELL, THAT MIGHT NOT WORK HERE.
NOTHING TELLS YOU HOW MANY PIECES THERE ARE!
WELL, I WOULD NOT EVEN BOTHER THEN!
I HATE IT WHEN YOU DO ALL THAT WORK SOLVING A PUZZLE AND THEN REALIZE THE REASON YOU GOT STUCK ON ONE PART IS BECAUSE SOMEBODY LOST SOME OF THE PIECES.
WELL, THAT'S KIND OF A GIVEN, GOING IN...
NOT THAT SOMEONE LOST PIECES, BUT THAT WE'VE NOT FOUND ALL THE PIECES YET.
THE PIECES ARE IDEAS, PRINCIPLES, AND TECHNIQUES.
THEY MIGHT BE ENTIRE FIELDS OF MATHEMATICS OR PHYSICS THAT WE'VE NOT THOUGHT OF YET.
YOU STILL CARRY ON, BECAUSE MAKING SENSE OF THE MANY GREAT PIECES WE DO HAVE MIGHT HELP FIND NEW PIECES.
OK, I SEE THAT.
YOU SEE, I'VE BEEN READING ABOUT RELATIVITY RECENTLY...
I GUESS EINSTEIN WOULD NOT HAVE COME TO **GENERAL RELATIVITY** WITHOUT FIRST DEVELOPING **SPECIAL RELATIVITY**.
HA! YES, THAT'S A GREAT EXAMPLE.
HMMM...
AND SPECIAL RELATIVITY WAS INSPIRED A LOT BY THE STUFF ON MY T-SHIRT, RIGHT?
ELECTROMAGNETISM?
IN PART, YES.
THE PHYSICS OF MAXWELL'S EQUATIONS WERE A GREAT PUZZLE PIECE, LEADING TO NEW PUZZLE PIECES.

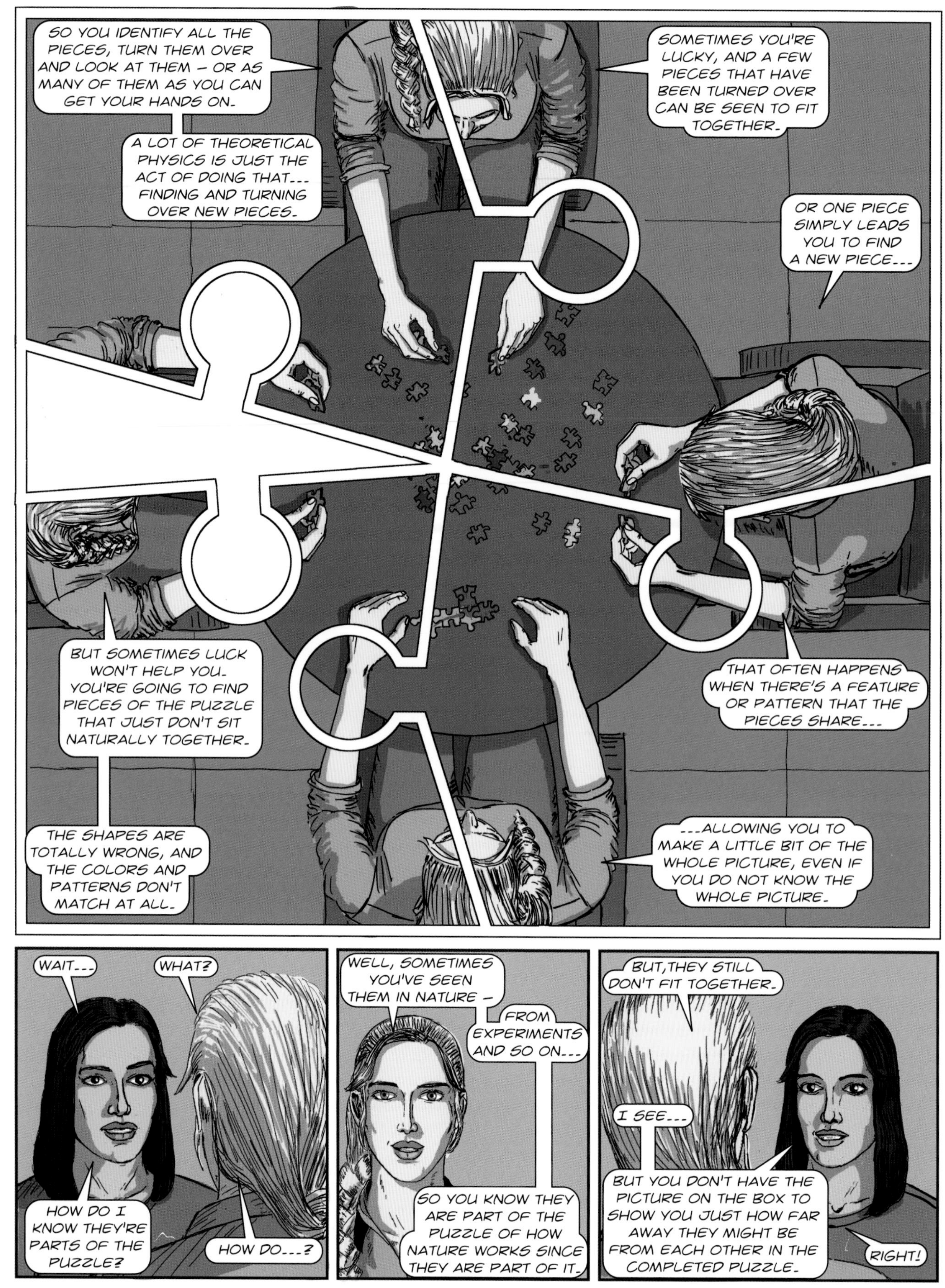
SO YOU IDENTIFY ALL THE PIECES, TURN THEM OVER AND LOOK AT THEM – OR AS MANY OF THEM AS YOU CAN GET YOUR HANDS ON.
A LOT OF THEORETICAL PHYSICS IS JUST THE ACT OF DOING THAT... FINDING AND TURNING OVER NEW PIECES.
SOMETIMES YOU'RE LUCKY, AND A FEW PIECES THAT HAVE BEEN TURNED OVER CAN BE SEEN TO FIT TOGETHER.
OR ONE PIECE SIMPLY LEADS YOU TO FIND A NEW PIECE...
THAT OFTEN HAPPENS WHEN THERE'S A FEATURE OR PATTERN THAT THE PIECES SHARE...
...ALLOWING YOU TO MAKE A LITTLE BIT OF THE WHOLE PICTURE, EVEN IF YOU DO NOT KNOW THE WHOLE PICTURE.
BUT SOMETIMES LUCK WON'T HELP YOU. YOU'RE GOING TO FIND PIECES OF THE PUZZLE THAT JUST DON'T SIT NATURALLY TOGETHER.
THE SHAPES ARE TOTALLY WRONG, AND THE COLORS AND PATTERNS DON'T MATCH AT ALL.
WAIT...
WHAT?
HOW DO I KNOW THEY'RE PARTS OF THE PUZZLE?
HOW DO...?
WELL, SOMETIMES YOU'VE SEEN THEM IN NATURE –
FROM EXPERIMENTS AND SO ON...
SO YOU KNOW THEY ARE PART OF THE PUZZLE OF HOW NATURE WORKS SINCE THEY ARE PART OF IT.
BUT, THEY STILL DON'T FIT TOGETHER.
I SEE...
BUT YOU DON'T HAVE THE PICTURE ON THE BOX TO SHOW YOU JUST HOW FAR AWAY THEY MIGHT BE FROM EACH OTHER IN THE COMPLETED PUZZLE.
RIGHT!

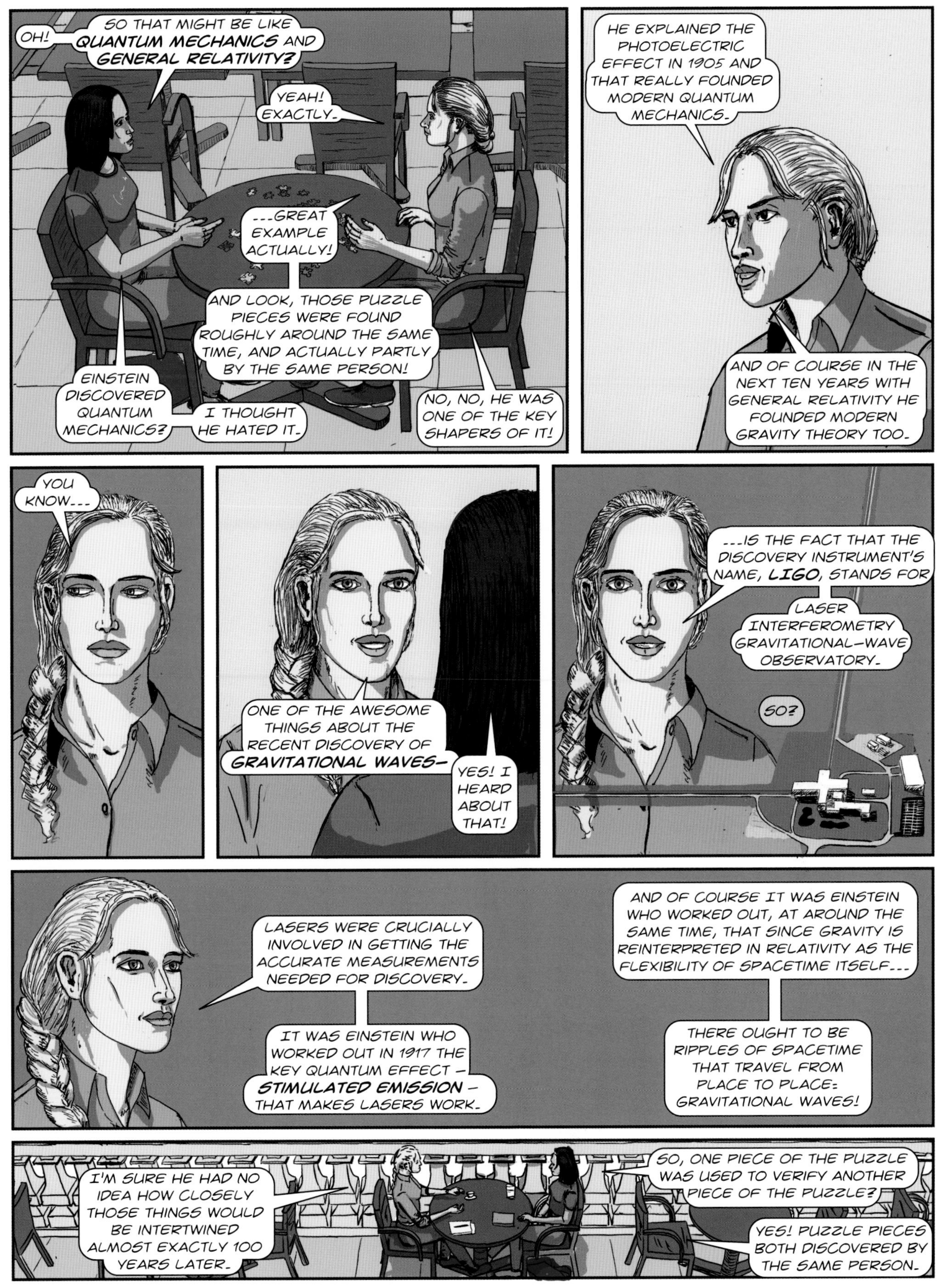
OH!
SO THAT MIGHT BE LIKE QUANTUM MECHANICS AND GENERAL RELATIVITY?
YEAH! EXACTLY.
...GREAT EXAMPLE ACTUALLY!
AND LOOK, THOSE PUZZLE PIECES WERE FOUND ROUGHLY AROUND THE SAME TIME, AND ACTUALLY PARTLY BY THE SAME PERSON!
EINSTEIN DISCOVERED QUANTUM MECHANICS?
I THOUGHT HE HATED IT.
NO, NO, HE WAS ONE OF THE KEY SHAPERS OF IT!
HE EXPLAINED THE PHOTOELECTRIC EFFECT IN 1905 AND THAT REALLY FOUNDED MODERN QUANTUM MECHANICS.
AND OF COURSE IN THE NEXT TEN YEARS WITH GENERAL RELATIVITY HE FOUNDED MODERN GRAVITY THEORY TOO.
YOU KNOW...
ONE OF THE AWESOME THINGS ABOUT THE RECENT DISCOVERY OF GRAVITATIONAL WAVES—
YES! I HEARD ABOUT THAT!
...IS THE FACT THAT THE DISCOVERY INSTRUMENT'S NAME, LIGO, STANDS FOR
LASER INTERFEROMETRY GRAVITATIONAL-WAVE OBSERVATORY.
SO?
LASERS WERE CRUCIALLY INVOLVED IN GETTING THE ACCURATE MEASUREMENTS NEEDED FOR DISCOVERY.
IT WAS EINSTEIN WHO WORKED OUT IN 1917 THE KEY QUANTUM EFFECT — STIMULATED EMISSION — THAT MAKES LASERS WORK.
AND OF COURSE IT WAS EINSTEIN WHO WORKED OUT, AT AROUND THE SAME TIME, THAT SINCE GRAVITY IS REINTERPRETED IN RELATIVITY AS THE FLEXIBILITY OF SPACETIME ITSELF...
THERE OUGHT TO BE RIPPLES OF SPACETIME THAT TRAVEL FROM PLACE TO PLACE: GRAVITATIONAL WAVES!
SO, ONE PIECE OF THE PUZZLE WAS USED TO VERIFY ANOTHER PIECE OF THE PUZZLE?
YES! PUZZLE PIECES BOTH DISCOVERED BY THE SAME PERSON.
I'M SURE HE HAD NO IDEA HOW CLOSELY THOSE THINGS WOULD BE INTERTWINED ALMOST EXACTLY 100 YEARS LATER.

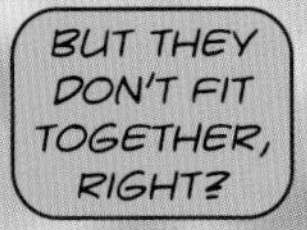
BUT THEY DON'T FIT TOGETHER, RIGHT?
RIGHT...
BUT THEY MUST, SOMEHOW, SINCE THEY ARE BOTH PART OF NATURE.
IN QUANTUM MECHANICS WE SEE MATTER, FORCE, AND ENERGY ALL BUILT OUT OF QUANTUM BUILDING BLOCKS, AT THE SMALLEST SCALES...
THIS QUANTUM DESCRIPTON UNDERLIES HOW THE STUFF WE'RE MADE OF AND THE THINGS WE SEE AROUND US – EVERYTHING WE KNOW – ARE PUT TOGETHER...
WE USE IT TO MOVE THINGS AROUND...
LIKE THE ELECTRONS IN THE ELECTRONICS OF YOUR PHONE...
...TO MANIPULATE HOW THE QUANTUM BLOCKS OF LIGHT – PHOTONS – INTERACT WITH THE ELECTRONS IN YOUR PHONE'S CAMERA TO MAKE PICTURES...
BUT NONE OF THIS WORKS FOR GRAVITY?
NOT SO FAR...
MOST OF THE THINGS THAT WORKED BEFORE FOR QUANTUM PHYSICS JUST FAIL FOR GRAVITY.
LIKE WHAT?
WELL, YOU CAN TRY TO DEFINE QUANTUM BUILDING BLOCKS OF THE GRAVITATIONAL FORCE – GRAVITONS – IN A SIMILAR WAY TO HOW PHOTONS ARE THE BUILDING BLOCKS OF LIGHT AND THE ELECTROMAGNETIC FORCE...
AND...?
...THAT FAILS.
OH.
WE UNDERSTAND GRAVITY ON THE LARGEST SCALES, GOVERNING THE MOTION OF PLANETS, STARS, GALAXIES – THE EVOLUTION OF THE WHOLE UNIVERSE...
IT'S DESCRIBED IN TERMS OF THE BENDING AND TWISTING OF SPACETIME – LIKE IT IS A SMOOTH FABRIC.
THAT DOES NOT SEEM EASY TO RECONCILE WITH LITTLE BASIC BUILDING BLOCKS... SO IT HAS LED TO A LOT OF CONFUSION.
SO YEAH, WE JUST DON'T KNOW HOW THEY FIT TOGETHER IN THE JIGSAW...

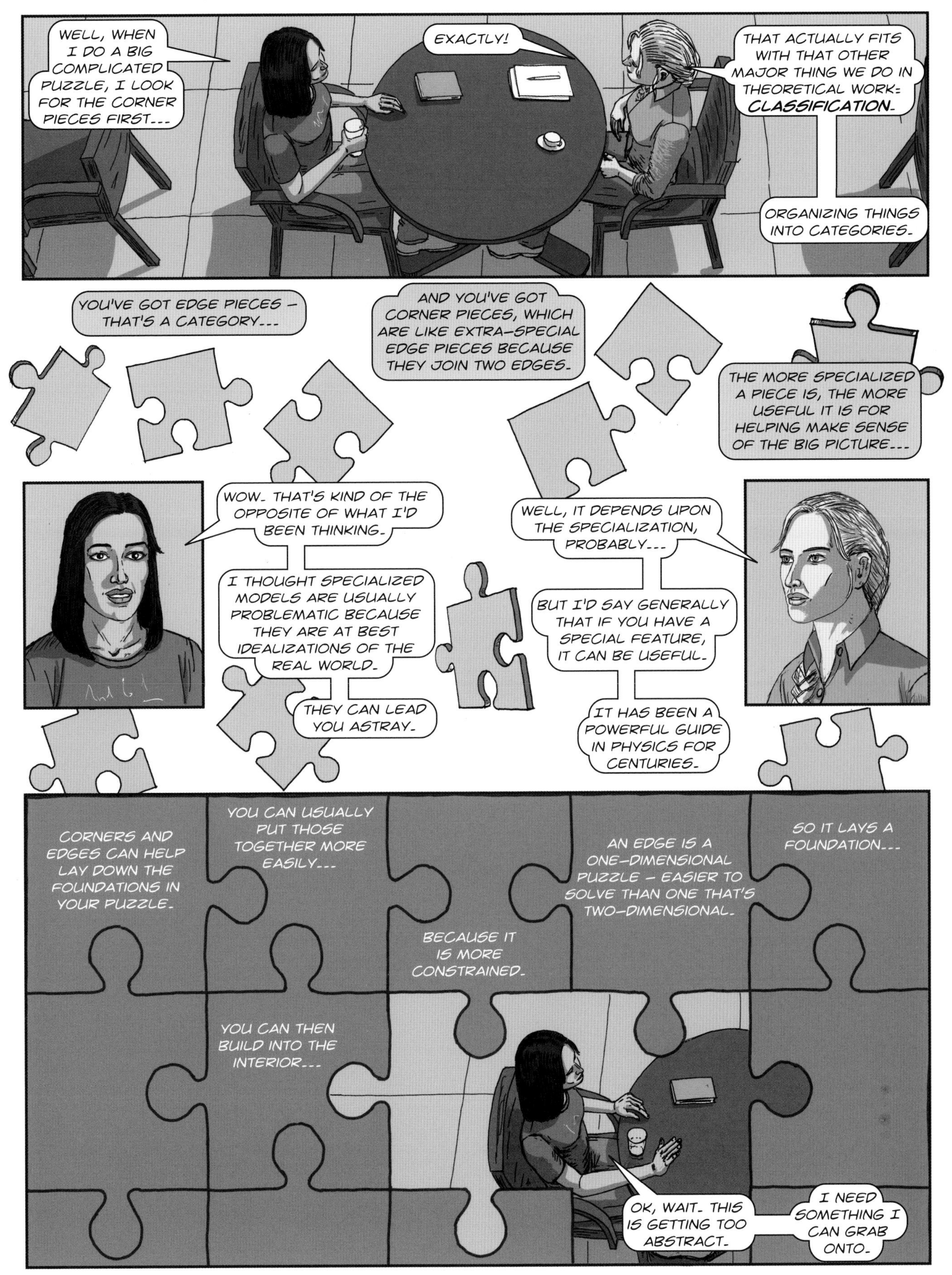
WELL, WHEN I DO A BIG COMPLICATED PUZZLE, I LOOK FOR THE CORNER PIECES FIRST...
EXACTLY!
THAT ACTUALLY FITS WITH THAT OTHER MAJOR THING WE DO IN THEORETICAL WORK: CLASSIFICATION.
ORGANIZING THINGS INTO CATEGORIES.
YOU'VE GOT EDGE PIECES – THAT'S A CATEGORY...
AND YOU'VE GOT CORNER PIECES, WHICH ARE LIKE EXTRA-SPECIAL EDGE PIECES BECAUSE THEY JOIN TWO EDGES.
THE MORE SPECIALIZED A PIECE IS, THE MORE USEFUL IT IS FOR HELPING MAKE SENSE OF THE BIG PICTURE...
WOW. THAT'S KIND OF THE OPPOSITE OF WHAT I'D BEEN THINKING.
I THOUGHT SPECIALIZED MODELS ARE USUALLY PROBLEMATIC BECAUSE THEY ARE AT BEST IDEALIZATIONS OF THE REAL WORLD.
THEY CAN LEAD YOU ASTRAY.
WELL, IT DEPENDS UPON THE SPECIALIZATION, PROBABLY...
BUT I'D SAY GENERALLY THAT IF YOU HAVE A SPECIAL FEATURE, IT CAN BE USEFUL.
IT HAS BEEN A POWERFUL GUIDE IN PHYSICS FOR CENTURIES.
CORNERS AND EDGES CAN HELP LAY DOWN THE FOUNDATIONS IN YOUR PUZZLE.
YOU CAN USUALLY PUT THOSE TOGETHER MORE EASILY...
BECAUSE IT IS MORE CONSTRAINED.
AN EDGE IS A ONE-DIMENSIONAL PUZZLE – EASIER TO SOLVE THAN ONE THAT'S TWO-DIMENSIONAL.
SO IT LAYS A FOUNDATION...
YOU CAN THEN BUILD INTO THE INTERIOR...
OK, WAIT. THIS IS GETTING TOO ABSTRACT.
I NEED SOMETHING I CAN GRAB ONTO.

OK...
MAYBE SOME OF THE THINGS ABOUT NATURE WE UNDERSTAND VERY WELL ARE LIKE EDGE PIECES.
NEWTON'S LAWS OF MOTION...
NEWTON'S LAW OF GRAVITY, ETC.
YOU KNOW...
NEWTONIAN MECHANICS MIGHT EVEN BE A CORNER PIECE...
REALLY?
BUT THEY ARE IN SO MUCH OF EVERYTHING WE DO!
YOU'RE SAYING THAT THEY'RE JUST THE EDGE OF WHAT THERE IS TO UNDERSTAND ABOUT THE UNIVERSE?
YES, THEY MIGHT WELL BE.
BOTTOM LINE IS THAT WE FOUND AND UNDERSTOOD THOSE ASPECTS OF NATURE FIRST BECAUSE THEY WERE THE MOST EASILY ACCESSIBLE.
THERE'S NO REASON THAT JUST BECAUSE THEY WERE EASY THEY MUST BE THE MOST IMPORTANT, OR EVEN THE MOST ABUNDANT.
WHAT IF NEWTON'S LAWS TURN OUT NOT TO BE EDGE PIECES AT ALL?
WOULD THAT AFFECT YOUR THINKING?
WELL, NOT TOO MUCH.
THE FOCUS ON CORNERS AND EDGES IS REALLY TO DO WITH THE FACT THAT THOSE ARE OFTEN EASIER TO PUT TOGETHER...
BECAUSE THEY SHARE FEATURES!
SO IT IS THE CLASSIFICATION THAT'S IMPORTANT?
IT'S NOT ABOUT WHAT THE CATEGORY IS – EDGE OR CORNER – YOU JUST WANT TO GROUP TOGETHER PIECES THAT HAVE A COMMON FEATURE...
EXACTLY!

SORTING INTO CATEGORIES IS ESPECIALLY VALUABLE FOR THE INTERIOR PIECES.
YOU GROUP THE PIECES THAT SHARE CERTAIN CHARACTERISTICS...
MAYBE COLOR, SHAPE, THE PRESENCE OF CERTAIN PATTERNS OR TEXTURES, ETC.
THIS CAN TAKE A LONG TIME, BUT I FIND THAT ONCE I'VE DONE THAT CLASSIFICATION PROCEDURE, IT SPEEDS UP THINGS A LOT.
IT MAKES IT EASY TO KNOW WHERE TO LOOK FOR CERTAIN KINDS OF PIECE WHEN I'M BUILDING OUT AN AREA IN THE PUZZLE.
ESPECIALLY WHEN I DON'T HAVE A PICTURE TO WORK WITH.
YEAH, I DO THAT TOO!
MY FAMILY AND I USED TO DO PUZZLES TOGETHER WHEN I WAS A KID.
ONCE A YEAR, AT CHRISTMAS, MY DAD WOULD BRING HOME A NEW ONE, AND WE'D WORK ON IT.
I USED TO GET ANNOYED WHEN PEOPLE WERE JUST TRYING OUT RANDOM PIECES INSTEAD OF USING A MORE SYSTEMATIC APPROACH!
HA HA!
YES, I REMEMBER MY OWN FRUSTRATIONS OF THAT KIND. IN THE END I PREFERRED TO DO THEM BY MYSELF!

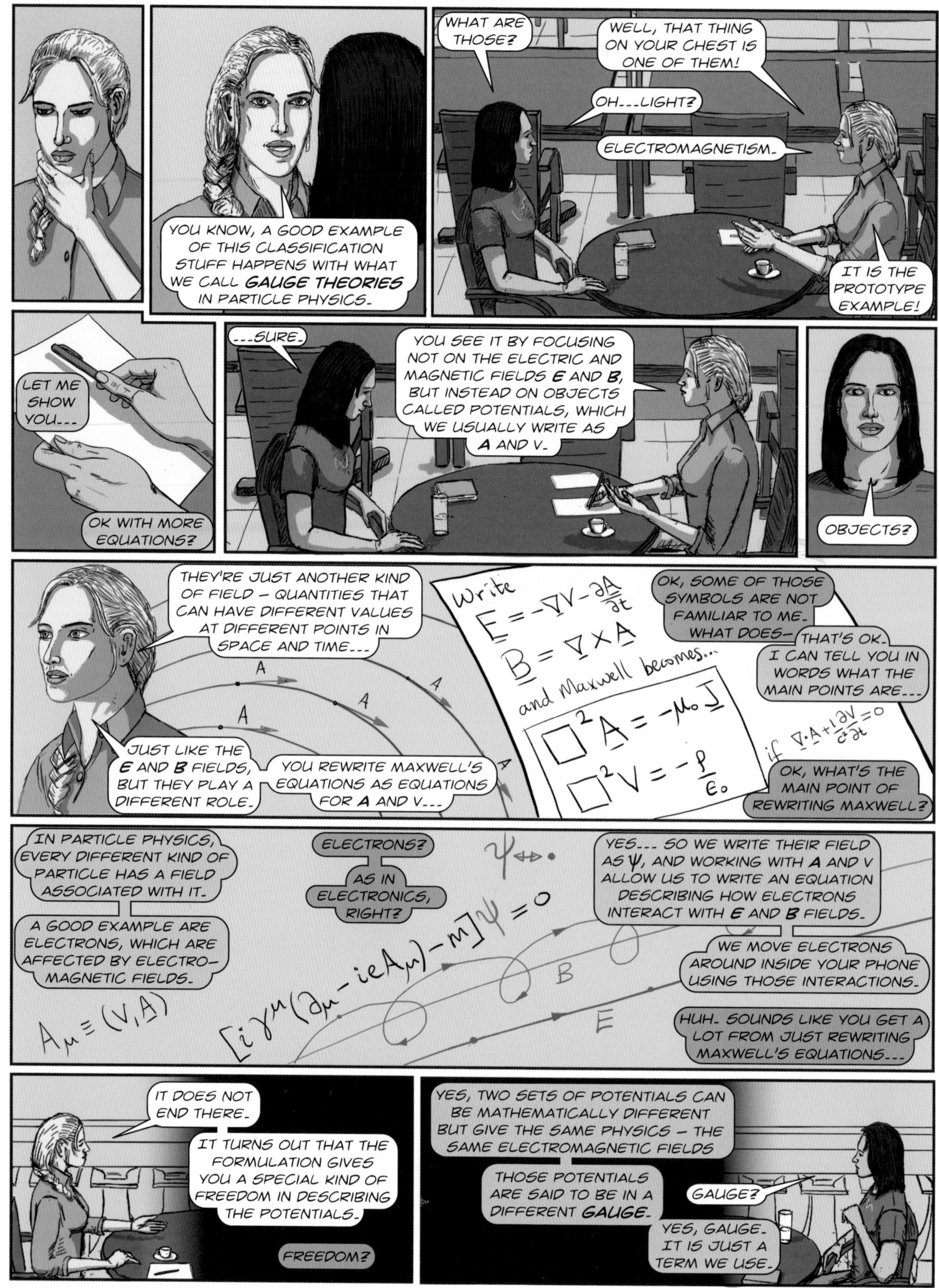

YOU KNOW, A GOOD EXAMPLE OF THIS CLASSIFICATION STUFF HAPPENS WITH WHAT WE CALL GAUGE THEORIES IN PARTICLE PHYSICS.
WHAT ARE THOSE?
WELL, THAT THING ON YOUR CHEST IS ONE OF THEM!
OH...LIGHT?
ELECTROMAGNETISM.
IT IS THE PROTOTYPE EXAMPLE!
LET ME SHOW YOU...
OK WITH MORE EQUATIONS?
...SURE.
YOU SEE IT BY FOCUSING NOT ON THE ELECTRIC AND MAGNETIC FIELDS E AND B, BUT INSTEAD ON OBJECTS CALLED POTENTIALS, WHICH WE USUALLY WRITE AS A AND V.
OBJECTS?
THEY'RE JUST ANOTHER KIND OF FIELD – QUANTITIES THAT CAN HAVE DIFFERENT VALUES AT DIFFERENT POINTS IN SPACE AND TIME...
JUST LIKE THE E AND B FIELDS, BUT THEY PLAY A DIFFERENT ROLE.
YOU REWRITE MAXWELL'S EQUATIONS AS EQUATIONS FOR A AND V...
Write
$E = -\nabla V - \frac{\partial A}{\partial t}$
$B = \nabla \times A$
and Maxwell becomes...
$\Box^2 A = -\mu_0 J$
$\Box^2 V = -\frac{\rho}{\epsilon_0}$
if $\nabla \cdot A + \frac{1}{c^2}\frac{\partial V}{\partial t} = 0$
OK, SOME OF THOSE SYMBOLS ARE NOT FAMILIAR TO ME. WHAT DOES—
THAT'S OK. I CAN TELL YOU IN WORDS WHAT THE MAIN POINTS ARE...
OK, WHAT'S THE MAIN POINT OF REWRITING MAXWELL?
IN PARTICLE PHYSICS, EVERY DIFFERENT KIND OF PARTICLE HAS A FIELD ASSOCIATED WITH IT.
A GOOD EXAMPLE ARE ELECTRONS, WHICH ARE AFFECTED BY ELECTRO-MAGNETIC FIELDS.
ELECTRONS?
AS IN ELECTRONICS, RIGHT?
YES... SO WE WRITE THEIR FIELD AS ψ, AND WORKING WITH A AND V ALLOW US TO WRITE AN EQUATION DESCRIBING HOW ELECTRONS INTERACT WITH E AND B FIELDS.
WE MOVE ELECTRONS AROUND INSIDE YOUR PHONE USING THOSE INTERACTIONS.
HUH. SOUNDS LIKE YOU GET A LOT FROM JUST REWRITING MAXWELL'S EQUATIONS...
$A_\mu \equiv (V, A)$
$[i\gamma^\mu(\partial_\mu - ieA_\mu) - m]\psi = 0$
IT DOES NOT END THERE.
IT TURNS OUT THAT THE FORMULATION GIVES YOU A SPECIAL KIND OF FREEDOM IN DESCRIBING THE POTENTIALS.
FREEDOM?
YES, TWO SETS OF POTENTIALS CAN BE MATHEMATICALLY DIFFERENT BUT GIVE THE SAME PHYSICS – THE SAME ELECTROMAGNETIC FIELDS
THOSE POTENTIALS ARE SAID TO BE IN A DIFFERENT GAUGE.
GAUGE?
YES, GAUGE. IT IS JUST A TERM WE USE.

THE TERM GAUGE ACTUALLY INVOKES A BIT OF ENGINEERING HISTORY, TO DO WITH TRAIN TRACKS AND SO FORTH, BUT NEVER MIND ABOUT THAT...
IT'S RELATED TO ANOTHER FREEDOM...
THE NUMBERS USED IN DESCRIBING AN ELECTRON'S FIELD DON'T ALL HAVE PHYSICAL MEANING.
REALLY?
THEN WHY USE THOSE NUMBERS?
SEEMS LIKE AN OBVIOUS BUG IN THE SYSTEM, RIGHT?
TURNS OUT THAT YOU MAKE PROGRESS BY STICKING WITH DESCRIBING THE FIELD IN THAT WAY, AND YOU TURN THAT BUG INTO A FEATURE.
HOW IS THAT A FEATURE...?
BECAUSE, FOR ONE THING, YOU GET TO WRITE DOWN SENSIBLE EQUATIONS ABOUT WELL-BEHAVED MATHEMATICAL OBJECTS.
SO BECAUSE THE MATH MAKES SENSE?
YES.
BUT HOW DO YOU KNOW YOU'RE STILL DOING PHYSICS?
GOOD QUESTION.
THE ULTIMATE TEST IS EXPERIMENT: IS NATURE DESCRIBED BY THE EQUATIONS YOU FIND?
IN THIS CASE, THE ANSWER IS YES:
THE EQUATIONS DESCRIBE HOW ELECTROMAGNETISM INTERACTS WITH CHARGED PARTICLES LIKE THE ELECTRON MORE ACCURATELY THAN ANY OTHER PHYSICAL THEORY EVER DEVISED TO DESCRIBE ANYTHING.
SO YEAH, GOOD MATH CAN'T BE THE ONLY GUIDE.

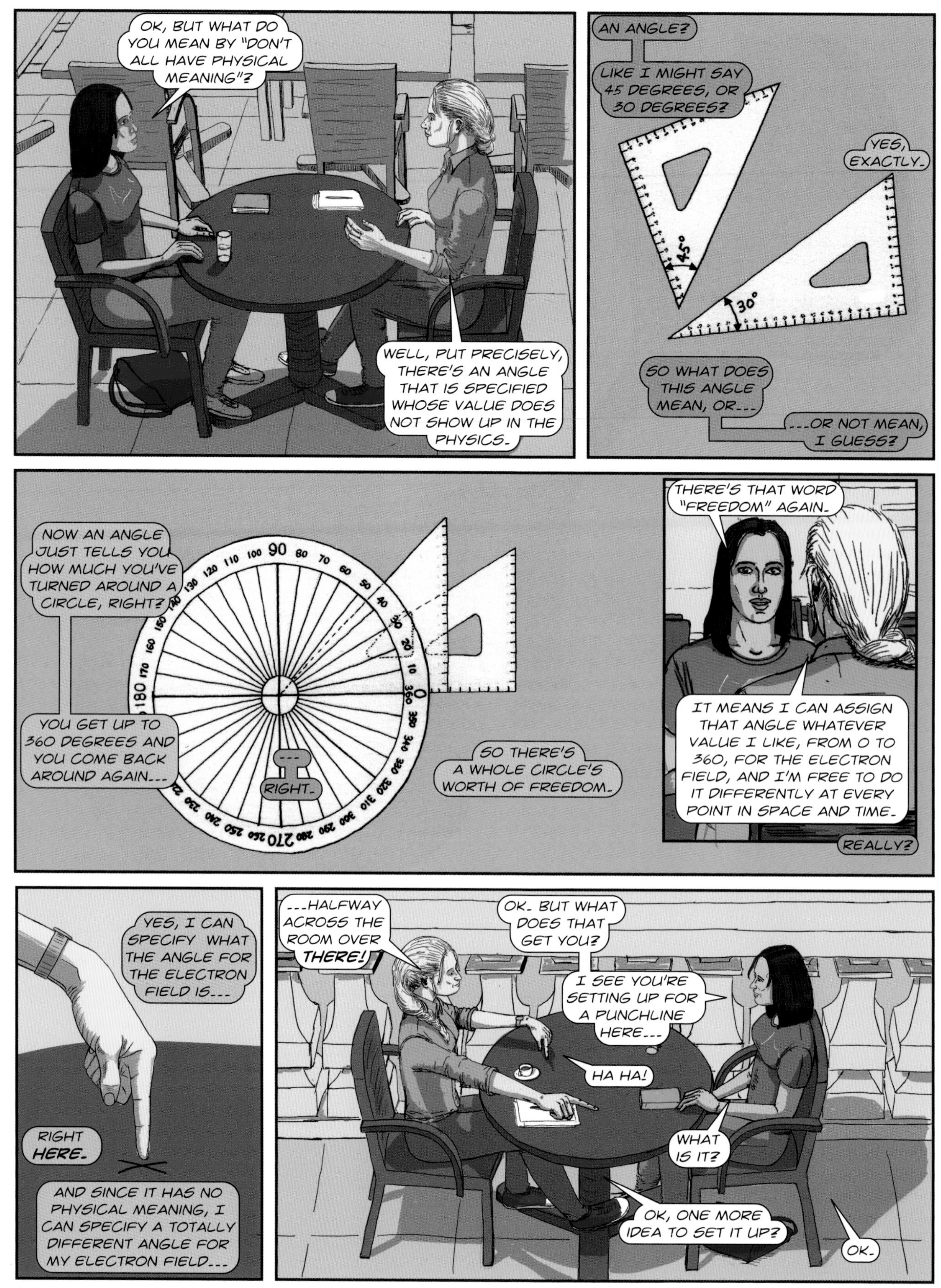

OK, BUT WHAT DO YOU MEAN BY "DON'T ALL HAVE PHYSICAL MEANING"?
WELL, PUT PRECISELY, THERE'S AN ANGLE THAT IS SPECIFIED WHOSE VALUE DOES NOT SHOW UP IN THE PHYSICS.
AN ANGLE?
LIKE I MIGHT SAY 45 DEGREES, OR 30 DEGREES?
45°
30°
YES, EXACTLY.
SO WHAT DOES THIS ANGLE MEAN, OR...
...OR NOT MEAN, I GUESS?
NOW AN ANGLE JUST TELLS YOU HOW MUCH YOU'VE TURNED AROUND A CIRCLE, RIGHT?
YOU GET UP TO 360 DEGREES AND YOU COME BACK AROUND AGAIN...
...
RIGHT.
SO THERE'S A WHOLE CIRCLE'S WORTH OF FREEDOM.
THERE'S THAT WORD "FREEDOM" AGAIN.
IT MEANS I CAN ASSIGN THAT ANGLE WHATEVER VALUE I LIKE, FROM 0 TO 360, FOR THE ELECTRON FIELD, AND I'M FREE TO DO IT DIFFERENTLY AT EVERY POINT IN SPACE AND TIME.
REALLY?
YES, I CAN SPECIFY WHAT THE ANGLE FOR THE ELECTRON FIELD IS...
RIGHT **HERE.**
AND SINCE IT HAS NO PHYSICAL MEANING, I CAN SPECIFY A TOTALLY DIFFERENT ANGLE FOR MY ELECTRON FIELD...
...HALFWAY ACROSS THE ROOM OVER **THERE!**
OK. BUT WHAT DOES THAT GET YOU?
I SEE YOU'RE SETTING UP FOR A PUNCHLINE HERE...
HA HA!
WHAT IS IT?
OK, ONE MORE IDEA TO SET IT UP?
OK.

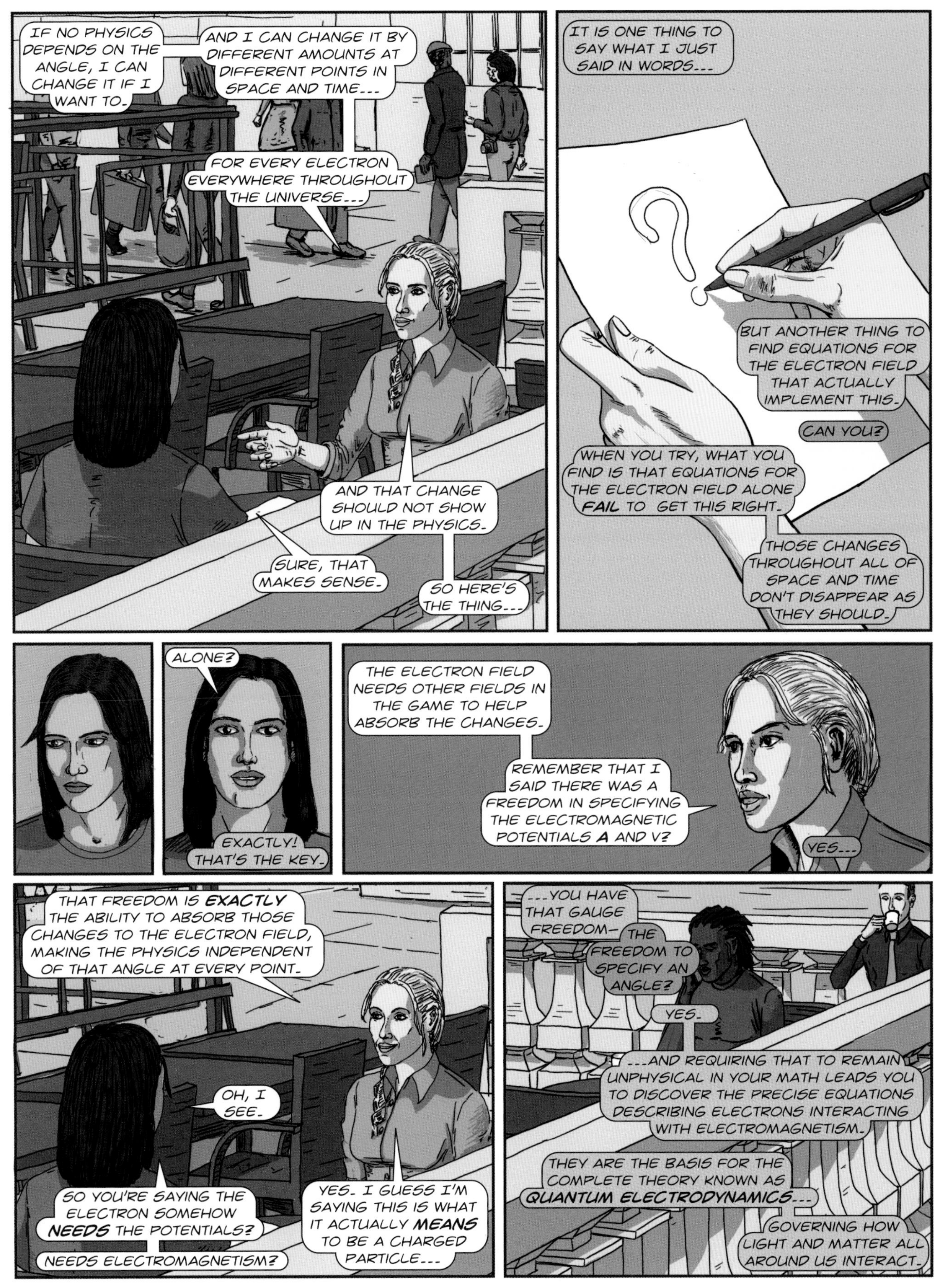
IF NO PHYSICS DEPENDS ON THE ANGLE, I CAN CHANGE IT IF I WANT TO.
AND I CAN CHANGE IT BY DIFFERENT AMOUNTS AT DIFFERENT POINTS IN SPACE AND TIME...
FOR EVERY ELECTRON EVERYWHERE THROUGHOUT THE UNIVERSE...
AND THAT CHANGE SHOULD NOT SHOW UP IN THE PHYSICS.
SURE, THAT MAKES SENSE.
SO HERE'S THE THING...
IT IS ONE THING TO SAY WHAT I JUST SAID IN WORDS...
BUT ANOTHER THING TO FIND EQUATIONS FOR THE ELECTRON FIELD THAT ACTUALLY IMPLEMENT THIS.
CAN YOU?
WHEN YOU TRY, WHAT YOU FIND IS THAT EQUATIONS FOR THE ELECTRON FIELD ALONE **FAIL** TO GET THIS RIGHT.
THOSE CHANGES THROUGHOUT ALL OF SPACE AND TIME DON'T DISAPPEAR AS THEY SHOULD.
ALONE?
EXACTLY! THAT'S THE KEY.
THE ELECTRON FIELD NEEDS OTHER FIELDS IN THE GAME TO HELP ABSORB THE CHANGES.
REMEMBER THAT I SAID THERE WAS A FREEDOM IN SPECIFYING THE ELECTROMAGNETIC POTENTIALS **A** AND V?
YES...
THAT FREEDOM IS **EXACTLY** THE ABILITY TO ABSORB THOSE CHANGES TO THE ELECTRON FIELD, MAKING THE PHYSICS INDEPENDENT OF THAT ANGLE AT EVERY POINT.
OH, I SEE.
SO YOU'RE SAYING THE ELECTRON SOMEHOW **NEEDS** THE POTENTIALS?
NEEDS ELECTROMAGNETISM?
YES. I GUESS I'M SAYING THIS IS WHAT IT ACTUALLY **MEANS** TO BE A CHARGED PARTICLE...
...YOU HAVE THAT GAUGE FREEDOM—
THE FREEDOM TO SPECIFY AN ANGLE?
YES.
...AND REQUIRING THAT TO REMAIN UNPHYSICAL IN YOUR MATH LEADS YOU TO DISCOVER THE PRECISE EQUATIONS DESCRIBING ELECTRONS INTERACTING WITH ELECTROMAGNETISM.
THEY ARE THE BASIS FOR THE COMPLETE THEORY KNOWN AS **QUANTUM ELECTRODYNAMICS**...
GOVERNING HOW LIGHT AND MATTER ALL AROUND US INTERACT.

THAT'S AWESOME. I SEE THAT...
BUT...
WHAT?
SO THERE ARE NO CIRCLES AT EVERY POINT IN SPACE AND TIME THROUGHOUT THE UNIVERSE, RIGHT?
RIGHT. NOT IN THIS SETUP.
THOSE CIRCLES ARE NOT REAL.
SO DON'T YOU FIND IT WEIRD THAT YOUR PHYSICS EQUATIONS ARE ALL CENTERED AROUND A FREEDOM TO DO SOMETHING THAT YOU WORK HARD TO MAKE INVISIBLE?
YES, IT IS STRANGE.
THE BASIC OBJECTS IN THE SETUP KNOW ABOUT THE CIRCLE MATHEMATICALLY, BUT THE ACTUAL PHYSICS EQUATIONS THAT THEY OBEY REMOVE THE CIRCLE FROM THE STORY ENTIRELY.
SO THE CIRCLE IS JUST A MATH TRICK TO GET YOU TO THE RIGHT ANSWER?
YOU COULD SAY THAT...
I PREFER TO THINK OF IT AS SCAFFOLDING.
SCAFFOLDING?
YES, YOU KNOW – A TEMPORARY STRUCTURE THAT'S ERECTED, AS A SORT OF GUIDE...
ONCE I'VE BUILT THE FINAL STRUCTURE – THE ACTUAL PHYSICS IN THIS CASE – I CAN TAKE THE SCAFFOLDING AWAY.
HUH. I SEE.
THIS HAPPENS OFTEN, ON THE WAY TO FINDING THE CORRECT WAY TO DESCRIBE SOME PIECE OF PHYSICS OR ANOTHER...
STILL WEIRD.
YES, I AGREE!
THE COOL THING IS THAT IT WORKS!
THAT'S THE REAL TEST OF ANY OF THIS REALLY. DOES IT AGREE WITH EXPERIMENT?

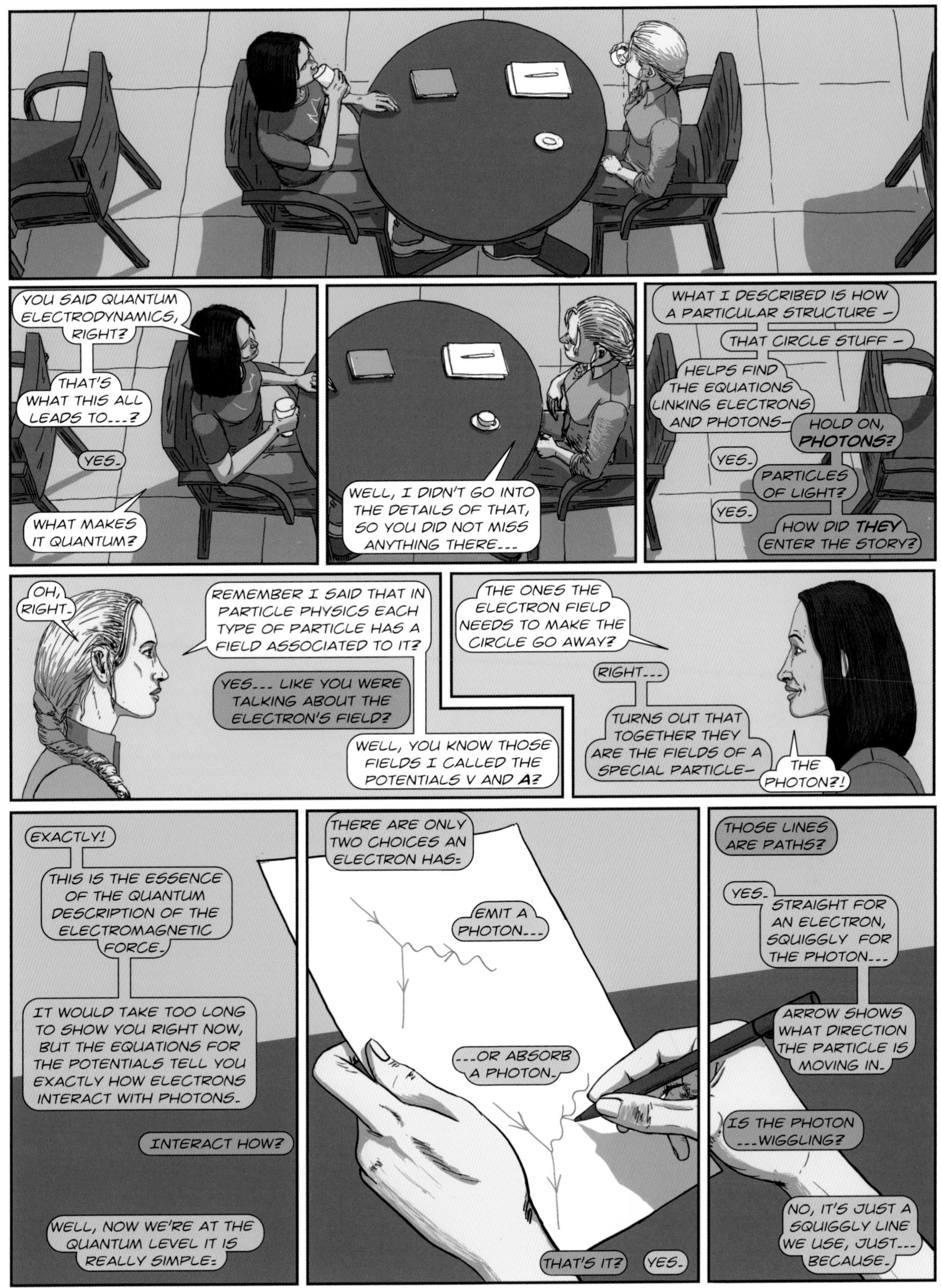
YOU SAID QUANTUM ELECTRODYNAMICS, RIGHT?
THAT'S WHAT THIS ALL LEADS TO...?
YES.
WHAT MAKES IT QUANTUM?
WELL, I DIDN'T GO INTO THE DETAILS OF THAT, SO YOU DID NOT MISS ANYTHING THERE...
WHAT I DESCRIBED IS HOW A PARTICULAR STRUCTURE –
THAT CIRCLE STUFF –
HELPS FIND THE EQUATIONS LINKING ELECTRONS AND PHOTONS–
HOLD ON, **PHOTONS?**
YES.
PARTICLES OF LIGHT?
YES.
HOW DID **THEY** ENTER THE STORY?
OH, RIGHT.
REMEMBER I SAID THAT IN PARTICLE PHYSICS EACH TYPE OF PARTICLE HAS A FIELD ASSOCIATED TO IT?
YES... LIKE YOU WERE TALKING ABOUT THE ELECTRON'S FIELD?
WELL, YOU KNOW THOSE FIELDS I CALLED THE POTENTIALS V AND **A**?
THE ONES THE ELECTRON FIELD NEEDS TO MAKE THE CIRCLE GO AWAY?
RIGHT...
TURNS OUT THAT TOGETHER THEY ARE THE FIELDS OF A SPECIAL PARTICLE–
THE PHOTON?!
EXACTLY!
THIS IS THE ESSENCE OF THE QUANTUM DESCRIPTION OF THE ELECTROMAGNETIC FORCE.
IT WOULD TAKE TOO LONG TO SHOW YOU RIGHT NOW, BUT THE EQUATIONS FOR THE POTENTIALS TELL YOU EXACTLY HOW ELECTRONS INTERACT WITH PHOTONS.
INTERACT HOW?
WELL, NOW WE'RE AT THE QUANTUM LEVEL IT IS REALLY SIMPLE:
THERE ARE ONLY TWO CHOICES AN ELECTRON HAS:
EMIT A PHOTON...
...OR ABSORB A PHOTON.
THAT'S IT?
YES.
THOSE LINES ARE PATHS?
YES.
STRAIGHT FOR AN ELECTRON, SQUIGGLY FOR THE PHOTON...
ARROW SHOWS WHAT DIRECTION THE PARTICLE IS MOVING IN.
IS THE PHOTON ...WIGGLING?
NO, IT'S JUST A SQUIGGLY LINE WE USE, JUST... BECAUSE.

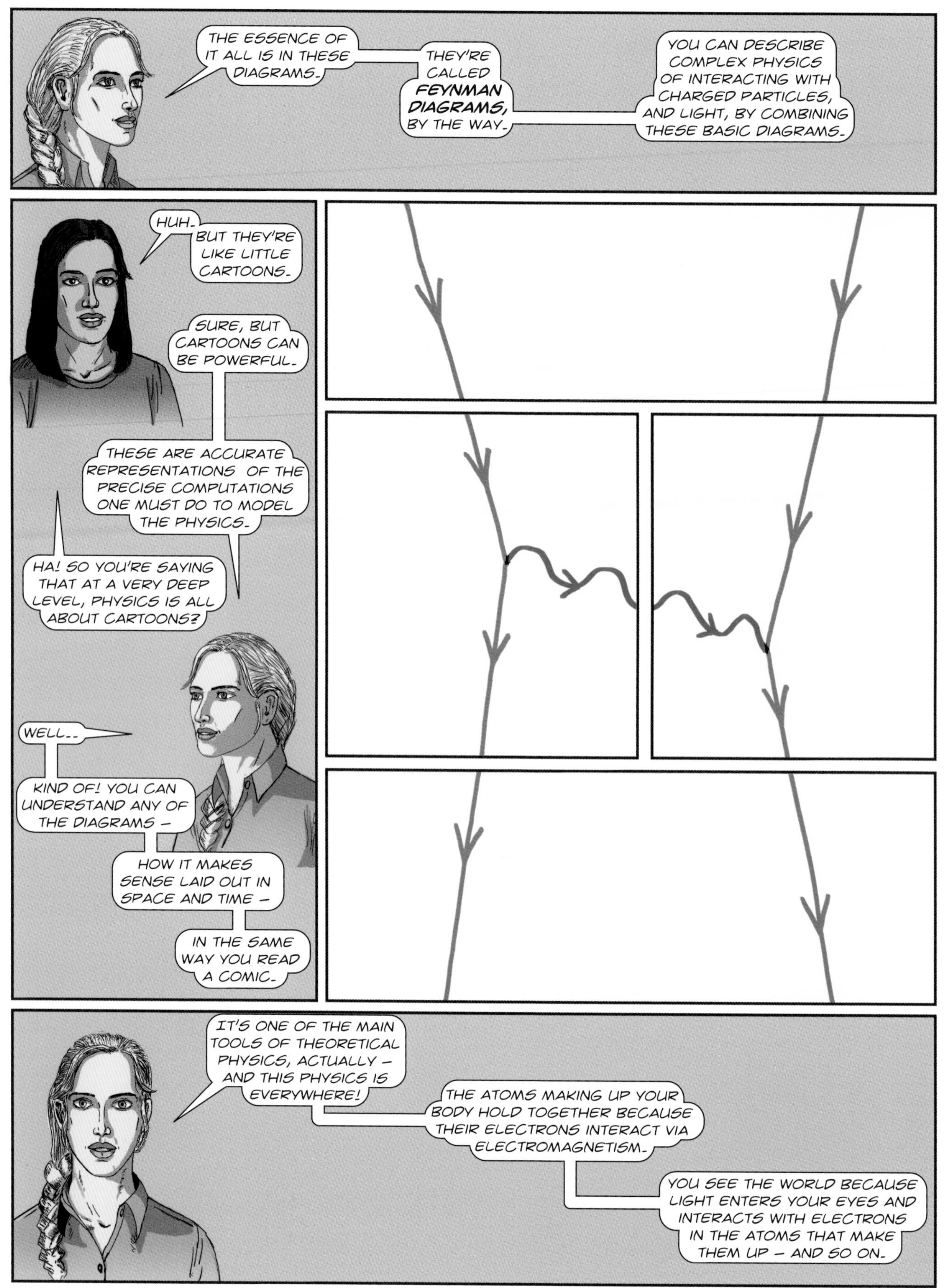

THE ESSENCE OF IT ALL IS IN THESE DIAGRAMS.
THEY'RE CALLED **FEYNMAN DIAGRAMS,** BY THE WAY.
YOU CAN DESCRIBE COMPLEX PHYSICS OF INTERACTING WITH CHARGED PARTICLES, AND LIGHT, BY COMBINING THESE BASIC DIAGRAMS.
HUH.
BUT THEY'RE LIKE LITTLE CARTOONS.
SURE, BUT CARTOONS CAN BE POWERFUL.
THESE ARE ACCURATE REPRESENTATIONS OF THE PRECISE COMPUTATIONS ONE MUST DO TO MODEL THE PHYSICS.
HA! SO YOU'RE SAYING THAT AT A VERY DEEP LEVEL, PHYSICS IS ALL ABOUT CARTOONS?
WELL...
KIND OF! YOU CAN UNDERSTAND ANY OF THE DIAGRAMS –
HOW IT MAKES SENSE LAID OUT IN SPACE AND TIME –
IN THE SAME WAY YOU READ A COMIC.
IT'S ONE OF THE MAIN TOOLS OF THEORETICAL PHYSICS, ACTUALLY – AND THIS PHYSICS IS EVERYWHERE!
THE ATOMS MAKING UP YOUR BODY HOLD TOGETHER BECAUSE THEIR ELECTRONS INTERACT VIA ELECTROMAGNETISM.
YOU SEE THE WORLD BECAUSE LIGHT ENTERS YOUR EYES AND INTERACTS WITH ELECTRONS IN THE ATOMS THAT MAKE THEM UP – AND SO ON.

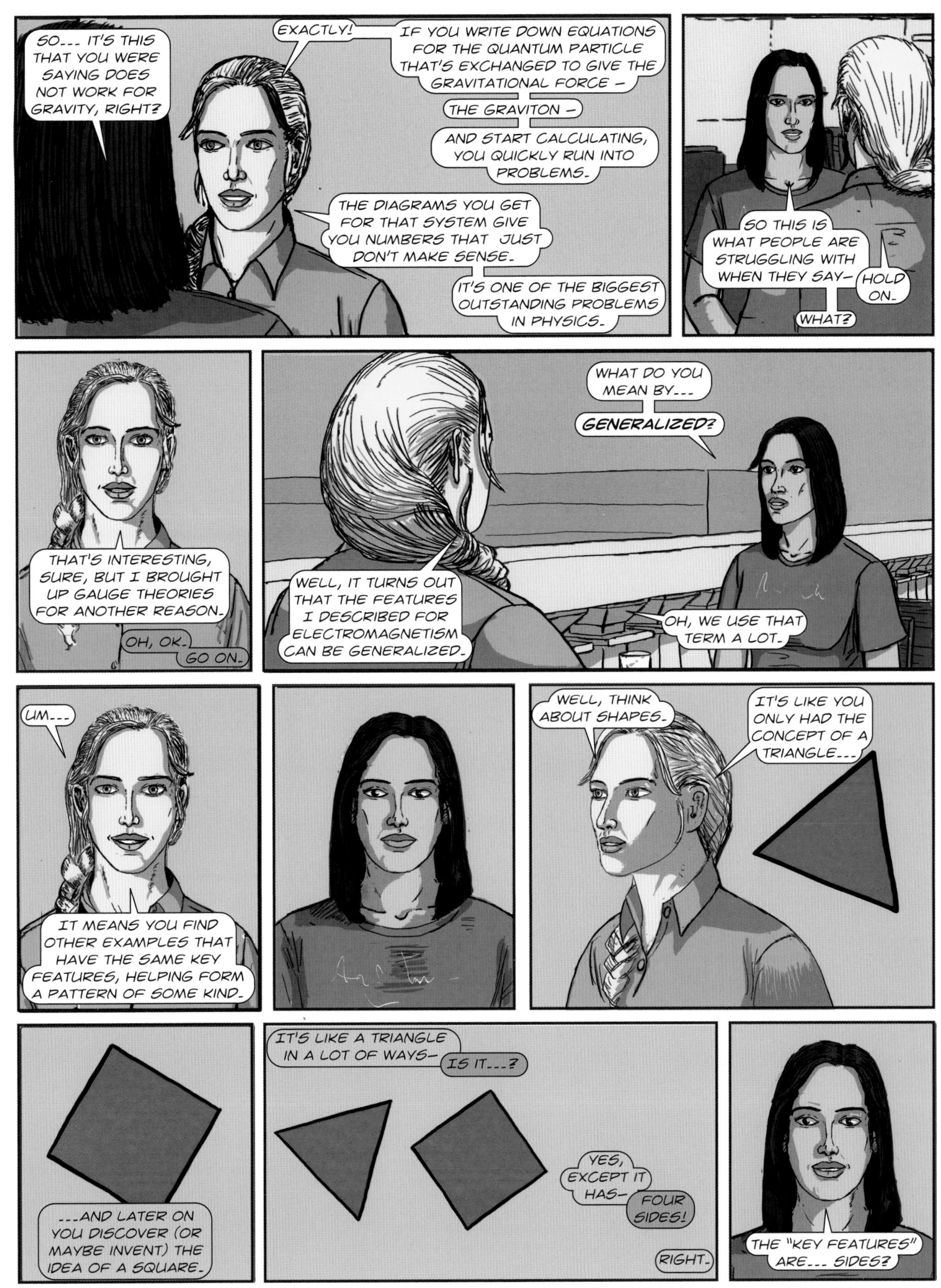

SO... IT'S THIS THAT YOU WERE SAYING DOES NOT WORK FOR GRAVITY, RIGHT?
EXACTLY!
IF YOU WRITE DOWN EQUATIONS FOR THE QUANTUM PARTICLE THAT'S EXCHANGED TO GIVE THE GRAVITATIONAL FORCE –
THE GRAVITON –
AND START CALCULATING, YOU QUICKLY RUN INTO PROBLEMS.
THE DIAGRAMS YOU GET FOR THAT SYSTEM GIVE YOU NUMBERS THAT JUST DON'T MAKE SENSE.
IT'S ONE OF THE BIGGEST OUTSTANDING PROBLEMS IN PHYSICS.
SO THIS IS WHAT PEOPLE ARE STRUGGLING WITH WHEN THEY SAY–
HOLD ON.
WHAT?
THAT'S INTERESTING, SURE, BUT I BROUGHT UP GAUGE THEORIES FOR ANOTHER REASON.
OH, OK.
GO ON.
WELL, IT TURNS OUT THAT THE FEATURES I DESCRIBED FOR ELECTROMAGNETISM CAN BE GENERALIZED.
OH, WE USE THAT TERM A LOT.
WHAT DO YOU MEAN BY...
GENERALIZED?
UM...
IT MEANS YOU FIND OTHER EXAMPLES THAT HAVE THE SAME KEY FEATURES, HELPING FORM A PATTERN OF SOME KIND.
WELL, THINK ABOUT SHAPES.
IT'S LIKE YOU ONLY HAD THE CONCEPT OF A TRIANGLE...
...AND LATER ON YOU DISCOVER (OR MAYBE INVENT) THE IDEA OF A SQUARE.
IT'S LIKE A TRIANGLE IN A LOT OF WAYS–
IS IT...?
YES, EXCEPT IT HAS–
FOUR SIDES!
RIGHT.
THE "KEY FEATURES" ARE... SIDES?

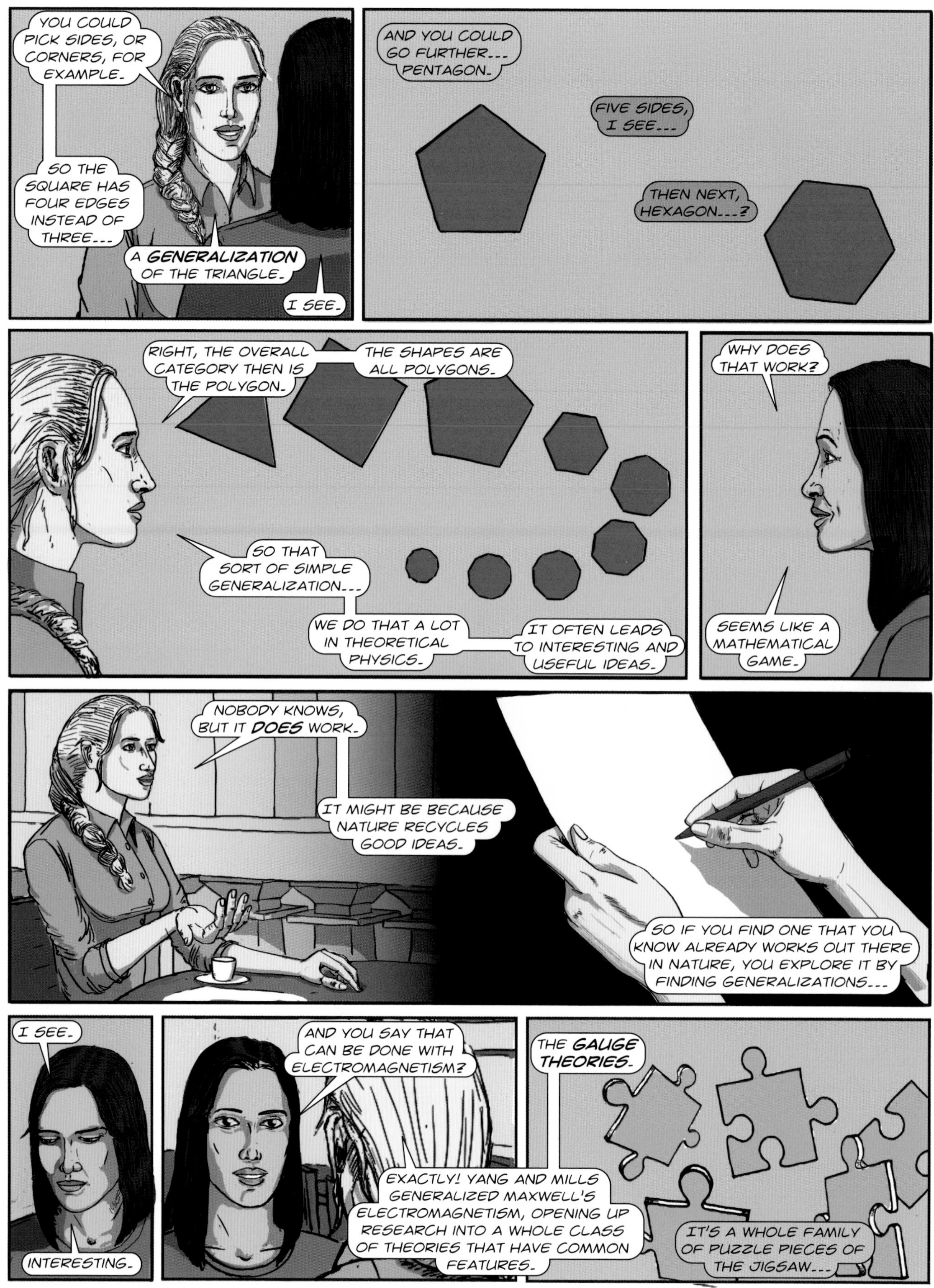

YOU COULD PICK SIDES, OR CORNERS, FOR EXAMPLE.
SO THE SQUARE HAS FOUR EDGES INSTEAD OF THREE...
A **GENERALIZATION** OF THE TRIANGLE.
I SEE.
AND YOU COULD GO FURTHER... PENTAGON.
FIVE SIDES, I SEE...
THEN NEXT, HEXAGON...?
RIGHT, THE OVERALL CATEGORY THEN IS THE POLYGON.
THE SHAPES ARE ALL POLYGONS.
SO THAT SORT OF SIMPLE GENERALIZATION...
WE DO THAT A LOT IN THEORETICAL PHYSICS.
IT OFTEN LEADS TO INTERESTING AND USEFUL IDEAS.
WHY DOES THAT WORK?
SEEMS LIKE A MATHEMATICAL GAME.
NOBODY KNOWS, BUT IT **DOES** WORK.
IT MIGHT BE BECAUSE NATURE RECYCLES GOOD IDEAS.
SO IF YOU FIND ONE THAT YOU KNOW ALREADY WORKS OUT THERE IN NATURE, YOU EXPLORE IT BY FINDING GENERALIZATIONS...
I SEE.
INTERESTING.
AND YOU SAY THAT CAN BE DONE WITH ELECTROMAGNETISM?
EXACTLY! YANG AND MILLS GENERALIZED MAXWELL'S ELECTROMAGNETISM, OPENING UP RESEARCH INTO A WHOLE CLASS OF THEORIES THAT HAVE COMMON FEATURES.
THE **GAUGE THEORIES**.
IT'S A WHOLE FAMILY OF PUZZLE PIECES OF THE JIGSAW...

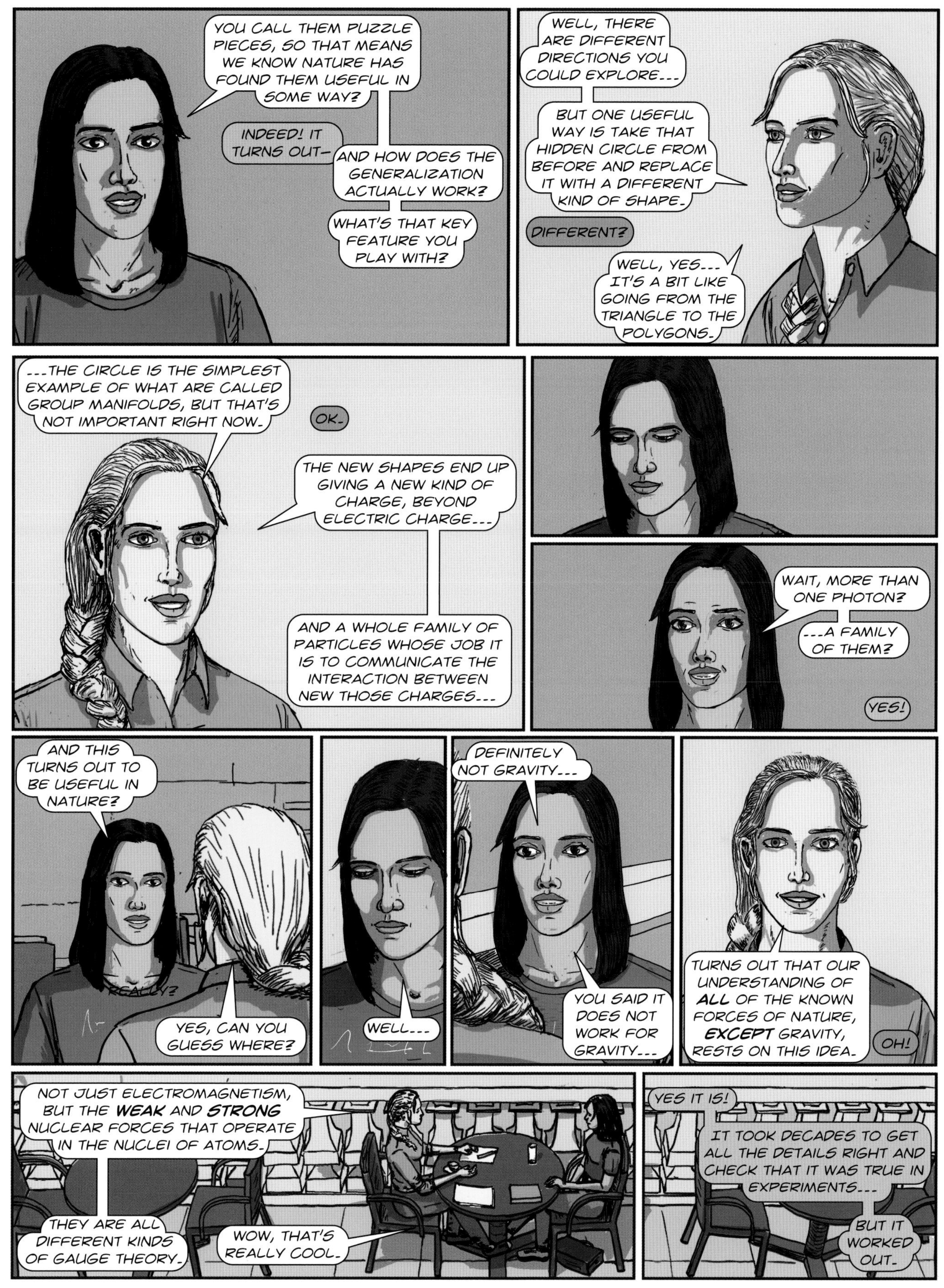

YOU CALL THEM PUZZLE PIECES, SO THAT MEANS WE KNOW NATURE HAS FOUND THEM USEFUL IN SOME WAY?
INDEED! IT TURNS OUT–
AND HOW DOES THE GENERALIZATION ACTUALLY WORK?
WHAT'S THAT KEY FEATURE YOU PLAY WITH?
WELL, THERE ARE DIFFERENT DIRECTIONS YOU COULD EXPLORE...
BUT ONE USEFUL WAY IS TAKE THAT HIDDEN CIRCLE FROM BEFORE AND REPLACE IT WITH A DIFFERENT KIND OF SHAPE.
DIFFERENT?
WELL, YES... IT'S A BIT LIKE GOING FROM THE TRIANGLE TO THE POLYGONS.
...THE CIRCLE IS THE SIMPLEST EXAMPLE OF WHAT ARE CALLED GROUP MANIFOLDS, BUT THAT'S NOT IMPORTANT RIGHT NOW.
OK.
THE NEW SHAPES END UP GIVING A NEW KIND OF CHARGE, BEYOND ELECTRIC CHARGE...
AND A WHOLE FAMILY OF PARTICLES WHOSE JOB IT IS TO COMMUNICATE THE INTERACTION BETWEEN NEW THOSE CHARGES...
WAIT, MORE THAN ONE PHOTON?
...A FAMILY OF THEM?
YES!
AND THIS TURNS OUT TO BE USEFUL IN NATURE?
YES, CAN YOU GUESS WHERE?
WELL...
DEFINITELY NOT GRAVITY...
YOU SAID IT DOES NOT WORK FOR GRAVITY...
TURNS OUT THAT OUR UNDERSTANDING OF **ALL** OF THE KNOWN FORCES OF NATURE, **EXCEPT** GRAVITY, RESTS ON THIS IDEA.
OH!
NOT JUST ELECTROMAGNETISM, BUT THE **WEAK** AND **STRONG** NUCLEAR FORCES THAT OPERATE IN THE NUCLEI OF ATOMS.
THEY ARE ALL DIFFERENT KINDS OF GAUGE THEORY.
WOW, THAT'S REALLY COOL.
YES IT IS!
IT TOOK DECADES TO GET ALL THE DETAILS RIGHT AND CHECK THAT IT WAS TRUE IN EXPERIMENTS...
BUT IT WORKED OUT.

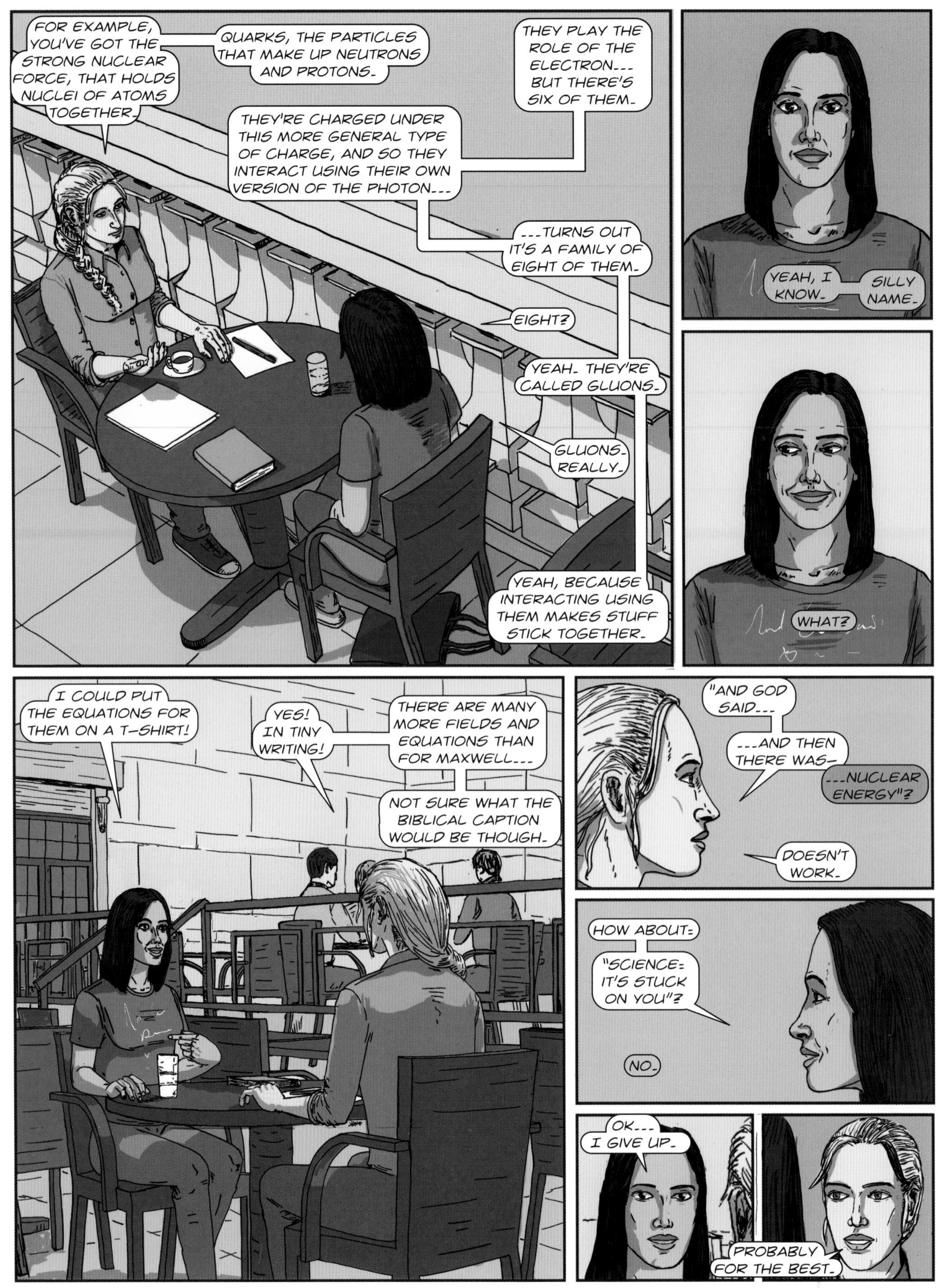
FOR EXAMPLE, YOU'VE GOT THE STRONG NUCLEAR FORCE, THAT HOLDS NUCLEI OF ATOMS TOGETHER.
QUARKS, THE PARTICLES THAT MAKE UP NEUTRONS AND PROTONS.
THEY PLAY THE ROLE OF THE ELECTRON... BUT THERE'S SIX OF THEM.
THEY'RE CHARGED UNDER THIS MORE GENERAL TYPE OF CHARGE, AND SO THEY INTERACT USING THEIR OWN VERSION OF THE PHOTON...
...TURNS OUT IT'S A FAMILY OF EIGHT OF THEM.
EIGHT?
YEAH. THEY'RE CALLED GLUONS.
GLUONS. REALLY.
YEAH, BECAUSE INTERACTING USING THEM MAKES STUFF STICK TOGETHER.
YEAH, I KNOW.
SILLY NAME.
WHAT?
I COULD PUT THE EQUATIONS FOR THEM ON A T-SHIRT!
YES! IN TINY WRITING!
THERE ARE MANY MORE FIELDS AND EQUATIONS THAN FOR MAXWELL...
NOT SURE WHAT THE BIBLICAL CAPTION WOULD BE THOUGH.
"AND GOD SAID...
...AND THEN THERE WAS-
...NUCLEAR ENERGY"?
DOESN'T WORK.
HOW ABOUT:
"SCIENCE: IT'S STUCK ON YOU"?
NO.
OK... I GIVE UP.
PROBABLY FOR THE BEST.

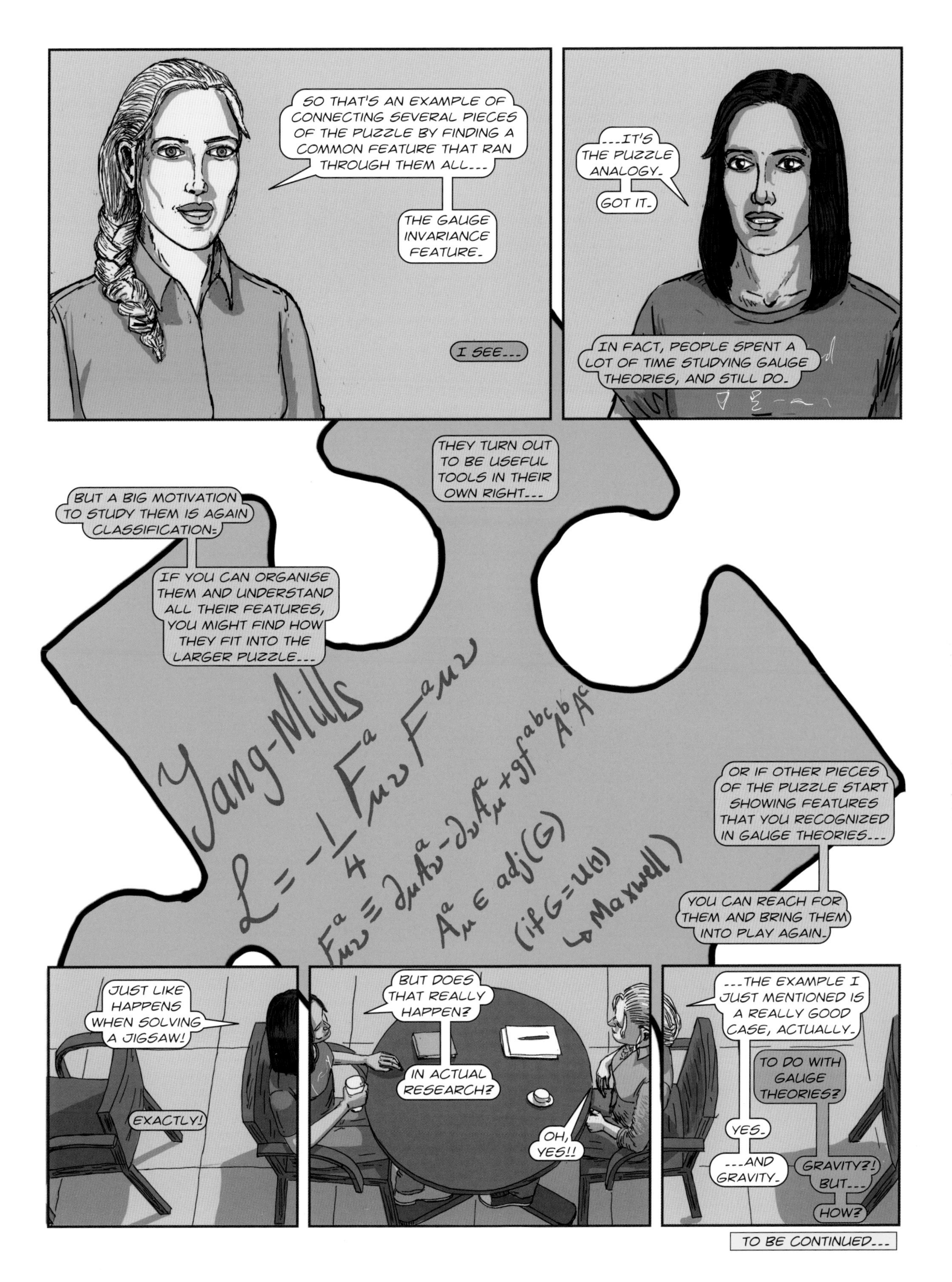

SO THAT'S AN EXAMPLE OF CONNECTING SEVERAL PIECES OF THE PUZZLE BY FINDING A COMMON FEATURE THAT RAN THROUGH THEM ALL...
THE GAUGE INVARIANCE FEATURE.
I SEE...
...IT'S THE PUZZLE ANALOGY.
GOT IT.
IN FACT, PEOPLE SPENT A LOT OF TIME STUDYING GAUGE THEORIES, AND STILL DO.
THEY TURN OUT TO BE USEFUL TOOLS IN THEIR OWN RIGHT...
BUT A BIG MOTIVATION TO STUDY THEM IS AGAIN CLASSIFICATION:
IF YOU CAN ORGANISE THEM AND UNDERSTAND ALL THEIR FEATURES, YOU MIGHT FIND HOW THEY FIT INTO THE LARGER PUZZLE...
Yang-Mills
$\mathcal{L} = -\frac{1}{4} F^a_{\mu\nu} F^{a\mu\nu}$
$F^a_{\mu\nu} \equiv \partial_\mu A^a_\nu - \partial_\nu A^a_\mu + g f^{abc} A^b A^c$
$A^a_\mu \in adj(G)$
(if $G=U(1)$ → Maxwell)
OR IF OTHER PIECES OF THE PUZZLE START SHOWING FEATURES THAT YOU RECOGNIZED IN GAUGE THEORIES...
YOU CAN REACH FOR THEM AND BRING THEM INTO PLAY AGAIN.
JUST LIKE HAPPENS WHEN SOLVING A JIGSAW!
EXACTLY!
BUT DOES THAT REALLY HAPPEN?
IN ACTUAL RESEARCH?
OH, YES!!
...THE EXAMPLE I JUST MENTIONED IS A REALLY GOOD CASE, ACTUALLY.
TO DO WITH GAUGE THEORIES?
YES.
...AND GRAVITY.
GRAVITY?! BUT... HOW?
TO BE CONTINUED...

## Notes

***Page 153*** – See chapters 1 and 3 for more on Maxwell's equations.

***Page 157*** – See the chapter 4 note for pages 79–82 and the chapter 6 note for pages 110–112 for readings about special and general relativity. Additionally, these two biographies unpack the physics journey Einstein makes from special to general very well: Abraham Pais, *Subtle Is the Lord: The Science and the Life of Albert Einstein* (Oxford: Oxford University Press, 1982); Walter Isaacson, *Einstein: His Life and Universe* (New York: Simon & Schuster, 2007).

***Pages 159–160*** – Chapters 6 and 7 had a lot of discussion about reconciling quantum mechanics and general relativity. See their notes for references.

***Page 159*** – For a fascinating and exhilarating inside look at the story of the gravitational wave discovery made at LIGO (Laser Interferometer Gravitational-Wave Observatory), see Janna Levin, *Black Hole Blues and Other Songs from Outer Space* (New York: Alfred A. Knopf, 2016).

***Pages 164–167*** – This is called the "potential formulation" of electromagnetism, and it is the starting point for most quantum treatments, and for the journey to generalizations of this approach to describe other forces. It culminates in what's called quantum field theory. A very accessible account of the workings of the prototype quantum theory of light and charged particles, which this discussion is leading to, is Richard P. Feynman, *QED: The Strange Theory of Light and Matter* (Princeton, NJ: Princeton University Press, 1985).

You've no doubt been thinking about quantum theory a lot already from what you've encountered in earlier conversations, and maybe you've done some follow-up reading. In case you want more, here's a short and accessible account of aspects of the quantum theory, with a particle physics feel to it: Brian Cox and Jeff Forshaw, *The Quantum Universe (And Why Anything that Can Happen, Does)* (Boston: Da Capo Press, 2012). For someone wanting to delve more deeply, it is hard to think of a better start than this: Leonard Susskind and Art Friedman, *Quantum Mechanics: The Theoretical Minimum* (New York: Basic Books, 2014). The Polkinghorne book mentioned in the chapter 7 notes for page 144 (panel 4) is also a good source.

***Page 168*** – Two good technical (but forgivingly and engagingly written) sources that delve into quantum field theory and say much more about things like gauge invariance are Anthony Zee, *Quantum Field Theory in a Nutshell* (Princeton: Princeton University Press, 2010); Tom Lancaster and Stephen J. Blundell, *Quantum Field Theory for the Gifted Amateur* (Oxford: Oxford University Press, 2014).

***Pages 169–170*** – The books mentioned above will have lots of discussion and further explanation about those wonderful tools called Feynman diagrams and their use in quantum field theory.

***Page 170*** – A note on the medium. Be sure to use your evident (because you've read this far) knowledge of how time works in contemporary sequential art to fully explore how Feynman diagrams work. Read those panels in the conventional order, with time ticking by as you move sequentially, and it will tell a story of the particles' interaction. But you can read in other orders too, and still there's a coherent story.

**Page 173** - These generalizations of electromagnetism are called Yang-Mills theories. The story of how they then get combined with the fields of additional particles to end up describing the strong and weak nuclear forces is a wonderful one. There's a nice essay by Christine Sutton in the collection of essays edited by Farmelo mentioned in the chapter 1 note to page 15, and a more intricate version of the tale can be found in this delightfully thorough book: Frank Close, *The Infinity Puzzle: Quantum Field Theory and the Hunt for an Orderly Universe* (New York: Basic Books, 2011).

I THOUGHT YOU SAID GAUGE THEORIES AND GRAVITY DID NOT FIT TOGETHER. WERE YOU JUST MAKING THAT UP, OR...?
IT'S AN INTERESTING AND SUBTLE STORY.
LET ME TRY TO EXPLAIN.

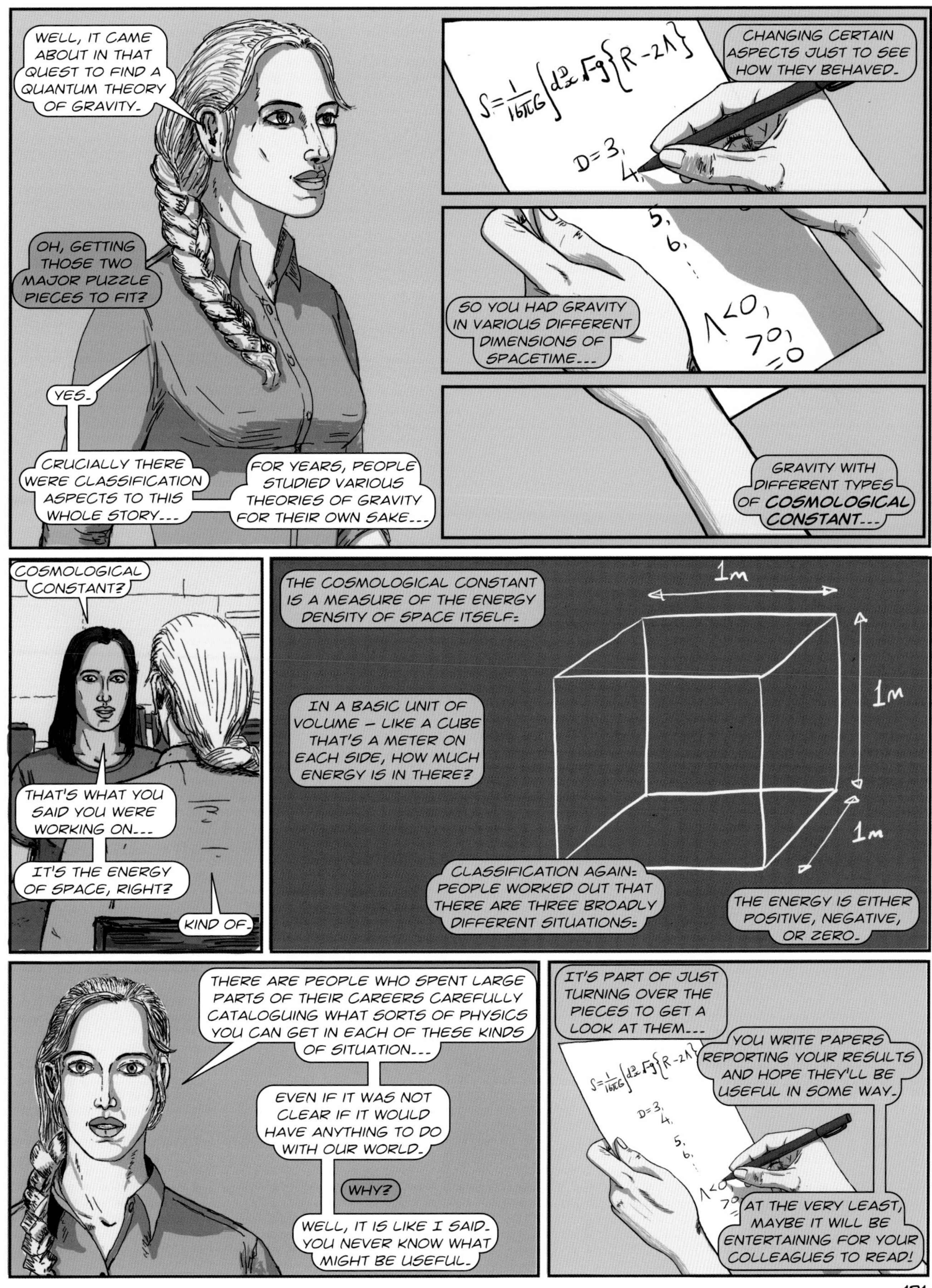
WELL, IT CAME ABOUT IN THAT QUEST TO FIND A QUANTUM THEORY OF GRAVITY.
OH, GETTING THOSE TWO MAJOR PUZZLE PIECES TO FIT?
YES.
CRUCIALLY THERE WERE CLASSIFICATION ASPECTS TO THIS WHOLE STORY...
FOR YEARS, PEOPLE STUDIED VARIOUS THEORIES OF GRAVITY FOR THEIR OWN SAKE...
CHANGING CERTAIN ASPECTS JUST TO SEE HOW THEY BEHAVED.
$S=\frac{1}{16\pi G}\int d^D x \sqrt{-g}\{R-2\Lambda\}$
D=3, 4, 5, 6, ...
SO YOU HAD GRAVITY IN VARIOUS DIFFERENT DIMENSIONS OF SPACETIME...
$\Lambda<0, >0, =0$
GRAVITY WITH DIFFERENT TYPES OF **COSMOLOGICAL CONSTANT**...
COSMOLOGICAL CONSTANT?
THAT'S WHAT YOU SAID YOU WERE WORKING ON...
IT'S THE ENERGY OF SPACE, RIGHT?
KIND OF.
THE COSMOLOGICAL CONSTANT IS A MEASURE OF THE ENERGY DENSITY OF SPACE ITSELF.
1m
1m
1m
IN A BASIC UNIT OF VOLUME – LIKE A CUBE THAT'S A METER ON EACH SIDE, HOW MUCH ENERGY IS IN THERE?
CLASSIFICATION AGAIN: PEOPLE WORKED OUT THAT THERE ARE THREE BROADLY DIFFERENT SITUATIONS:
THE ENERGY IS EITHER POSITIVE, NEGATIVE, OR ZERO.
THERE ARE PEOPLE WHO SPENT LARGE PARTS OF THEIR CAREERS CAREFULLY CATALOGUING WHAT SORTS OF PHYSICS YOU CAN GET IN EACH OF THESE KINDS OF SITUATION...
EVEN IF IT WAS NOT CLEAR IF IT WOULD HAVE ANYTHING TO DO WITH OUR WORLD.
WHY?
WELL, IT IS LIKE I SAID. YOU NEVER KNOW WHAT MIGHT BE USEFUL.
IT'S PART OF JUST TURNING OVER THE PIECES TO GET A LOOK AT THEM...
YOU WRITE PAPERS REPORTING YOUR RESULTS AND HOPE THEY'LL BE USEFUL IN SOME WAY.
AT THE VERY LEAST, MAYBE IT WILL BE ENTERTAINING FOR YOUR COLLEAGUES TO READ!

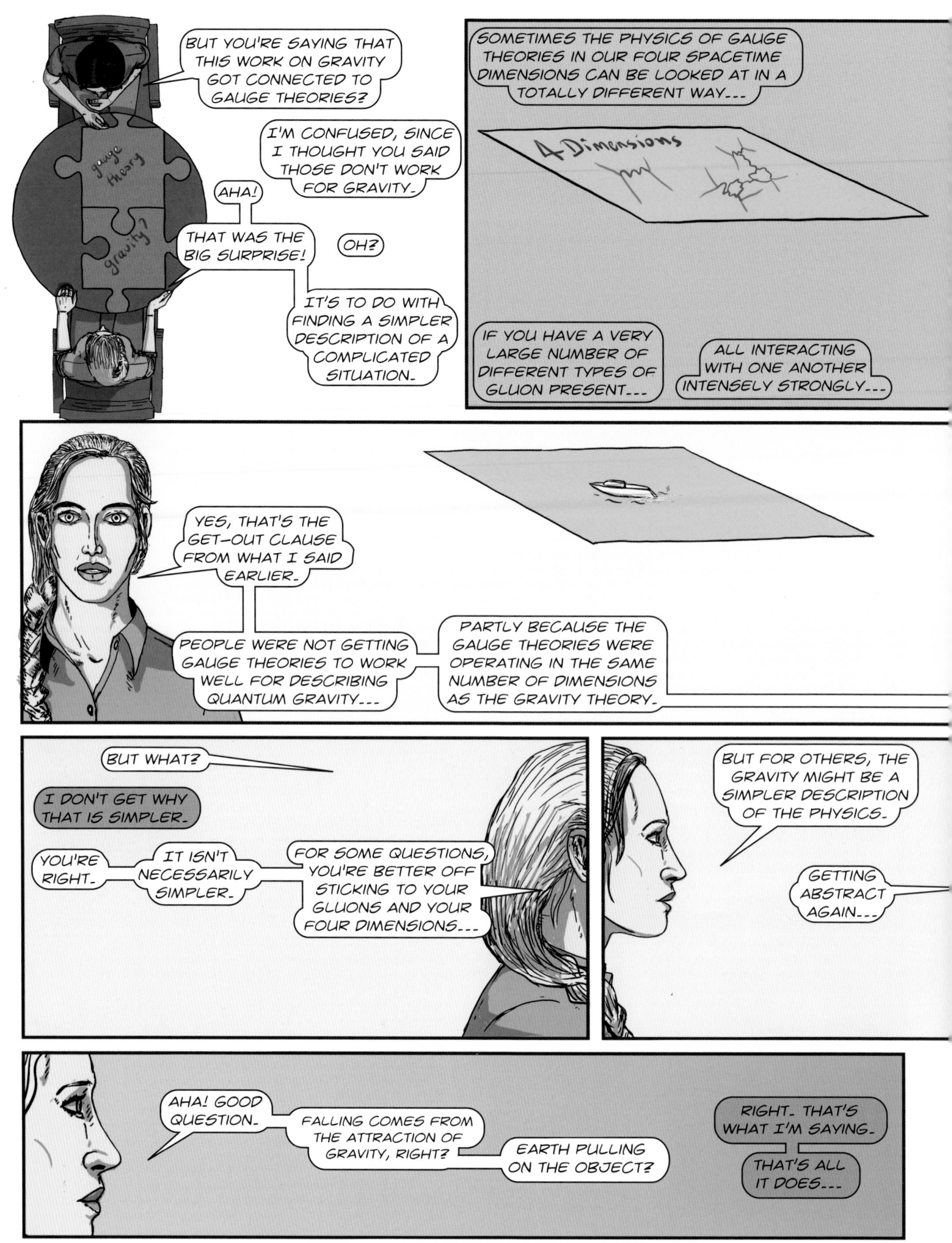
BUT YOU'RE SAYING THAT THIS WORK ON GRAVITY GOT CONNECTED TO GAUGE THEORIES?
I'M CONFUSED, SINCE I THOUGHT YOU SAID THOSE DON'T WORK FOR GRAVITY.
AHA!
THAT WAS THE BIG SURPRISE!
OH?
IT'S TO DO WITH FINDING A SIMPLER DESCRIPTION OF A COMPLICATED SITUATION.
gauge theory
gravity?
SOMETIMES THE PHYSICS OF GAUGE THEORIES IN OUR FOUR SPACETIME DIMENSIONS CAN BE LOOKED AT IN A TOTALLY DIFFERENT WAY...
4 Dimensions
IF YOU HAVE A VERY LARGE NUMBER OF DIFFERENT TYPES OF GLUON PRESENT...
ALL INTERACTING WITH ONE ANOTHER INTENSELY STRONGLY...
YES, THAT'S THE GET-OUT CLAUSE FROM WHAT I SAID EARLIER.
PEOPLE WERE NOT GETTING GAUGE THEORIES TO WORK WELL FOR DESCRIBING QUANTUM GRAVITY...
PARTLY BECAUSE THE GAUGE THEORIES WERE OPERATING IN THE SAME NUMBER OF DIMENSIONS AS THE GRAVITY THEORY.
BUT WHAT?
I DON'T GET WHY THAT IS SIMPLER.
YOU'RE RIGHT.
IT ISN'T NECESSARILY SIMPLER.
FOR SOME QUESTIONS, YOU'RE BETTER OFF STICKING TO YOUR GLUONS AND YOUR FOUR DIMENSIONS...
BUT FOR OTHERS, THE GRAVITY MIGHT BE A SIMPLER DESCRIPTION OF THE PHYSICS.
GETTING ABSTRACT AGAIN...
AHA! GOOD QUESTION.
FALLING COMES FROM THE ATTRACTION OF GRAVITY, RIGHT?
EARTH PULLING ON THE OBJECT?
RIGHT. THAT'S WHAT I'M SAYING.
THAT'S ALL IT DOES...

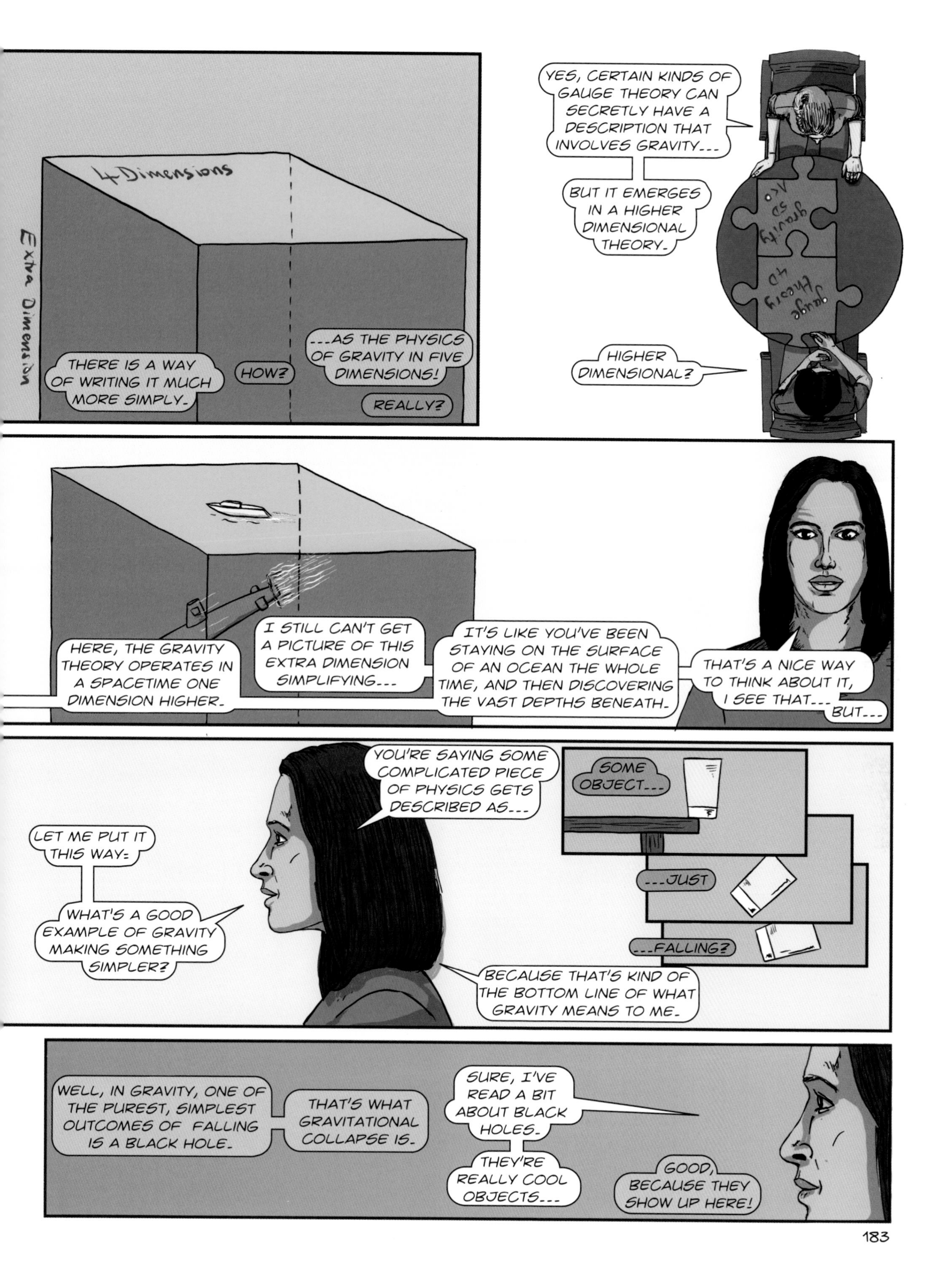
4 Dimensions
Extra Dimension
THERE IS A WAY OF WRITING IT MUCH MORE SIMPLY.
HOW?
...AS THE PHYSICS OF GRAVITY IN FIVE DIMENSIONS!
REALLY?
YES, CERTAIN KINDS OF GAUGE THEORY CAN SECRETLY HAVE A DESCRIPTION THAT INVOLVES GRAVITY...
BUT IT EMERGES IN A HIGHER DIMENSIONAL THEORY.
gravity 5D
gauge theory 4D
HIGHER DIMENSIONAL?
HERE, THE GRAVITY THEORY OPERATES IN A SPACETIME ONE DIMENSION HIGHER.
I STILL CAN'T GET A PICTURE OF THIS EXTRA DIMENSION SIMPLIFYING...
IT'S LIKE YOU'VE BEEN STAYING ON THE SURFACE OF AN OCEAN THE WHOLE TIME, AND THEN DISCOVERING THE VAST DEPTHS BENEATH.
THAT'S A NICE WAY TO THINK ABOUT IT, I SEE THAT...
BUT...
YOU'RE SAYING SOME COMPLICATED PIECE OF PHYSICS GETS DESCRIBED AS...
SOME OBJECT...
...JUST
...FALLING?
LET ME PUT IT THIS WAY:
WHAT'S A GOOD EXAMPLE OF GRAVITY MAKING SOMETHING SIMPLER?
BECAUSE THAT'S KIND OF THE BOTTOM LINE OF WHAT GRAVITY MEANS TO ME.
WELL, IN GRAVITY, ONE OF THE PUREST, SIMPLEST OUTCOMES OF FALLING IS A BLACK HOLE.
THAT'S WHAT GRAVITATIONAL COLLAPSE IS.
SURE, I'VE READ A BIT ABOUT BLACK HOLES.
THEY'RE REALLY COOL OBJECTS...
GOOD, BECAUSE THEY SHOW UP HERE!

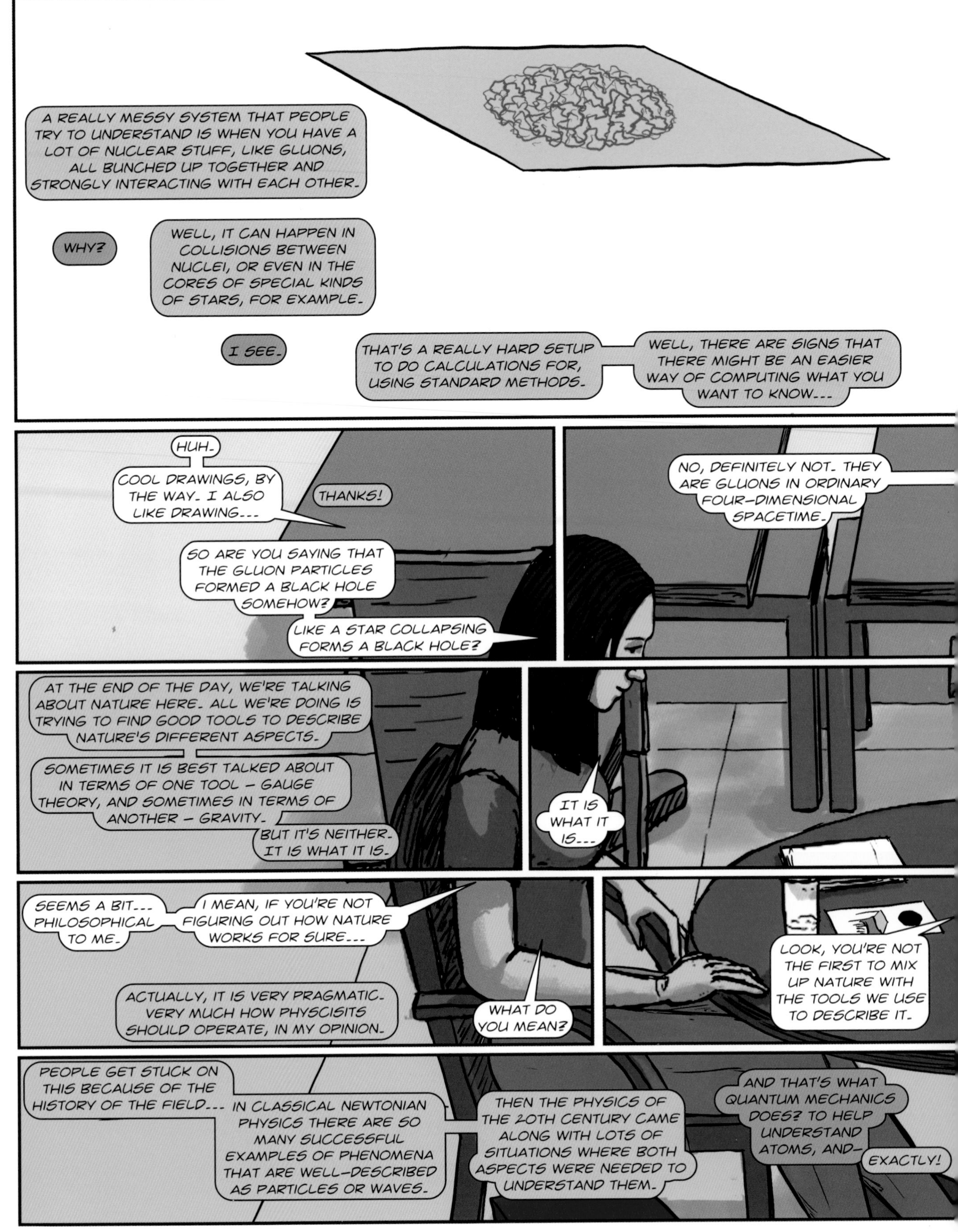
A REALLY MESSY SYSTEM THAT PEOPLE TRY TO UNDERSTAND IS WHEN YOU HAVE A LOT OF NUCLEAR STUFF, LIKE GLUONS, ALL BUNCHED UP TOGETHER AND STRONGLY INTERACTING WITH EACH OTHER.
WHY?
WELL, IT CAN HAPPEN IN COLLISIONS BETWEEN NUCLEI, OR EVEN IN THE CORES OF SPECIAL KINDS OF STARS, FOR EXAMPLE.
I SEE.
THAT'S A REALLY HARD SETUP TO DO CALCULATIONS FOR, USING STANDARD METHODS.
WELL, THERE ARE SIGNS THAT THERE MIGHT BE AN EASIER WAY OF COMPUTING WHAT YOU WANT TO KNOW...
HUH.
COOL DRAWINGS, BY THE WAY. I ALSO LIKE DRAWING...
THANKS!
SO ARE YOU SAYING THAT THE GLUON PARTICLES FORMED A BLACK HOLE SOMEHOW?
LIKE A STAR COLLAPSING FORMS A BLACK HOLE?
NO, DEFINITELY NOT. THEY ARE GLUONS IN ORDINARY FOUR-DIMENSIONAL SPACETIME.
AT THE END OF THE DAY, WE'RE TALKING ABOUT NATURE HERE. ALL WE'RE DOING IS TRYING TO FIND GOOD TOOLS TO DESCRIBE NATURE'S DIFFERENT ASPECTS.
SOMETIMES IT IS BEST TALKED ABOUT IN TERMS OF ONE TOOL – GAUGE THEORY, AND SOMETIMES IN TERMS OF ANOTHER – GRAVITY.
BUT IT'S NEITHER. IT IS WHAT IT IS.
IT IS WHAT IT IS...
SEEMS A BIT... PHILOSOPHICAL TO ME.
I MEAN, IF YOU'RE NOT FIGURING OUT HOW NATURE WORKS FOR SURE...
ACTUALLY, IT IS VERY PRAGMATIC. VERY MUCH HOW PHYSCISITS SHOULD OPERATE, IN MY OPINION.
WHAT DO YOU MEAN?
LOOK, YOU'RE NOT THE FIRST TO MIX UP NATURE WITH THE TOOLS WE USE TO DESCRIBE IT.
PEOPLE GET STUCK ON THIS BECAUSE OF THE HISTORY OF THE FIELD...
IN CLASSICAL NEWTONIAN PHYSICS THERE ARE SO MANY SUCCESSFUL EXAMPLES OF PHENOMENA THAT ARE WELL-DESCRIBED AS PARTICLES OR WAVES.
THEN THE PHYSICS OF THE 20TH CENTURY CAME ALONG WITH LOTS OF SITUATIONS WHERE BOTH ASPECTS WERE NEEDED TO UNDERSTAND THEM.
AND THAT'S WHAT QUANTUM MECHANICS DOES? TO HELP UNDERSTAND ATOMS, AND–
EXACTLY!

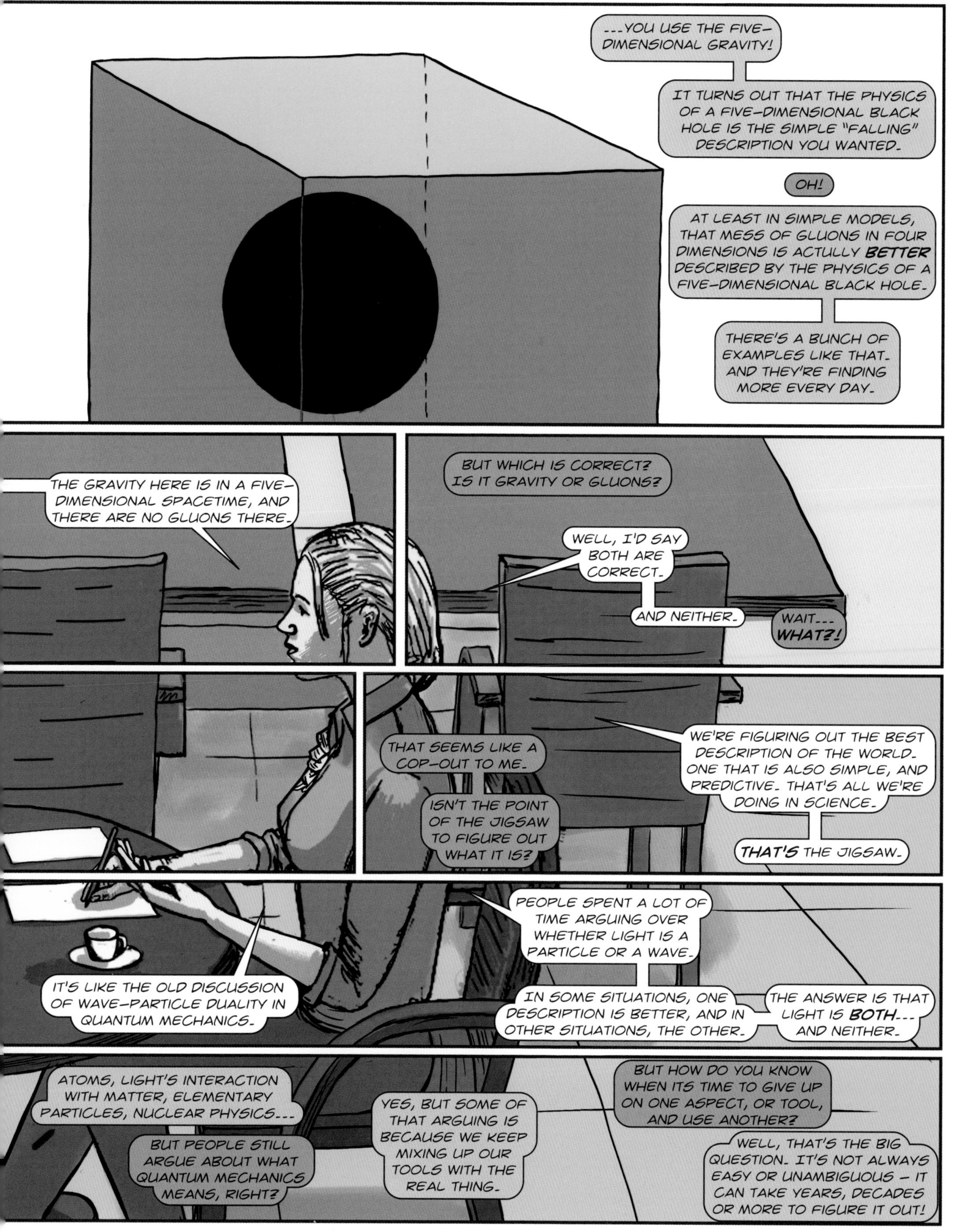
...YOU USE THE FIVE-DIMENSIONAL GRAVITY!
IT TURNS OUT THAT THE PHYSICS OF A FIVE-DIMENSIONAL BLACK HOLE IS THE SIMPLE "FALLING" DESCRIPTION YOU WANTED.
OH!
AT LEAST IN SIMPLE MODELS, THAT MESS OF GLUONS IN FOUR DIMENSIONS IS ACTULLY **BETTER** DESCRIBED BY THE PHYSICS OF A FIVE-DIMENSIONAL BLACK HOLE.
THERE'S A BUNCH OF EXAMPLES LIKE THAT. AND THEY'RE FINDING MORE EVERY DAY.
THE GRAVITY HERE IS IN A FIVE-DIMENSIONAL SPACETIME, AND THERE ARE NO GLUONS THERE.
BUT WHICH IS CORRECT? IS IT GRAVITY OR GLUONS?
WELL, I'D SAY BOTH ARE CORRECT.
AND NEITHER.
WAIT... **WHAT?!**
THAT SEEMS LIKE A COP-OUT TO ME.
ISN'T THE POINT OF THE JIGSAW TO FIGURE OUT WHAT IT IS?
WE'RE FIGURING OUT THE BEST DESCRIPTION OF THE WORLD. ONE THAT IS ALSO SIMPLE, AND PREDICTIVE. THAT'S ALL WE'RE DOING IN SCIENCE.
**THAT'S** THE JIGSAW.
IT'S LIKE THE OLD DISCUSSION OF WAVE-PARTICLE DUALITY IN QUANTUM MECHANICS.
PEOPLE SPENT A LOT OF TIME ARGUING OVER WHETHER LIGHT IS A PARTICLE OR A WAVE.
IN SOME SITUATIONS, ONE DESCRIPTION IS BETTER, AND IN OTHER SITUATIONS, THE OTHER.
THE ANSWER IS THAT LIGHT IS **BOTH**... AND NEITHER.
ATOMS, LIGHT'S INTERACTION WITH MATTER, ELEMENTARY PARTICLES, NUCLEAR PHYSICS...
BUT PEOPLE STILL ARGUE ABOUT WHAT QUANTUM MECHANICS MEANS, RIGHT?
YES, BUT SOME OF THAT ARGUING IS BECAUSE WE KEEP MIXING UP OUR TOOLS WITH THE REAL THING.
BUT HOW DO YOU KNOW WHEN ITS TIME TO GIVE UP ON ONE ASPECT, OR TOOL, AND USE ANOTHER?
WELL, THAT'S THE BIG QUESTION. IT'S NOT ALWAYS EASY OR UNAMBIGUOUS – IT CAN TAKE YEARS, DECADES OR MORE TO FIGURE IT OUT!

I SEE.
AND YOU THINK THAT'S WHAT'S GOING ON HERE WITH GAUGE THEORY GLUONS AND GRAVITY?
YES... IT SEEMS TO BE ONE OF THE SUBTLE WAYS THAT SOME OF OUR PUZZLE PIECES FIT TOGETHER!
BUT IT IS STILL EARLY ON IN THIS JOURNEY OF DISCOVERY.
THESE ARE SIMPLIFIED MODELS OF GLUONS... AND SIMPLIFIED MODELS OF GRAVITY.
SIMPLIFIED? SO THEY MIGHT NOT HAVE ANYTHING TO DO WITH THE REAL WORLD?
I THINK THAT THEY TEACH US THINGS ABOUT IT...POINT THE RIGHT WAY... BUT THERE'S A LOT TO DO.
FOR EXAMPLE, WE'VE STILL MUCH TO LEARN ABOUT THE GRAVITY OF OUR WORLD. THAT HIGHER DIMENSIONAL GRAVITY THAT SHOWS UP FORM THE GLUONS IS NOT THE GRAVITY OF OUR WORLD.
WHY NOT?
WELL, FOR A START IT IS FIVE DIMENSIONAL!
OK, TRUE!
FOR ANOTHER THING, IT IS GRAVITY WITH NEGATIVE COSMOLOGICAL CONSTANT. OUR UNIVERSE HAS, AS FAR AS WE CAN TELL, A POSITIVE COSMOLOGICAL CONSTANT.
OH!
CAN OUR GRAVITY IN FOUR DIMENSIONS BE THOUGHT OF AS A THEORY OF GLUONS OR WHATEVER IN SOME OTHER DIMENSIONS?!
GOOD QUESTION!
THAT'S EXACTLY WHAT A LOT OF PEOPLE HAVE BEEN PUZZLING OVER FOR A WHILE.
WELL, CAN IT?
NOBODY'S FOUND A COMPELLING ANSWER YET.
SO WHAT HAS ALL THIS TAUGHT US ABOUT OUR WORLD THEN?
WHAT DOES THAT MEAN FOR THE PUZZLE?
WELL, PROBABLY THAT WE'VE BEEN THINKING ABOUT GRAVITY – MAYBE SPACETIME ITSELF – AS MORE FUNDAMENTAL THAN IT REALLY IS...
WHAT DO YOU MEAN?
WELL, MANY OF ITS PROPERTIES, LIKE DEFINITE LOCATIONS, NUMBER OF DIMENSIONS, ARE REALLY ONLY PROPERTIES THAT EMERGE IN THE RIGHT SORT OF SITUATION...
HOW DOES THAT MAKE IT LESS FUNDAMENTAL?
YOU'RE RIGHT. THOSE ARE NOT GOOD WORDS.
THAT'S PART OF THE PROBLEM. THE LANGUAGE WE TYPICALLY USE IN THE FIELD IS SOMEWHAT POORLY ADAPTED TO THE DISCOVERIES WE'RE MAKING.
LET ME GIVE YOU AN ANALOGY.
OK!

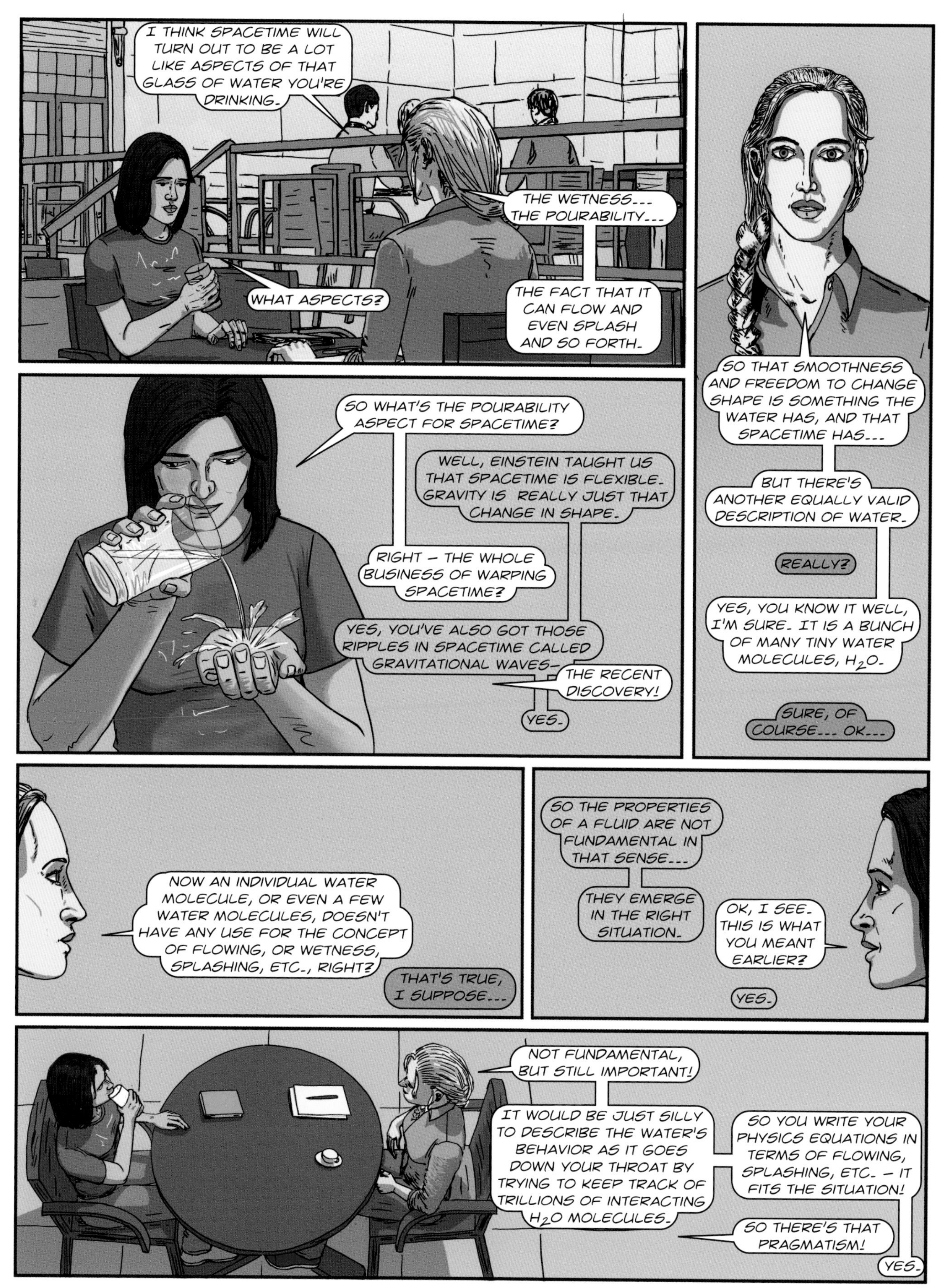

I THINK SPACETIME WILL TURN OUT TO BE A LOT LIKE ASPECTS OF THAT GLASS OF WATER YOU'RE DRINKING.
WHAT ASPECTS?
THE WETNESS... THE POURABILITY...
THE FACT THAT IT CAN FLOW AND EVEN SPLASH AND SO FORTH.
SO THAT SMOOTHNESS AND FREEDOM TO CHANGE SHAPE IS SOMETHING THE WATER HAS, AND THAT SPACETIME HAS...
BUT THERE'S ANOTHER EQUALLY VALID DESCRIPTION OF WATER.
REALLY?
YES, YOU KNOW IT WELL, I'M SURE. IT IS A BUNCH OF MANY TINY WATER MOLECULES, $H_2O$.
SURE, OF COURSE... OK...
SO WHAT'S THE POURABILITY ASPECT FOR SPACETIME?
WELL, EINSTEIN TAUGHT US THAT SPACETIME IS FLEXIBLE. GRAVITY IS REALLY JUST THAT CHANGE IN SHAPE.
RIGHT – THE WHOLE BUSINESS OF WARPING SPACETIME?
YES, YOU'VE ALSO GOT THOSE RIPPLES IN SPACETIME CALLED GRAVITATIONAL WAVES.
THE RECENT DISCOVERY!
YES.
NOW AN INDIVIDUAL WATER MOLECULE, OR EVEN A FEW WATER MOLECULES, DOESN'T HAVE ANY USE FOR THE CONCEPT OF FLOWING, OR WETNESS, SPLASHING, ETC., RIGHT?
THAT'S TRUE, I SUPPOSE...
SO THE PROPERTIES OF A FLUID ARE NOT FUNDAMENTAL IN THAT SENSE...
THEY EMERGE IN THE RIGHT SITUATION.
OK, I SEE. THIS IS WHAT YOU MEANT EARLIER?
YES.
NOT FUNDAMENTAL, BUT STILL IMPORTANT!
IT WOULD BE JUST SILLY TO DESCRIBE THE WATER'S BEHAVIOR AS IT GOES DOWN YOUR THROAT BY TRYING TO KEEP TRACK OF TRILLIONS OF INTERACTING $H_2O$ MOLECULES.
SO YOU WRITE YOUR PHYSICS EQUATIONS IN TERMS OF FLOWING, SPLASHING, ETC. – IT FITS THE SITUATION!
SO THERE'S THAT PRAGMATISM!
YES.

ON THE OTHER HAND, THE MOLECULAR DESCRIPTION, WHERE, BY THE WAY, THE QUANTUM PHYSICS TURNS ON, IS USEFUL FOR HANDLING OTHER ISSUES...
LIKE WATER'S INTERACTION WITH OTHER CHEMICALS, OR HOW IT EVAPORATES AND TURNS TO STEAM, OR FREEZES AND TURNS TO ICE, AND SO FORTH.
SO WHAT ARE THE WATER MOLECULES?
HMM? OH, YOU KNOW, $H_2O$ – TWO HYDROGEN ATOMS BONDED TO AN OXY–
NO.
YOU'RE IMPLYING THAT THERE'S SOME OTHER ASPECT OF SPACETIME – THE MOLECULE ASPECT, LIKE WATER HAS...
WHAT IS IT?
OH I SEE!
WELL, THAT'S WHAT MANY ARE WORKING ON, USING MANY DIFFERENT APPROACHES.
BUT THERE'VE BEEN SOME INTERESTING CLUES...
AND THAT CONNECTION BETWEEN GLUONS AND FIVE-DIMENSIONAL GRAVITY IS ONE CLUE, I THINK.
THE GLUONS ARE THE MOLECULES?
HMMM, KIND OF. THEY'RE PART OF A LARGER STORY COMING FROM A SITUATION THAT TURNS UP IN STRING THEORY, WHERE THE MOLECULES OF SPACETIME ARE THINGS CALLED D-BRANES...
BUT IT'S A LONGER STORY TO GO INTO STRING THEORY, D-BRANES, AND ALL THAT...
IT'S OK, I CAN WRITE IT DOWN AND LOOK IT UP ONLINE LATER...
YEAH, THAT'S A GOOD IDEA. THERE'S A LOT ABOUT THAT OUT THERE...
...AND THE SPELLING IS THE LETTER D, THEN A DASH, THEN B-R-A-N-E-S.
Deebrains
-D-branes
-quantum gravity
-Cosmological constant
OK. THANKS!
SO IF YOU HAD A HANDLE ON THESE MOLECULES OF SPACETIME... QUANTUM BUILDING BLOCKS, I THINK I'VE HEARD THEM CALLED...
WHAT WOULD THEY BE GOOD FOR?
RIGHT.
FIRST OF ALL, WHEN THEY'RE IN THE RIGHT SITUATION... MAYBE A LARGE NUMBER OF THEM INTERACTING...
...THINGS WE CLASSICALLY ASSOCIATE WITH SPACETIME, LIKE NUMBER OF DIMENSIONS, DEFINITE POSITIONS, ETC., WILL EMERGE AS THE SMART TERMS TO USE...
LIKE THE WETNESS OF WATER...?
EXACTLY!
BUT THEN AWAY FROM THAT REGIME WE HOPE THEY DESCRIBE PHYSICS IN SITUATIONS WHERE OUR CURRENT UNDERSTANDING OF GRAVITY AND SPACETIME BREAK DOWN...
LIKE WHERE?

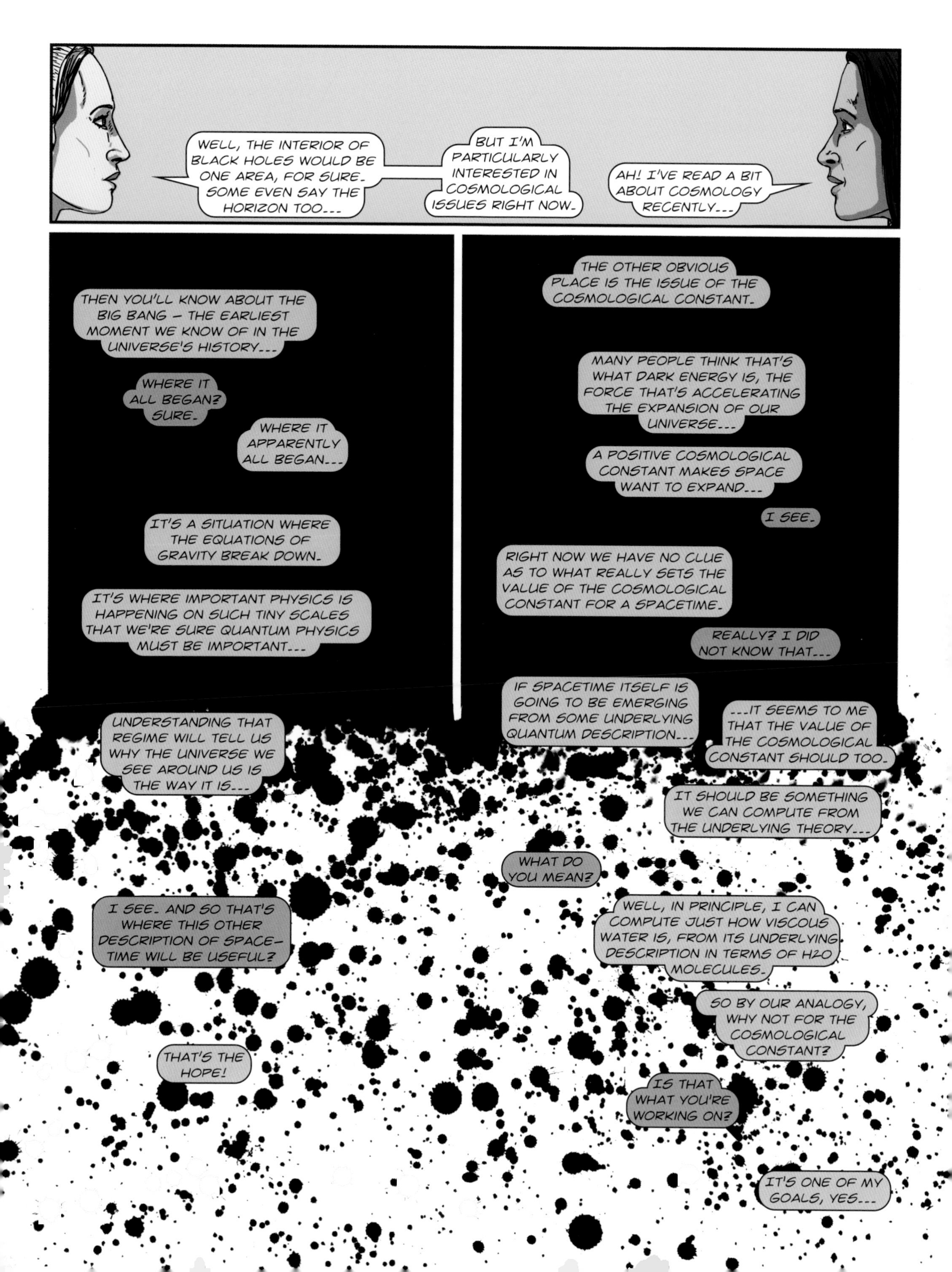
WELL, THE INTERIOR OF BLACK HOLES WOULD BE ONE AREA, FOR SURE. SOME EVEN SAY THE HORIZON TOO...
BUT I'M PARTICULARLY INTERESTED IN COSMOLOGICAL ISSUES RIGHT NOW.
AH! I'VE READ A BIT ABOUT COSMOLOGY RECENTLY...
THEN YOU'LL KNOW ABOUT THE BIG BANG – THE EARLIEST MOMENT WE KNOW OF IN THE UNIVERSE'S HISTORY...
WHERE IT ALL BEGAN? SURE.
WHERE IT APPARENTLY ALL BEGAN...
IT'S A SITUATION WHERE THE EQUATIONS OF GRAVITY BREAK DOWN.
IT'S WHERE IMPORTANT PHYSICS IS HAPPENING ON SUCH TINY SCALES THAT WE'RE SURE QUANTUM PHYSICS MUST BE IMPORTANT...
UNDERSTANDING THAT REGIME WILL TELL US WHY THE UNIVERSE WE SEE AROUND US IS THE WAY IT IS...
I SEE. AND SO THAT'S WHERE THIS OTHER DESCRIPTION OF SPACE-TIME WILL BE USEFUL?
THAT'S THE HOPE!
THE OTHER OBVIOUS PLACE IS THE ISSUE OF THE COSMOLOGICAL CONSTANT.
MANY PEOPLE THINK THAT'S WHAT DARK ENERGY IS, THE FORCE THAT'S ACCELERATING THE EXPANSION OF OUR UNIVERSE...
A POSITIVE COSMOLOGICAL CONSTANT MAKES SPACE WANT TO EXPAND...
I SEE.
RIGHT NOW WE HAVE NO CLUE AS TO WHAT REALLY SETS THE VALUE OF THE COSMOLOGICAL CONSTANT FOR A SPACETIME.
REALLY? I DID NOT KNOW THAT...
IF SPACETIME ITSELF IS GOING TO BE EMERGING FROM SOME UNDERLYING QUANTUM DESCRIPTION...
...IT SEEMS TO ME THAT THE VALUE OF THE COSMOLOGICAL CONSTANT SHOULD TOO.
IT SHOULD BE SOMETHING WE CAN COMPUTE FROM THE UNDERLYING THEORY...
WHAT DO YOU MEAN?
WELL, IN PRINCIPLE, I CAN COMPUTE JUST HOW VISCOUS WATER IS, FROM ITS UNDERLYING DESCRIPTION IN TERMS OF H2O MOLECULES.
SO BY OUR ANALOGY, WHY NOT FOR THE COSMOLOGICAL CONSTANT?
IS THAT WHAT YOU'RE WORKING ON?
IT'S ONE OF MY GOALS, YES...

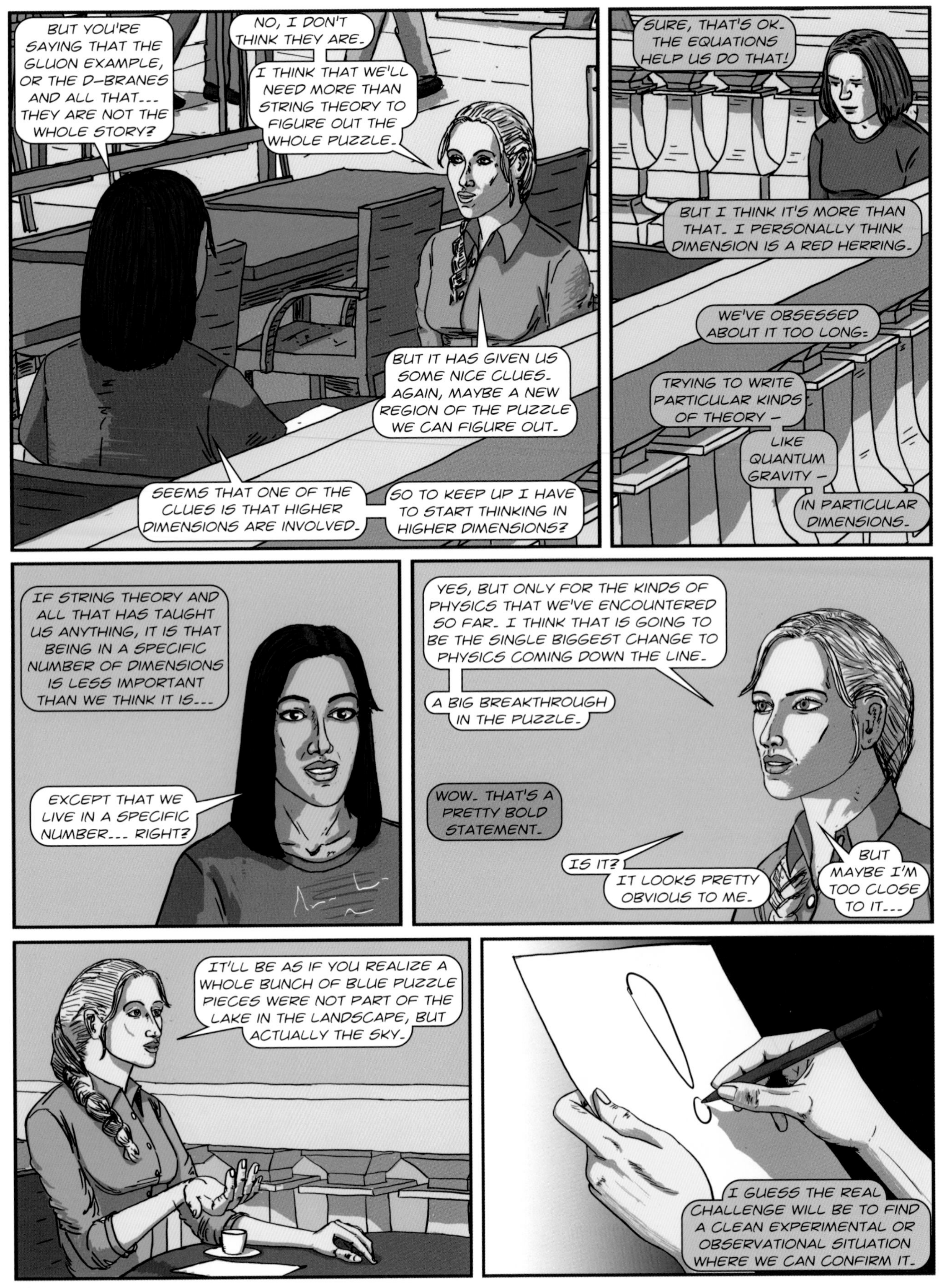

BUT YOU'RE SAYING THAT THE GLUON EXAMPLE, OR THE D-BRANES AND ALL THAT... THEY ARE NOT THE WHOLE STORY?
NO, I DON'T THINK THEY ARE.
I THINK THAT WE'LL NEED MORE THAN STRING THEORY TO FIGURE OUT THE WHOLE PUZZLE.
BUT IT HAS GIVEN US SOME NICE CLUES. AGAIN, MAYBE A NEW REGION OF THE PUZZLE WE CAN FIGURE OUT.
SEEMS THAT ONE OF THE CLUES IS THAT HIGHER DIMENSIONS ARE INVOLVED.
SO TO KEEP UP I HAVE TO START THINKING IN HIGHER DIMENSIONS?
SURE, THAT'S OK. THE EQUATIONS HELP US DO THAT!
BUT I THINK IT'S MORE THAN THAT. I PERSONALLY THINK DIMENSION IS A RED HERRING.
WE'VE OBSESSED ABOUT IT TOO LONG:
TRYING TO WRITE PARTICULAR KINDS OF THEORY —
LIKE QUANTUM GRAVITY —
IN PARTICULAR DIMENSIONS.
IF STRING THEORY AND ALL THAT HAS TAUGHT US ANYTHING, IT IS THAT BEING IN A SPECIFIC NUMBER OF DIMENSIONS IS LESS IMPORTANT THAN WE THINK IT IS...
EXCEPT THAT WE LIVE IN A SPECIFIC NUMBER... RIGHT?
YES, BUT ONLY FOR THE KINDS OF PHYSICS THAT WE'VE ENCOUNTERED SO FAR. I THINK THAT IS GOING TO BE THE SINGLE BIGGEST CHANGE TO PHYSICS COMING DOWN THE LINE.
A BIG BREAKTHROUGH IN THE PUZZLE.
WOW. THAT'S A PRETTY BOLD STATEMENT.
IS IT?
IT LOOKS PRETTY OBVIOUS TO ME.
BUT MAYBE I'M TOO CLOSE TO IT...
IT'LL BE AS IF YOU REALIZE A WHOLE BUNCH OF BLUE PUZZLE PIECES WERE NOT PART OF THE LAKE IN THE LANDSCAPE, BUT ACTUALLY THE SKY.
I GUESS THE REAL CHALLENGE WILL BE TO FIND A CLEAN EXPERIMENTAL OR OBSERVATIONAL SITUATION WHERE WE CAN CONFIRM IT.

FROM THE WAY YOU'VE DESCRIBED THINGS, MAYBE THERE'LL NEVER BE A DEFINITE STATEMENT ABOUT IT...
ABOUT WHAT...?
NOT BEING IN OUR REGULAR FOUR DIMENSIONS.
WHAT DO YOU MEAN?
WELL, SOMEONE COULD ALWAYS JUST SAY YOU'RE NOT LOOKING AT IT THE RIGHT WAY...
LIKE WITH YOUR LARGE NUMBER OF GLUONS SETUP...
ON THE ONE HAND, IT'S FOUR-DIMENSIONAL...
WHILE ON THE OTHER HAND IT'S FIVE-DIMENSIONAL, AND THERE'S GRAVITY INVOLVED.
BUT ACCORDING TO WHAT YOU SAID, NOBODY LIVING IN OUR FOUR DIMENSIONS IS GOING TO FEEL THAT FIVE-DIMENSIONAL GRAVITY...
TRUE...
THEY'LL JUST SAY WE LIVE IN FOUR DIMENSIONS AND YOU'VE FOUND A NICE TRICK FOR MAKING THE EQUATIONS LOOK FIVE-DIMENSIONAL.
YEAH. YOU'RE RIGHT, I SUPPOSE. BUT-
DOES THAT NOT BOTHER YOU?
NO. IT'LL COME DOWN TO PRAGMATISM AGAIN.
HOW SO?

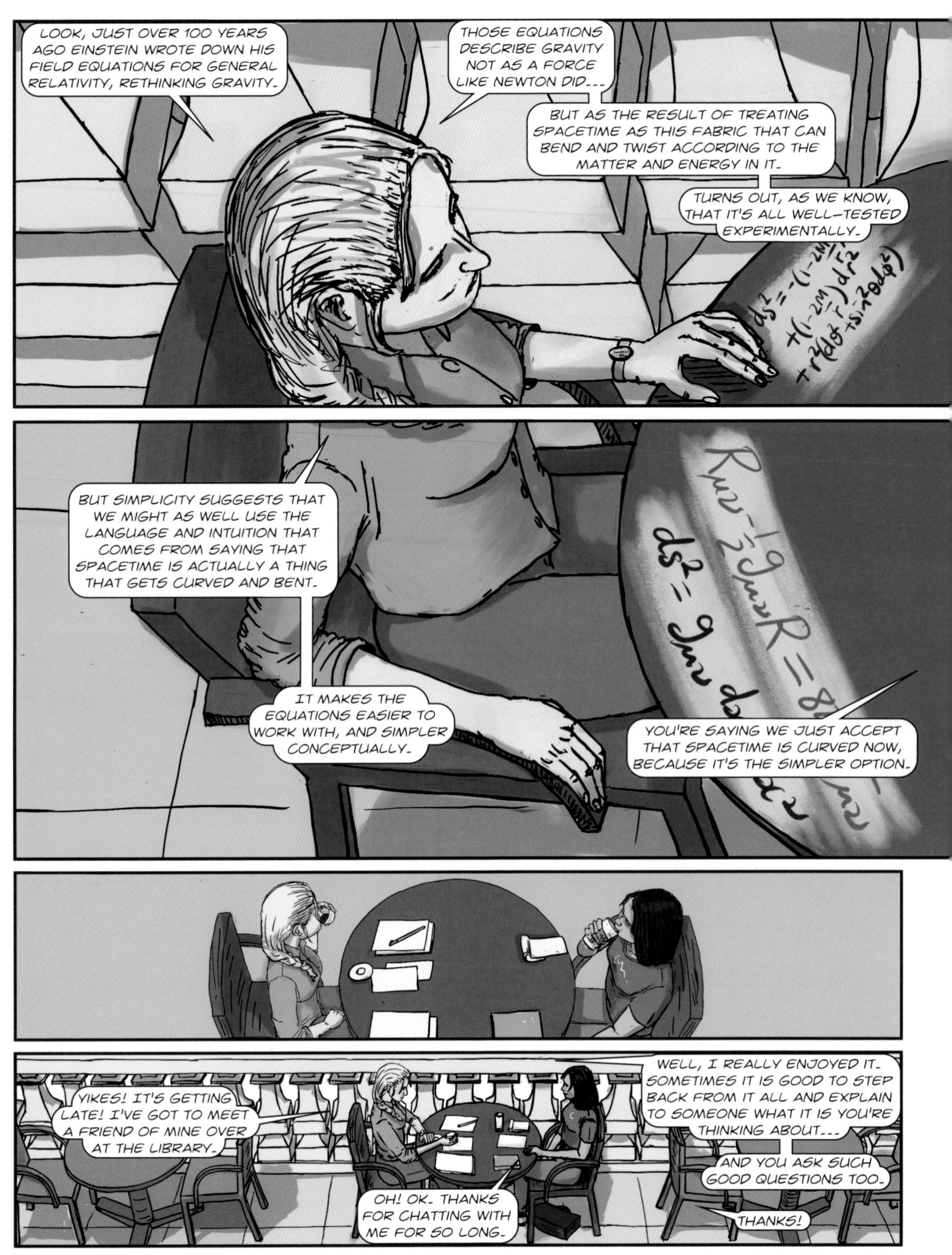
LOOK, JUST OVER 100 YEARS AGO EINSTEIN WROTE DOWN HIS FIELD EQUATIONS FOR GENERAL RELATIVITY, RETHINKING GRAVITY.
THOSE EQUATIONS DESCRIBE GRAVITY NOT AS A FORCE LIKE NEWTON DID...
BUT AS THE RESULT OF TREATING SPACETIME AS THIS FABRIC THAT CAN BEND AND TWIST ACCORDING TO THE MATTER AND ENERGY IN IT.
TURNS OUT, AS WE KNOW, THAT IT'S ALL WELL-TESTED EXPERIMENTALLY.
BUT SIMPLICITY SUGGESTS THAT WE MIGHT AS WELL USE THE LANGUAGE AND INTUITION THAT COMES FROM SAYING THAT SPACETIME IS ACTUALLY A THING THAT GETS CURVED AND BENT.
IT MAKES THE EQUATIONS EASIER TO WORK WITH, AND SIMPLER CONCEPTUALLY.
YOU'RE SAYING WE JUST ACCEPT THAT SPACETIME IS CURVED NOW, BECAUSE IT'S THE SIMPLER OPTION.
YIKES! IT'S GETTING LATE! I'VE GOT TO MEET A FRIEND OF MINE OVER AT THE LIBRARY.
OH! OK. THANKS FOR CHATTING WITH ME FOR SO LONG.
WELL, I REALLY ENJOYED IT. SOMETIMES IT IS GOOD TO STEP BACK FROM IT ALL AND EXPLAIN TO SOMEONE WHAT IT IS YOU'RE THINKING ABOUT...
AND YOU ASK SUCH GOOD QUESTIONS TOO.
THANKS!

SURE... LIGHT BENT BY THE SUN, GRAVITATIONAL WAVES, AND SO FORTH... BUT WHAT ARE YOU GETTING AT?
WE ALL TALK ABOUT BENDING AND CURVING SPACETIME NOW... BUT YOU COULD SIMPLY THINK OF GRAVITY AS A FORCE BEING DESCRIBED BY THOSE EQUATIONS.
YOU CAN FORGET ABOUT THE BENDING AND THE CURVATURE OF SPACETIME AND CHOOSE NOT TO BELIEVE IN IT AT ALL. JUST TAKE THE EQUATIONS AND DO YOUR PHYSICS.
YES. NOT JUST SIMPLER, BUT CONCEPTUALLY IT IS HE BETTER GUIDE.
SO I THINK THE SAME THING WILL HAPPEN WITH DIMENSION TOO, ONCE WE FIND MANY GOOD PRAGMATIC REASONS TO DESCRIBE THE NUMBER OF DIMENSIONS AS DIFFERENT FROM FOUR, AS I AM SURE WE WILL ONE DAY.
GOOD REASONS HERE MEANS EXPERIMENTS?
EXPERIMENTS, YES, AND ALSO OBSERVATIONS PROBABLY IN ASTROPHYSICS AND COSMOLOGY...
AND THEY'D WORK ALONGSIDE WHAT YOU MIGHT CALL INDIRECT GOOD REASONS: EXPLANATIONS OF THINGS WE ALREADY KNOW THAT ARE BETTER DONE IN TERMS OF DIFFERENT NUMBERS OF DIMENSIONS.
HUH...
I GUESS WE'LL SEE.
PERHAPS.
CAN WE STAY IN TOUCH?
SURE...
HERE'S MY EMAIL...
'BYE!
GOOD-BYE!

## Notes

***Page 182*** - A large number of different types of gluon. This is what you'll see called "large N" in other sources. Typically, N is a measure of the size of the structure that generalizes the circle (the "rank" of the "gauge group," for those interested) and the number of different types of gluon that generalize the photon grows as $N^2$.

***Pages 182–185*** - This dimension-crossing connection between the physics of gauge theory and physics of gravity is called the AdS/CFT correspondence, and it is an example of a broader phenomenon in string theory called gauge/gravity duality.* The change in dimension is part of the reason it is referred to as "holographic."** AdS/CFT can be phrased in such a way that needs no reference to its string theory origins, and it is so robust in its structure that it has become a powerful tool for tackling physics problems that (once upon a time) seemed very far from the concerns of string theory, used by scientists in a variety of fields such as condensed matter physics and nuclear physics to organize certain kinds of physics. To date, few accessible nonexpert accounts of AdS/CFT exist in book form.*** Here is an excellent article by some of its pioneers: Igor Klebanov and Juan Maldacena, "Solving Quantum Field Theories via Curved Spacetimes," *Physics Today* 62 (2009): 28. (An open access version is here: https://arxiv.org/abs/gr-qc/9310026/.)

*The best understood version of the correspondence involves extra fields that gives the whole model an extra symmetry called "supersymmetry," but that's well beyond the scope of this conversation. See the reading.

**The idea that gravity is fundamentally able to be captured using physics in one dimension fewer than it operates in (i.e., it is "holographic") is due to Gerard 't Hooft. See his essay in A. Ali, J. Ellis, and S. Randjbar-Daemi, eds., *Conference on Highlights of Particle and Condensed Matter Physics (SALAMFEST)* (River Edge, NJ: World Scientific, 1993).

***That short book by another of the AdS/CFT pioneers, Gubser (mentioned in the chapter 3 notes for pages 52–53) has some core aspects of AdS/CFT in it, with applications to ideas in heavy-ion physics, and it is written for nonexperts, but otherwise not much exists. Some more technical books have begun to appear as researcher handbooks for applying AdS/CFT to various types of physics. Browsing the Cambridge University Press catalog yields several excellent recent ones. These sorts of applications—connecting very different fields, and very different kinds of physicist—has been one of the remarkable recent examples of the benefits of the "jigsaw puzzle" nature of research described in the conversations in chapters 8 and 9. For a flavor of all this (if you are not an expert, it is worth looking at the introductory sections before it gets too technical), see, for example, Makoto Natsuume, *AdS/CFT Duality User Guide* (Tokyo: Springer, 2015). (An online version is at https://arxiv.org/abs/1409.3575/.)

***Pages 186–190*** - The conversations in chapters 6 and 7 touched upon the idea of the geometrical description of spacetime breaking down and being replaced by something else in a quantum theory of gravity, and it is being revisited here. See the notes in those chapters for further reading. Hints of how it works appear in various ways in string theory (through various kinds of string duality), and this AdS/CFT context is one of them. For sample discussions of to what extent the AdS/CFT example constitutes an example of emergent spacetime, see the discussion (and several of the references) in

Dean Rickles, "AdS/CFT Duality and the Emergence of Spacetime," *Studies in History and Philosophy of Science* B44 (2013): 312–320.

A more recent set of techniques building on the AdS/CFT approach is even more explicitly quantum in nature, showing that spacetime can be reconstructed or inferred from the quantum entanglement between objects in the dual spacetime. See the (advanced-level) lectures by one of the pioneers of that approach, here: Mark Van Raamsdonk, "Lectures on Gravity and Entanglement," in *New Frontiers in Fields and Strings*, ed. Joseph Polchinski, Pedro Vieira, and Oliver DeWolf (River Edge, NJ: World Scientific, 2017). (There is an open access online version here: http://arxiv.org/abs/1609.00026/.)

Clearly, our understanding of spacetime is evolving very quickly in this area. It would be nice if sharply testable predictions for our own universe were to come out of this. As a start in this direction, Erik Verlinde has made a proposal about how the gravity of our own universe might be emergent, including implications for the origin of the dark matter problem. See, for example, Erik Verlinde, "Emergent Gravity and the Dark Universe," *SciPost Physics* 2 (2017): 16. (There is an online version here: http://arxiv.org/abs/1611.02269/.) That's an advanced-level paper, so an alternative is this article: Natalie Wolchover, "The Case against Dark Matter," *Quanta Magazine* (November 2016), http://www.quantamagazine.org/20161129-verlinde-gravity-dark-matter/.

***Pages 188–189*** – Note that the language of sequential narrative art on the page is coming into play here to illustrate the breakdown of spacetime. Panels are the fabric of spacetime in comics (see the preface and the chapter 7 note for page 145 [panels 1–4]), and their arrangement relative to each other, including the spaces between them, conveys time. Their dissolving on the page does not just disrupt the spacetime structure here, but renders it entirely meaningless.

***Page 188*** – D-branes are introduced in the AdS/CFT references given above, but here's an advanced-level book devoted entirely to them: Clifford V. Johnson, *D-Branes* (Cambridge: Cambridge University Press, 2003). It also contains some chapters on AdS/CFT, including some early suggestions about how the idea might be used as a tool for applications.

***Page 190*** – The book by Randall (see the chapter 3 notes for page 45) describes some ideas and approaches to the puzzle of how extra dimensions, if they exist, might reveal themselves experimentally one day. Perhaps our interlocutor has some of these in mind. Or maybe ideas from some of the AdS/CFT types of scenarios just described. Or perhaps others?

***Page 193*** – For more about LIGO and gravitational waves, see the book by Levin in the chapter 8 notes for page 159.

10

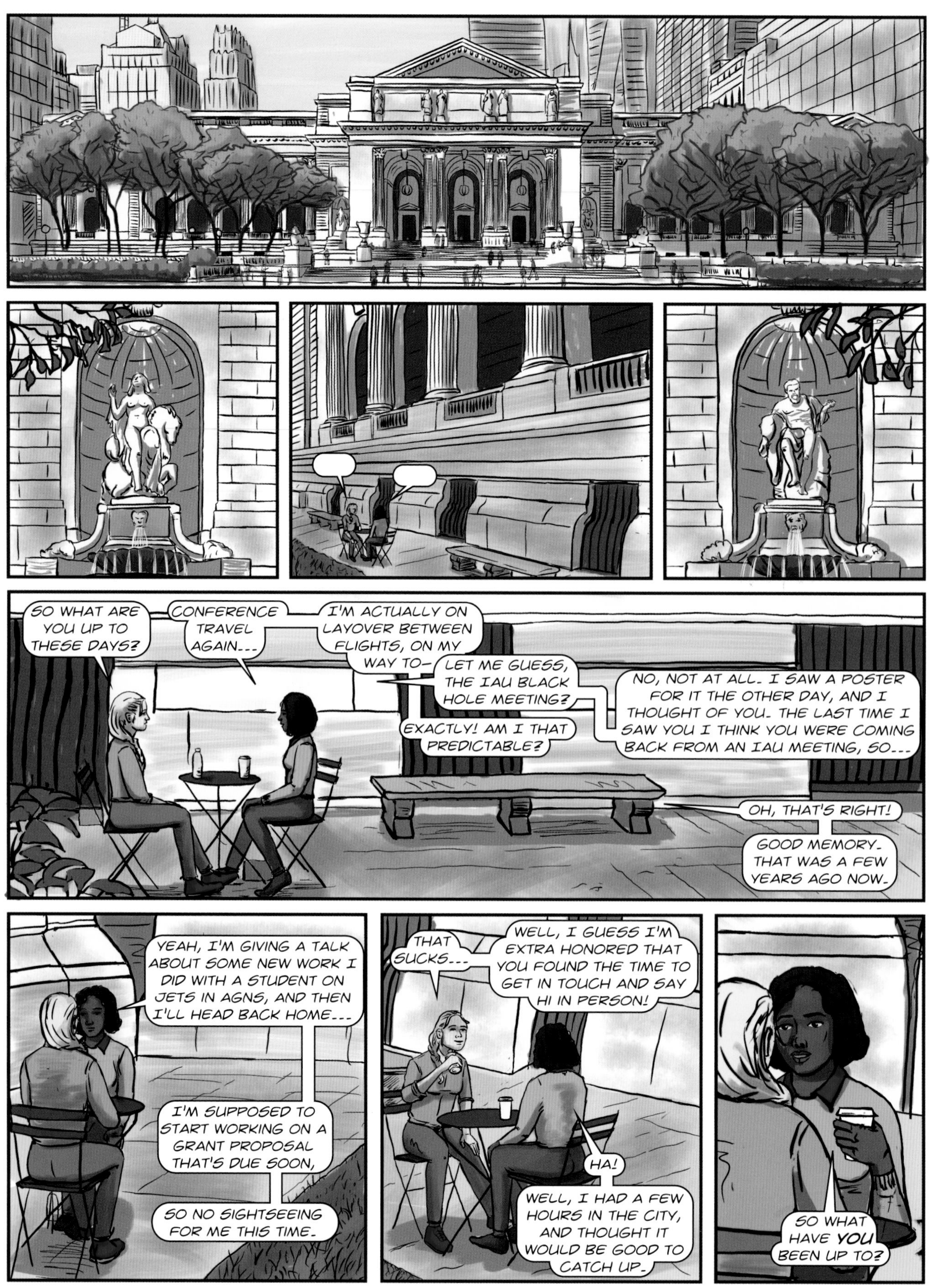
SO WHAT ARE YOU UP TO THESE DAYS?
CONFERENCE TRAVEL AGAIN...
I'M ACTUALLY ON LAYOVER BETWEEN FLIGHTS, ON MY WAY TO–
LET ME GUESS, THE IAU BLACK HOLE MEETING?
EXACTLY! AM I THAT PREDICTABLE?
NO, NOT AT ALL. I SAW A POSTER FOR IT THE OTHER DAY, AND I THOUGHT OF YOU. THE LAST TIME I SAW YOU I THINK YOU WERE COMING BACK FROM AN IAU MEETING, SO...
OH, THAT'S RIGHT!
GOOD MEMORY. THAT WAS A FEW YEARS AGO NOW.
YEAH, I'M GIVING A TALK ABOUT SOME NEW WORK I DID WITH A STUDENT ON JETS IN AGNS, AND THEN I'LL HEAD BACK HOME...
I'M SUPPOSED TO START WORKING ON A GRANT PROPOSAL THAT'S DUE SOON,
SO NO SIGHTSEEING FOR ME THIS TIME.
THAT SUCKS...
WELL, I GUESS I'M EXTRA HONORED THAT YOU FOUND THE TIME TO GET IN TOUCH AND SAY HI IN PERSON!
HA!
WELL, I HAD A FEW HOURS IN THE CITY, AND THOUGHT IT WOULD BE GOOD TO CATCH UP.
SO WHAT HAVE **YOU** BEEN UP TO?

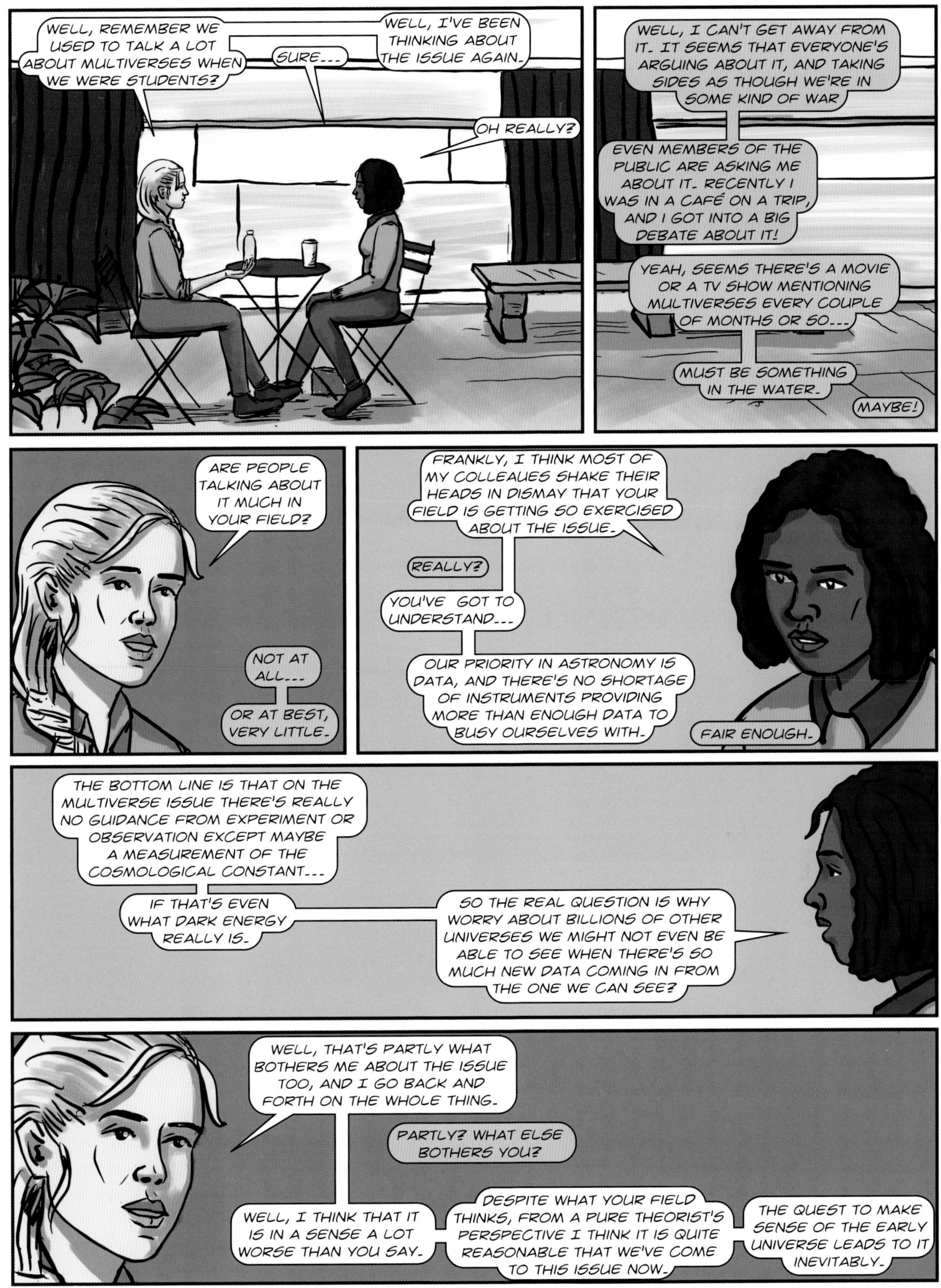
WELL, REMEMBER WE USED TO TALK A LOT ABOUT MULTIVERSES WHEN WE WERE STUDENTS?
SURE...
WELL, I'VE BEEN THINKING ABOUT THE ISSUE AGAIN.
OH REALLY?
WELL, I CAN'T GET AWAY FROM IT. IT SEEMS THAT EVERYONE'S ARGUING ABOUT IT, AND TAKING SIDES AS THOUGH WE'RE IN SOME KIND OF WAR
EVEN MEMBERS OF THE PUBLIC ARE ASKING ME ABOUT IT. RECENTLY I WAS IN A CAFÉ ON A TRIP, AND I GOT INTO A BIG DEBATE ABOUT IT!
YEAH, SEEMS THERE'S A MOVIE OR A TV SHOW MENTIONING MULTIVERSES EVERY COUPLE OF MONTHS OR SO...
MUST BE SOMETHING IN THE WATER.
MAYBE!
ARE PEOPLE TALKING ABOUT IT MUCH IN YOUR FIELD?
NOT AT ALL...
OR AT BEST, VERY LITTLE.
FRANKLY, I THINK MOST OF MY COLLEAUES SHAKE THEIR HEADS IN DISMAY THAT YOUR FIELD IS GETTING SO EXERCISED ABOUT THE ISSUE.
REALLY?
YOU'VE GOT TO UNDERSTAND...
OUR PRIORITY IN ASTRONOMY IS DATA, AND THERE'S NO SHORTAGE OF INSTRUMENTS PROVIDING MORE THAN ENOUGH DATA TO BUSY OURSELVES WITH.
FAIR ENOUGH.
THE BOTTOM LINE IS THAT ON THE MULTIVERSE ISSUE THERE'S REALLY NO GUIDANCE FROM EXPERIMENT OR OBSERVATION EXCEPT MAYBE A MEASUREMENT OF THE COSMOLOGICAL CONSTANT...
IF THAT'S EVEN WHAT DARK ENERGY REALLY IS.
SO THE REAL QUESTION IS WHY WORRY ABOUT BILLIONS OF OTHER UNIVERSES WE MIGHT NOT EVEN BE ABLE TO SEE WHEN THERE'S SO MUCH NEW DATA COMING IN FROM THE ONE WE CAN SEE?
WELL, THAT'S PARTLY WHAT BOTHERS ME ABOUT THE ISSUE TOO, AND I GO BACK AND FORTH ON THE WHOLE THING.
PARTLY? WHAT ELSE BOTHERS YOU?
WELL, I THINK THAT IT IS IN A SENSE A LOT WORSE THAN YOU SAY.
DESPITE WHAT YOUR FIELD THINKS, FROM A PURE THEORIST'S PERSPECTIVE I THINK IT IS QUITE REASONABLE THAT WE'VE COME TO THIS ISSUE NOW.
THE QUEST TO MAKE SENSE OF THE EARLY UNIVERSE LEADS TO IT INEVITABLY.

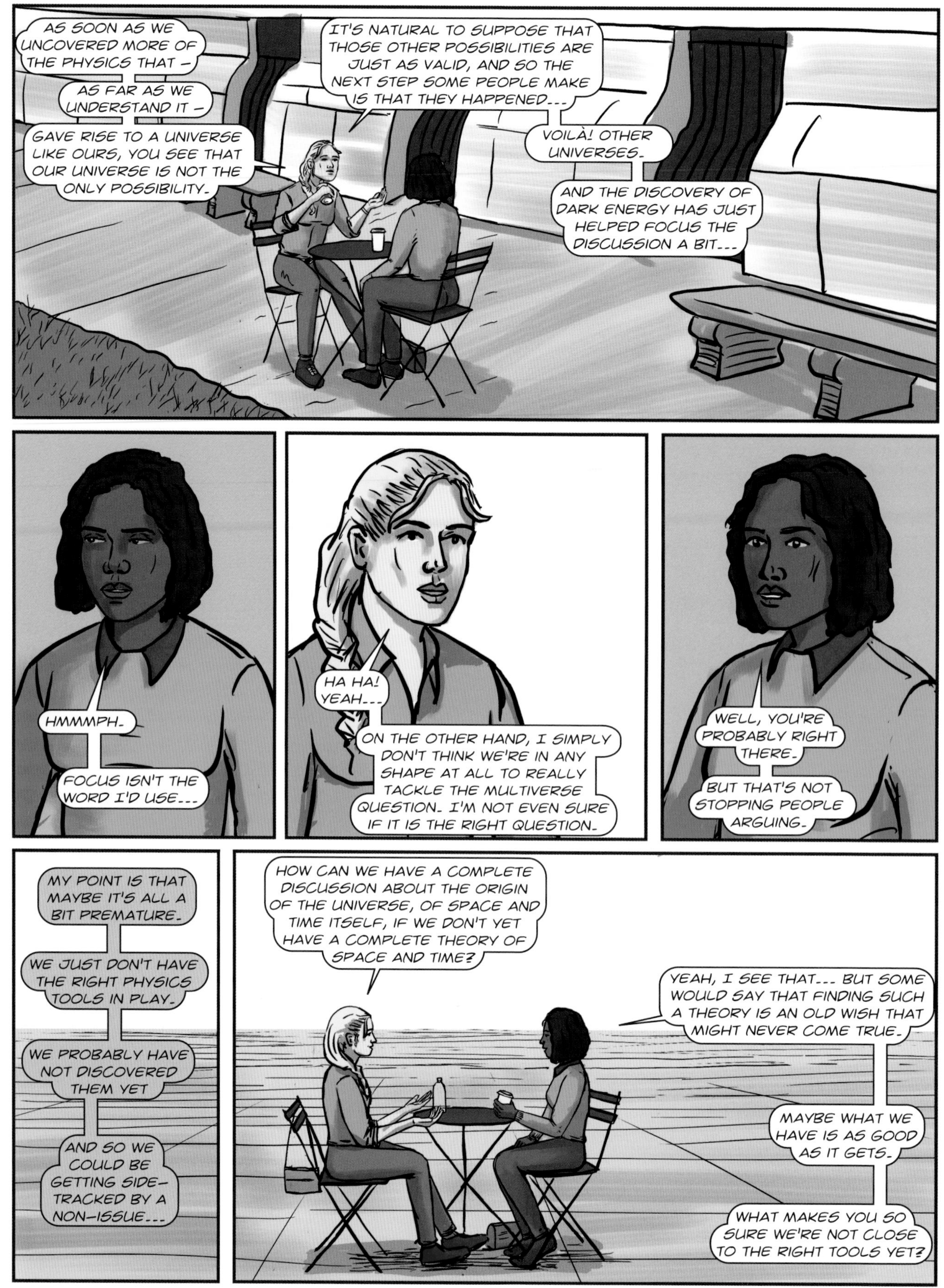
AS SOON AS WE UNCOVERED MORE OF THE PHYSICS THAT —
AS FAR AS WE UNDERSTAND IT —
GAVE RISE TO A UNIVERSE LIKE OURS, YOU SEE THAT OUR UNIVERSE IS NOT THE ONLY POSSIBILITY.
IT'S NATURAL TO SUPPOSE THAT THOSE OTHER POSSIBILITIES ARE JUST AS VALID, AND SO THE NEXT STEP SOME PEOPLE MAKE IS THAT THEY HAPPENED...
VOILÀ! OTHER UNIVERSES.
AND THE DISCOVERY OF DARK ENERGY HAS JUST HELPED FOCUS THE DISCUSSION A BIT...
HMMMPH.
FOCUS ISN'T THE WORD I'D USE...
HA HA! YEAH...
ON THE OTHER HAND, I SIMPLY DON'T THINK WE'RE IN ANY SHAPE AT ALL TO REALLY TACKLE THE MULTIVERSE QUESTION. I'M NOT EVEN SURE IF IT IS THE RIGHT QUESTION.
WELL, YOU'RE PROBABLY RIGHT THERE.
BUT THAT'S NOT STOPPING PEOPLE ARGUING.
MY POINT IS THAT MAYBE IT'S ALL A BIT PREMATURE.
WE JUST DON'T HAVE THE RIGHT PHYSICS TOOLS IN PLAY.
WE PROBABLY HAVE NOT DISCOVERED THEM YET
AND SO WE COULD BE GETTING SIDE-TRACKED BY A NON-ISSUE...
HOW CAN WE HAVE A COMPLETE DISCUSSION ABOUT THE ORIGIN OF THE UNIVERSE, OF SPACE AND TIME ITSELF, IF WE DON'T YET HAVE A COMPLETE THEORY OF SPACE AND TIME?
YEAH, I SEE THAT... BUT SOME WOULD SAY THAT FINDING SUCH A THEORY IS AN OLD WISH THAT MIGHT NEVER COME TRUE.
MAYBE WHAT WE HAVE IS AS GOOD AS IT GETS.
WHAT MAKES YOU SO SURE WE'RE NOT CLOSE TO THE RIGHT TOOLS YET?

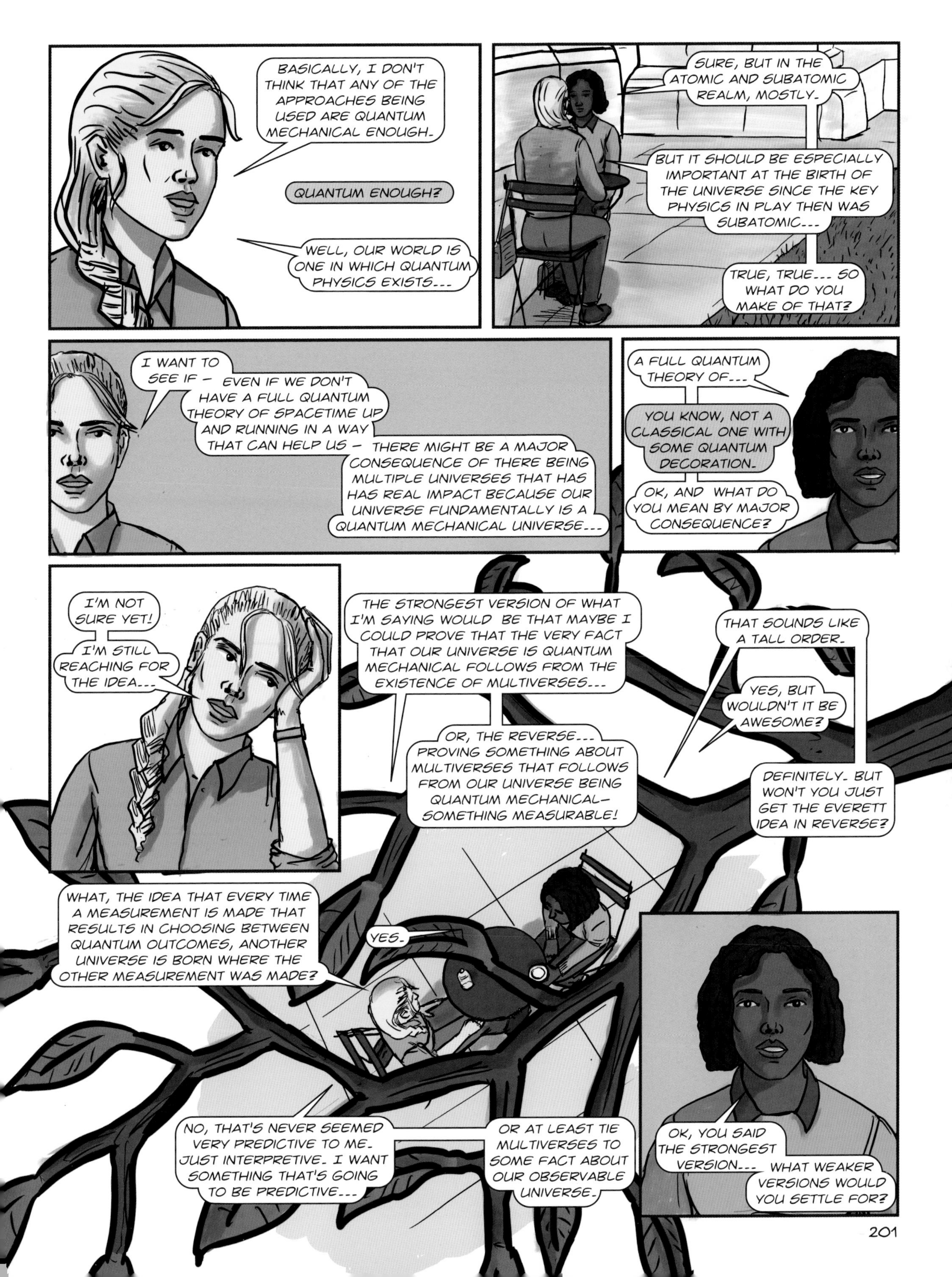
BASICALLY, I DON'T THINK THAT ANY OF THE APPROACHES BEING USED ARE QUANTUM MECHANICAL ENOUGH.
QUANTUM ENOUGH?
WELL, OUR WORLD IS ONE IN WHICH QUANTUM PHYSICS EXISTS...
SURE, BUT IN THE ATOMIC AND SUBATOMIC REALM, MOSTLY.
BUT IT SHOULD BE ESPECIALLY IMPORTANT AT THE BIRTH OF THE UNIVERSE SINCE THE KEY PHYSICS IN PLAY THEN WAS SUBATOMIC...
TRUE, TRUE... SO WHAT DO YOU MAKE OF THAT?
I WANT TO SEE IF –
EVEN IF WE DON'T HAVE A FULL QUANTUM THEORY OF SPACETIME UP AND RUNNING IN A WAY THAT CAN HELP US –
THERE MIGHT BE A MAJOR CONSEQUENCE OF THERE BEING MULTIPLE UNIVERSES THAT HAS HAS REAL IMPACT BECAUSE OUR UNIVERSE FUNDAMENTALLY IS A QUANTUM MECHANICAL UNIVERSE...
A FULL QUANTUM THEORY OF...
YOU KNOW, NOT A CLASSICAL ONE WITH SOME QUANTUM DECORATION.
OK, AND WHAT DO YOU MEAN BY MAJOR CONSEQUENCE?
I'M NOT SURE YET!
I'M STILL REACHING FOR THE IDEA...
THE STRONGEST VERSION OF WHAT I'M SAYING WOULD BE THAT MAYBE I COULD PROVE THAT THE VERY FACT THAT OUR UNIVERSE IS QUANTUM MECHANICAL FOLLOWS FROM THE EXISTENCE OF MULTIVERSES...
OR, THE REVERSE... PROVING SOMETHING ABOUT MULTIVERSES THAT FOLLOWS FROM OUR UNIVERSE BEING QUANTUM MECHANICAL– SOMETHING MEASURABLE!
THAT SOUNDS LIKE A TALL ORDER.
YES, BUT WOULDN'T IT BE AWESOME?
DEFINITELY. BUT WON'T YOU JUST GET THE EVERETT IDEA IN REVERSE?
WHAT, THE IDEA THAT EVERY TIME A MEASUREMENT IS MADE THAT RESULTS IN CHOOSING BETWEEN QUANTUM OUTCOMES, ANOTHER UNIVERSE IS BORN WHERE THE OTHER MEASUREMENT WAS MADE?
YES.
NO, THAT'S NEVER SEEMED VERY PREDICTIVE TO ME. JUST INTERPRETIVE. I WANT SOMETHING THAT'S GOING TO BE PREDICTIVE...
OR AT LEAST TIE MULTIVERSES TO SOME FACT ABOUT OUR OBSERVABLE UNIVERSE.
OK, YOU SAID THE STRONGEST VERSION...
WHAT WEAKER VERSIONS WOULD YOU SETTLE FOR?

AH...
WELL, I WAS THINKING THAT IT WOULD BE SOMETHING LIKE DIRAC'S ARGUMENT, BUT WITH A COSMOLOGICAL FLAVOR.
DIRAC'S ARGUMENT? WHICH ONE?
WELL, REMEMBER THAT HE PROVED, QUITE BEAUTIFULLY, THAT IF YOU INTRODUCE POINT MAGNETIC CHARGES INTO ELECTROMAGNETISM...
THE FACT THAT WE LIVE IN A QUANTUM MECHANICAL WORLD FORCES ELECTRIC CHARGE TO COME IN ONLY DISCRETE UNITS.
OH, QUANTIZATION OF CHARGE, RIGHT...
BUT NOBODY'S YET FOUND SUCH A MAGNETIC MONOPOLE.
TRUE... BUT THE FACT THAT ELECTRICALLY CHARGED PARTICLES **DO** ONLY CARRY DISCRETE UNITS OF CHANGE HAS NEVER BEEN EXPLAINED.
SO YOU'RE SAYING MONOPOLES ***MUST*** EXIST?
I'M SAYING THAT SINCE YOU NEED ONLY ONE MONOPOLE TO EXIST SOMEWHERE IN THE UNIVERSE TO EXPLAIN CHARGE QUANTIZATION, IT'S A PRETTY COMPELLING SETUP.
OK, SPOKEN LIKE A TRUE THEORIST...
BUT WHAT DO YOU WANT TO DO WITH THAT?
I'M SAYING THAT I WANT TO FORMULATE SOMETHING LIKE THAT, BUT FOR MULTIVERSES.
SOMETHING STRONG AND COMPELLING THAT USES THE QUANTUM NATURE OF OUR UNIVERSE HEAD-ON TO EXPRESS THE CONSEQUENCES OF MULTIVERSES.

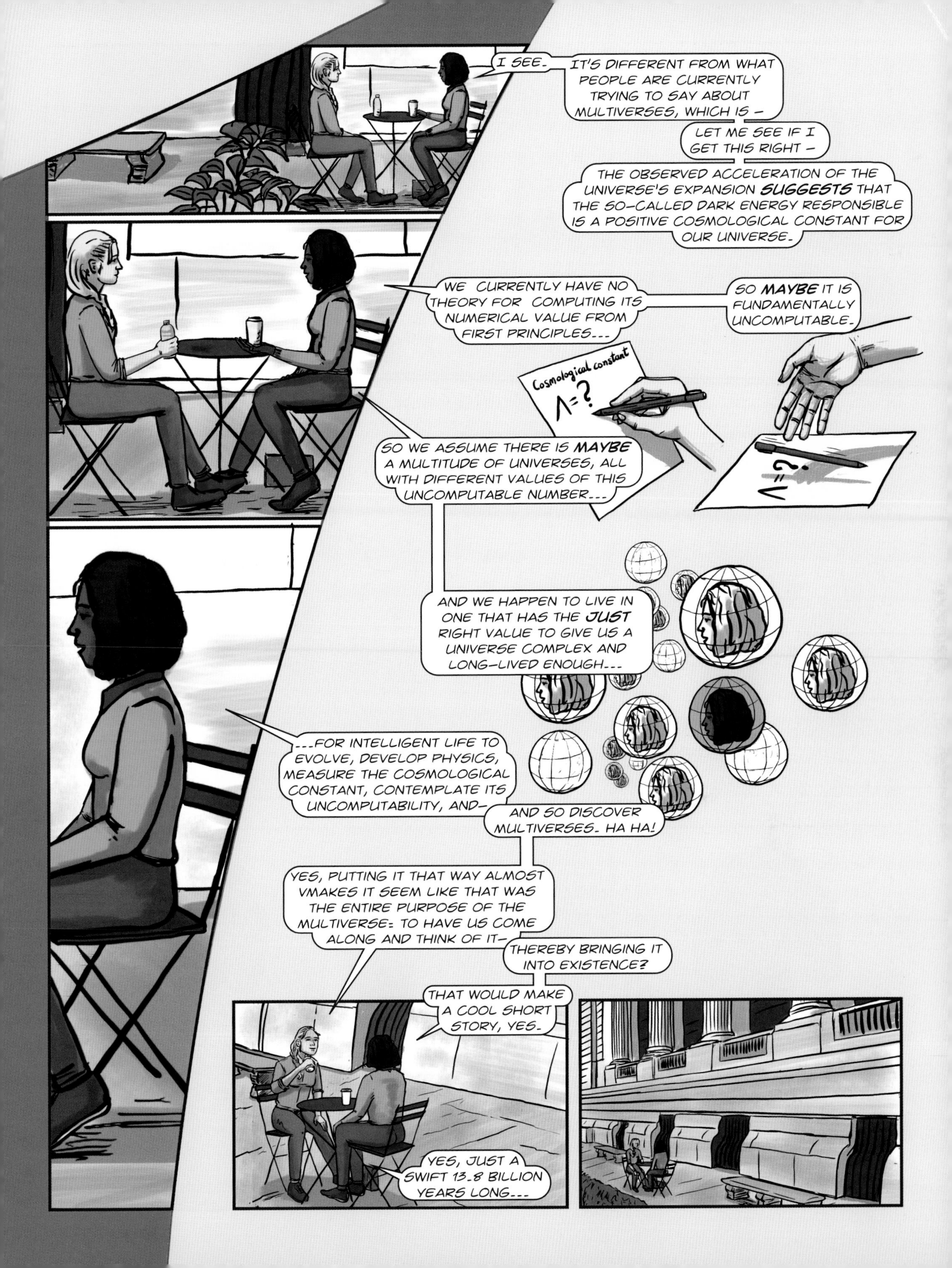
I SEE.
IT'S DIFFERENT FROM WHAT PEOPLE ARE CURRENTLY TRYING TO SAY ABOUT MULTIVERSES, WHICH IS –
LET ME SEE IF I GET THIS RIGHT –
THE OBSERVED ACCELERATION OF THE UNIVERSE'S EXPANSION SUGGESTS THAT THE SO-CALLED DARK ENERGY RESPONSIBLE IS A POSITIVE COSMOLOGICAL CONSTANT FOR OUR UNIVERSE.
WE CURRENTLY HAVE NO THEORY FOR COMPUTING ITS NUMERICAL VALUE FROM FIRST PRINCIPLES...
SO MAYBE IT IS FUNDAMENTALLY UNCOMPUTABLE.
Cosmological constant
Λ=?
Λ=?
SO WE ASSUME THERE IS MAYBE A MULTITUDE OF UNIVERSES, ALL WITH DIFFERENT VALUES OF THIS UNCOMPUTABLE NUMBER...
AND WE HAPPEN TO LIVE IN ONE THAT HAS THE JUST RIGHT VALUE TO GIVE US A UNIVERSE COMPLEX AND LONG-LIVED ENOUGH...
...FOR INTELLIGENT LIFE TO EVOLVE, DEVELOP PHYSICS, MEASURE THE COSMOLOGICAL CONSTANT, CONTEMPLATE ITS UNCOMPUTABILITY, AND–
AND SO DISCOVER MULTIVERSES. HA HA!
YES, PUTTING IT THAT WAY ALMOST VMAKES IT SEEM LIKE THAT WAS THE ENTIRE PURPOSE OF THE MULTIVERSE: TO HAVE US COME ALONG AND THINK OF IT–
THEREBY BRINGING IT INTO EXISTENCE?
THAT WOULD MAKE A COOL SHORT STORY, YES.
YES, JUST A SWIFT 13.8 BILLION YEARS LONG...

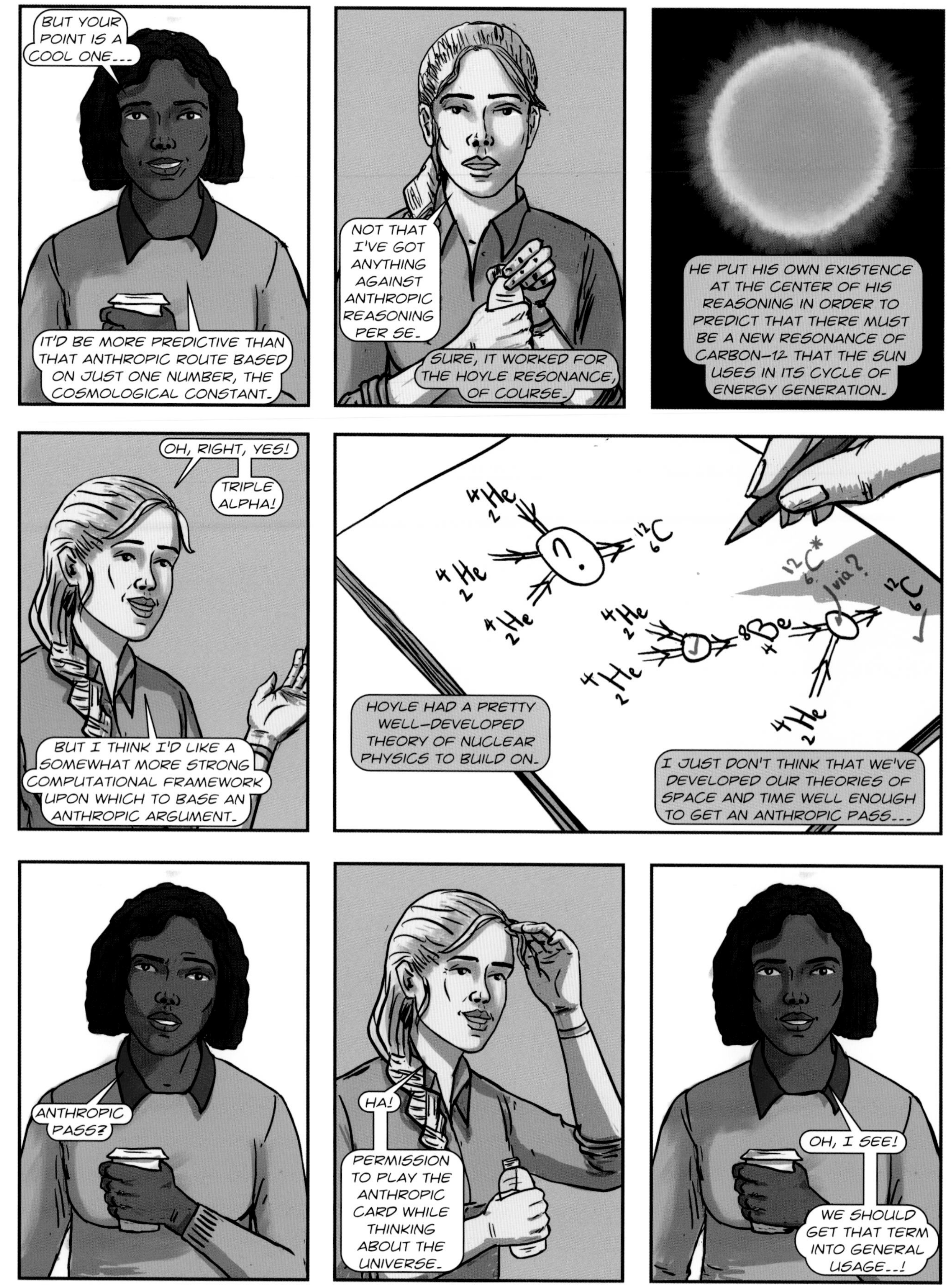
BUT YOUR POINT IS A COOL ONE...
IT'D BE MORE PREDICTIVE THAN THAT ANTHROPIC ROUTE BASED ON JUST ONE NUMBER, THE COSMOLOGICAL CONSTANT.
NOT THAT I'VE GOT ANYTHING AGAINST ANTHROPIC REASONING PER SE.
SURE, IT WORKED FOR THE HOYLE RESONANCE, OF COURSE.
HE PUT HIS OWN EXISTENCE AT THE CENTER OF HIS REASONING IN ORDER TO PREDICT THAT THERE MUST BE A NEW RESONANCE OF CARBON-12 THAT THE SUN USES IN ITS CYCLE OF ENERGY GENERATION.
OH, RIGHT, YES!
TRIPLE ALPHA!
BUT I THINK I'D LIKE A SOMEWHAT MORE STRONG COMPUTATIONAL FRAMEWORK UPON WHICH TO BASE AN ANTHROPIC ARGUMENT.
$^4_2He$
$^4_2He$
$^4_2He$
?
$^{12}_6C$
$^4_2He$
$^4_2He$
$^8_4Be$
$^{12}_6C^*$
via?
$^{12}_6C$
$^4_2He$
HOYLE HAD A PRETTY WELL-DEVELOPED THEORY OF NUCLEAR PHYSICS TO BUILD ON.
I JUST DON'T THINK THAT WE'VE DEVELOPED OUR THEORIES OF SPACE AND TIME WELL ENOUGH TO GET AN ANTHROPIC PASS...
ANTHROPIC PASS?
HA!
PERMISSION TO PLAY THE ANTHROPIC CARD WHILE THINKING ABOUT THE UNIVERSE.
OH, I SEE!
WE SHOULD GET THAT TERM INTO GENERAL USAGE..!

SO LET ME MAKE SURE I UNDERSTAND...
YOU WANT SOME SORT OF OBSERVATIONAL PROPERTY OF OUR UNIVERSE THAT IS A CONSEQUENCE OF THE EXISTENCE OF A MULTIVERSE.
I WANT TO SAY IT STRONGER.
I WANT THE EXISTENCE OF A MULTIVERSE TO IMPLY –
GIVEN THAT WE KNOW OUR UNIVERSE OBEYS THE LAWS OF QUANTUM MECHANICS –
...TO IMPLY SOMETHING VERY SPECIFIC AND OBSERVABLE.
AND BY OBSERVABLE YOU MEAN SOMETHING AN ASTRONOMER MIGHT SEE?
THAT WOULD BE MY PREFERENCE...
IT'S A COSMOLOGICAL THING SO I WANT IT TO HAVE COSMOLOGICAL CONSEQUENCES.
CONSEQUENCES WRITTEN IN THE SKY THAT WE'VE ALREADY SEEN AND HAVEN'T EXPLAINED, OR MAYBE WILL SEE ONE DAY SOON.
OR, CONSEQUENCES, THAT IF **NOT** SEEN, WOULD RULE OUT THE MULTIVERSE.
OBSERVATIONAL CONSEQUENCES...
LOOK, I'D SETTLE FOR IT TO BE SOME COOL FEATURE OF THE SPECTRUM OF ELEMENTARY PARTICLES...
BUT I FEEL THAT MIGHT JUST GIVE LOCAL INFORMATION ABOUT OUR OWN UNIVERSE'S SPACETIME...
I'M NOT SURE.
I SEE...
OH, AND IT CAN'T BE THE SAME OLD OBSERVATIONS THAT LED TO INFLATION...

...THAT'S TOO VAGUE.
OK, OK. WOULD YOU LIKE FRIES WITH THAT?

WHAT?
JUST KIDDING...

IT WAS GETTING TO BE A LONG LIST OF REQUESTS OF THE UNIVERSE..
OH! HA HA! I SEE.
I GUESS MY QUESTION FOR YOU IS WHAT DO YOU THINK?

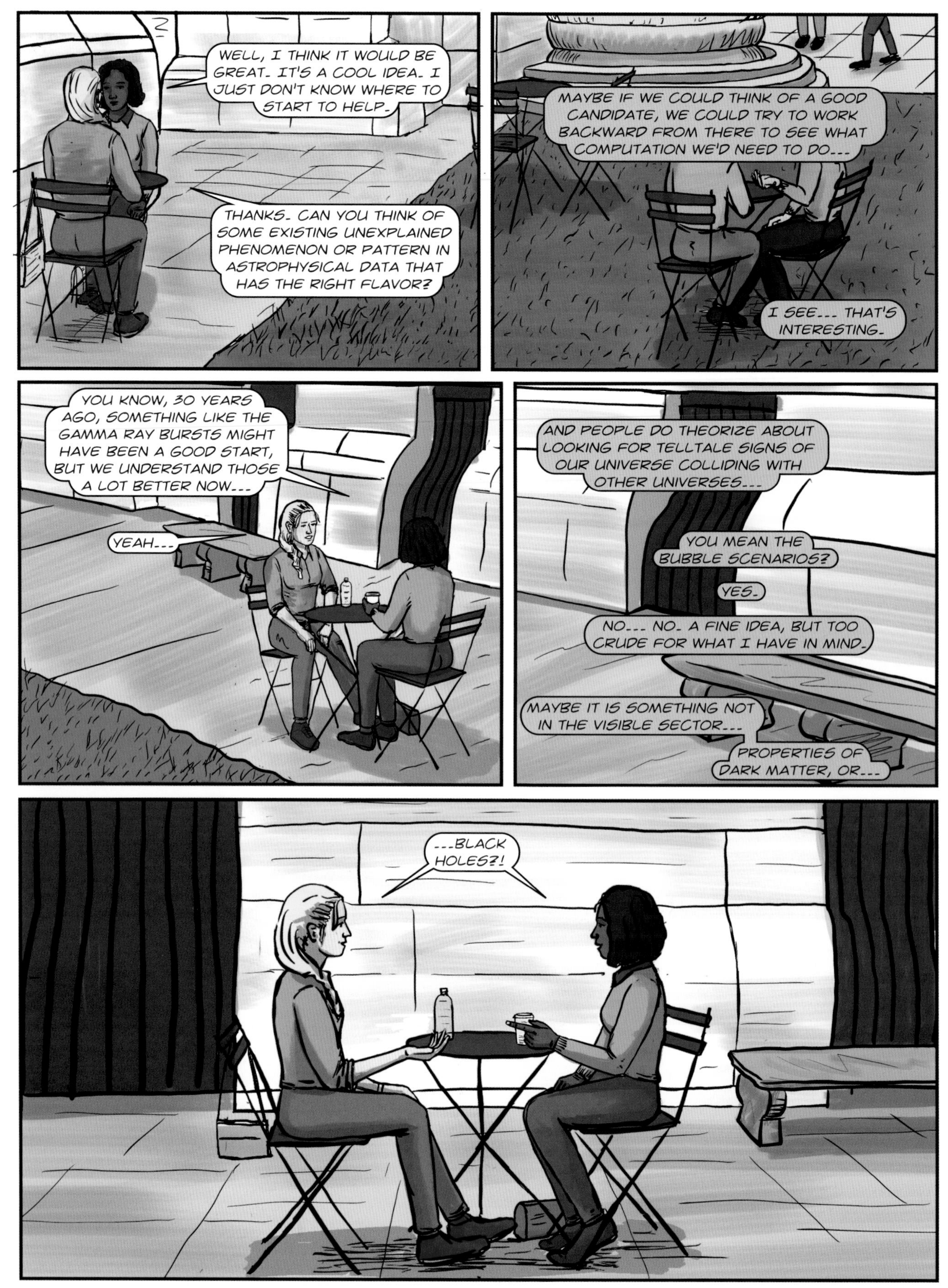
WELL, I THINK IT WOULD BE GREAT. IT'S A COOL IDEA. I JUST DON'T KNOW WHERE TO START TO HELP.
THANKS. CAN YOU THINK OF SOME EXISTING UNEXPLAINED PHENOMENON OR PATTERN IN ASTROPHYSICAL DATA THAT HAS THE RIGHT FLAVOR?
MAYBE IF WE COULD THINK OF A GOOD CANDIDATE, WE COULD TRY TO WORK BACKWARD FROM THERE TO SEE WHAT COMPUTATION WE'D NEED TO DO...
I SEE... THAT'S INTERESTING.
YOU KNOW, 30 YEARS AGO, SOMETHING LIKE THE GAMMA RAY BURSTS MIGHT HAVE BEEN A GOOD START, BUT WE UNDERSTAND THOSE A LOT BETTER NOW...
YEAH...
AND PEOPLE DO THEORIZE ABOUT LOOKING FOR TELLTALE SIGNS OF OUR UNIVERSE COLLIDING WITH OTHER UNIVERSES...
YOU MEAN THE BUBBLE SCENARIOS?
YES.
NO... NO. A FINE IDEA, BUT TOO CRUDE FOR WHAT I HAVE IN MIND.
MAYBE IT IS SOMETHING NOT IN THE VISIBLE SECTOR...
PROPERTIES OF DARK MATTER, OR...
...BLACK HOLES?!

GREAT MINDS THINK ALIKE!
YOU KNOW, MY GRANT PROPOSAL IS GOING TO BE ALL ABOUT THE COLLIDING BLACK HOLE DATA FROM GRAVITATIONAL WAVES...
DATA ACCUMULATED SINCE THE LIGO ANNOUNCEMENTS AND OVER THE NEXT FIVE YEARS...
MAYBE THERE'LL BE SOME STRIKING OBSERVATIONAL FACT FROM THAT?
YEAH...
IT'LL PROBABLY TAKE YEARS FOR A PATTERN TO DEVELOP THOUGH.
TRUE. WELL, I WON'T BE PUTTING ALL MY EGGS IN ONE BASKET.
AND BESIDES, IT MIGHT TAKE ME YEARS TO DEVELOP THE RIGHT KIND OF COMPUTATION TO USE TO MAKE THE QUANTUM MECHANICAL PREDICTION.
I'M HOPING IT WON'T BE AS HARD AS COMPUTING THE COSMOLOGICAL CONSTANT FROM SCRATCH.
YOU REALLY WON'T NEED A COMPLETE THEORY OF QUANTUM GRAVITY?
I HOPE NOT!
IF SO WE'D BE IN FOR A LONG WAIT...
WHAT I'M HOPING IS THAT I MIGHT BE ABLE TO FORMULATE A GENERAL ARGUMENT POWERFUL ENOUGH THAT IT DOES NOT NEED THE DETAILS OF HOW QUANTUM GRAVITY WORKS...
...JUST KEY INGREDIENTS.
I SEE, LIKE DIRAC DID...
AS YOU SAID...
EXACTLY!
HIS ARGUMENT ENDED UP APPLYING TO ALL KINDS OF QUANTUM THEORIES WITH CHARGED OBJECTS THAT HE WOULD NOT HAVE DREAMED OF BACK IN 1931!
OR, LIKE HAWKING DID, SHOWING THAT BLACK HOLES MUST RADIATE BECAUSE OF QUANTUM MECHANICS.
YEAH. BUT NO BLACK HOLE IN ASTROPHYSICS HAS MEASURABLE HAWKING RADIATION, SO...THERE'S THAT.
FAIR ENOIUGH, BUT THE PREDICITON IS ROBUST ENOUGH THAT WE THINK ANY QUANTUM THEORY OF GRAVITY MUST CONTAIN IT...
HMMMM...

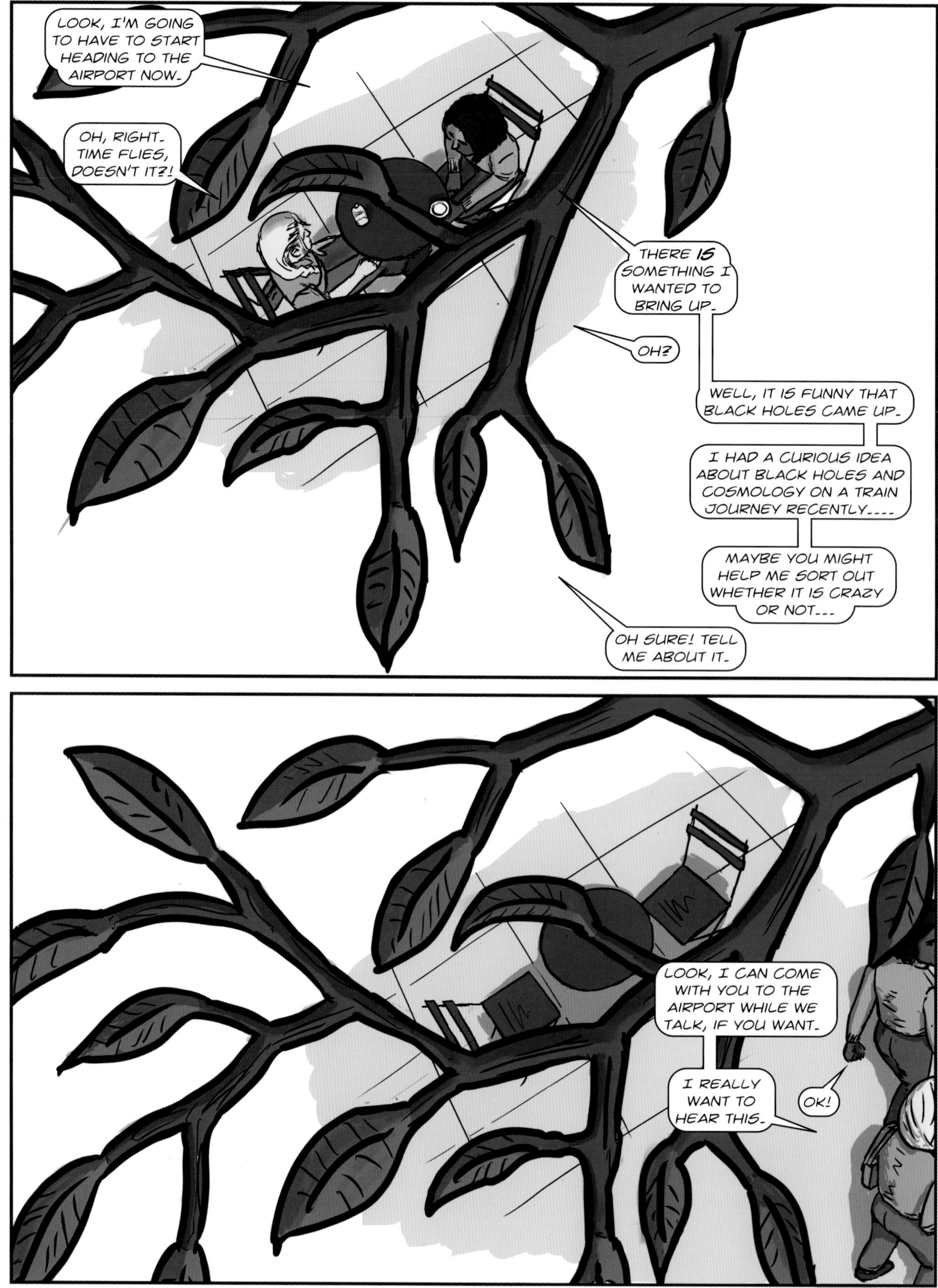

LOOK, I'M GOING TO HAVE TO START HEADING TO THE AIRPORT NOW.
OH, RIGHT. TIME FLIES, DOESN'T IT?!
THERE **IS** SOMETHING I WANTED TO BRING UP.
OH?
WELL, IT IS FUNNY THAT BLACK HOLES CAME UP.
I HAD A CURIOUS IDEA ABOUT BLACK HOLES AND COSMOLOGY ON A TRAIN JOURNEY RECENTLY....
MAYBE YOU MIGHT HELP ME SORT OUT WHETHER IT IS CRAZY OR NOT...
OH SURE! TELL ME ABOUT IT.
LOOK, I CAN COME WITH YOU TO THE AIRPORT WHILE WE TALK, IF YOU WANT.
I REALLY WANT TO HEAR THIS.
OK!

BEAUTY
OLD YET EVER NEW
ETERNAL VOICE
AND INWARD WORD

on DETOUR

M104 ON DETOUR
...AND SO ALL THAT WEIRD PHYSICS OF THE FLUID IN OUR FOUR-DIMENSIONAL WORLD IS THE SAME AS THE REGULAR PHYSICS OF A BLACK HOLE IN FIVE DIMENSIONS.
FIVE DIMENSIONS?
YES, IF YOU LOOK AT IT THE RIGHT WAY.
THE RIGHT WAY?
M104 ON DETOUR

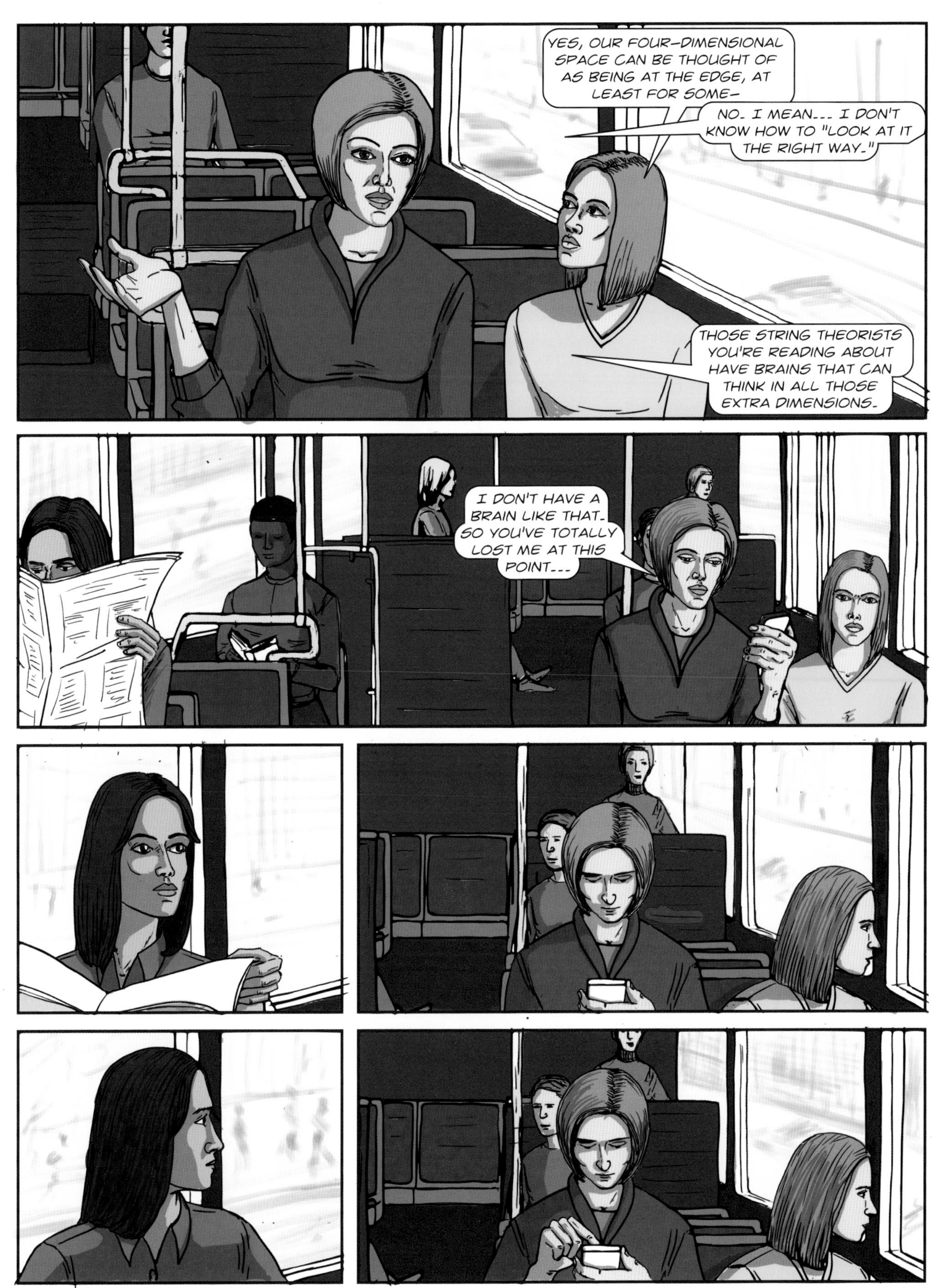
YES, OUR FOUR–DIMENSIONAL SPACE CAN BE THOUGHT OF AS BEING AT THE EDGE, AT LEAST FOR SOME–
NO. I MEAN... I DON'T KNOW HOW TO "LOOK AT IT THE RIGHT WAY."
THOSE STRING THEORISTS YOU'RE READING ABOUT HAVE BRAINS THAT CAN THINK IN ALL THOSE EXTRA DIMENSIONS.
I DON'T HAVE A BRAIN LIKE THAT. SO YOU'VE TOTALLY LOST ME AT THIS POINT...

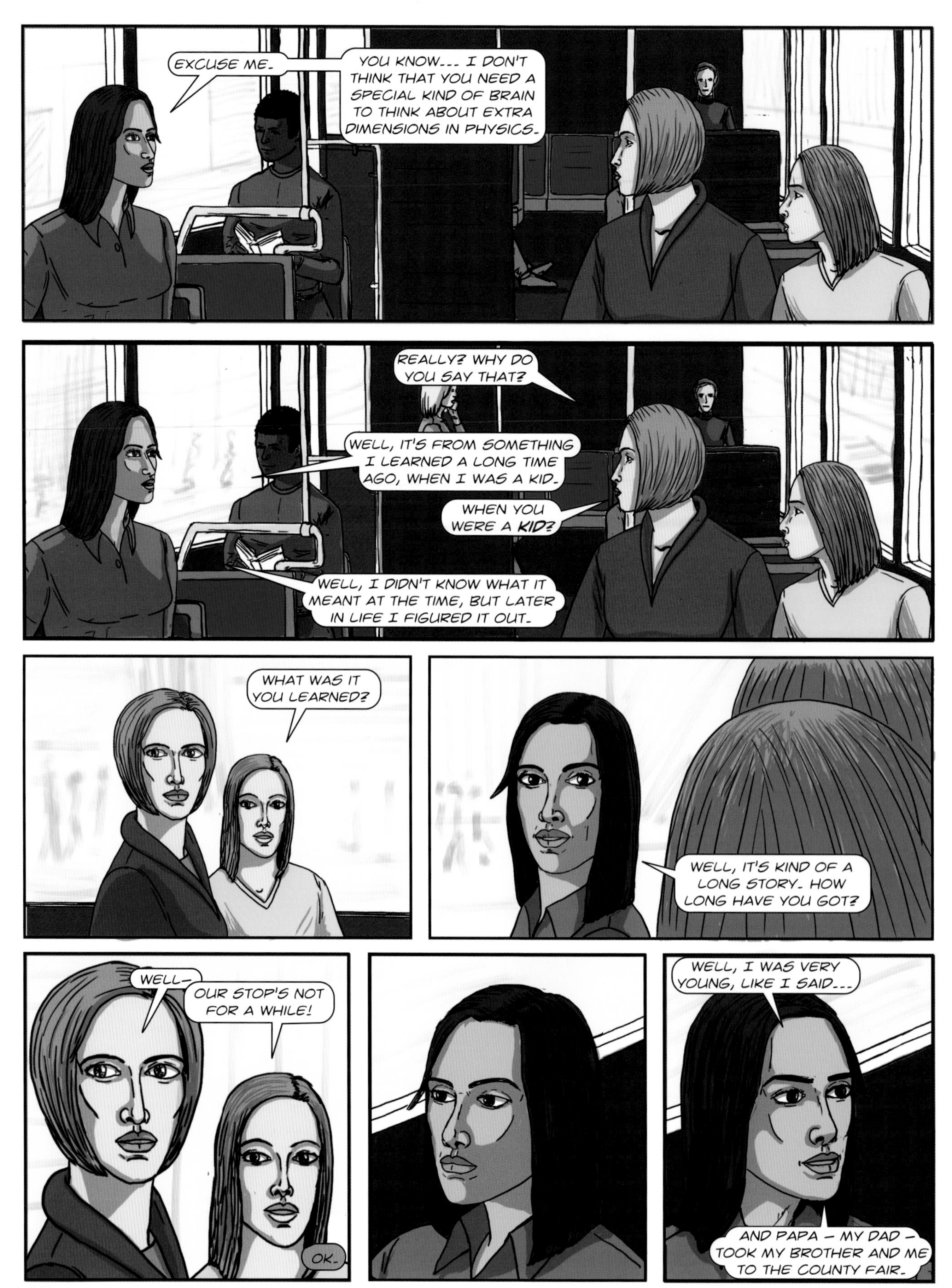
EXCUSE ME.
YOU KNOW... I DON'T THINK THAT YOU NEED A SPECIAL KIND OF BRAIN TO THINK ABOUT EXTRA DIMENSIONS IN PHYSICS.
REALLY? WHY DO YOU SAY THAT?
WELL, IT'S FROM SOMETHING I LEARNED A LONG TIME AGO, WHEN I WAS A KID.
WHEN YOU WERE A **KID**?
WELL, I DIDN'T KNOW WHAT IT MEANT AT THE TIME, BUT LATER IN LIFE I FIGURED IT OUT.
WHAT WAS IT YOU LEARNED?
WELL, IT'S KIND OF A LONG STORY. HOW LONG HAVE YOU GOT?
WELL–
OUR STOP'S NOT FOR A WHILE!
OK.
WELL, I WAS VERY YOUNG, LIKE I SAID...
AND PAPA – MY DAD – TOOK MY BROTHER AND ME TO THE COUNTY FAIR.

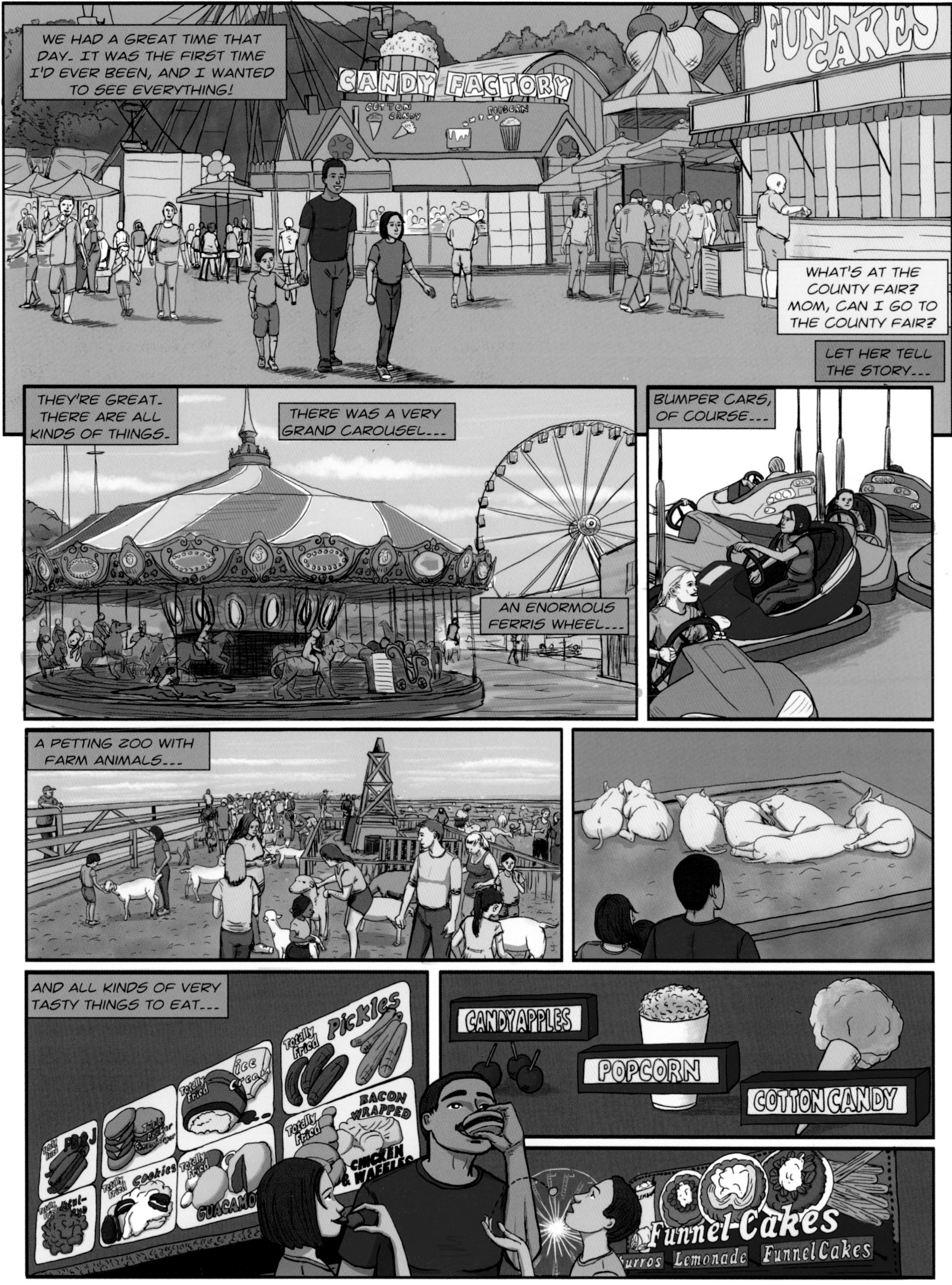
WE HAD A GREAT TIME THAT DAY. IT WAS THE FIRST TIME I'D EVER BEEN, AND I WANTED TO SEE EVERYTHING!
CANDY FACTORY
COTTON CANDY
POPCORN
FUNNEL CAKES
WHAT'S AT THE COUNTY FAIR? MOM, CAN I GO TO THE COUNTY FAIR?
LET HER TELL THE STORY...
THEY'RE GREAT. THERE ARE ALL KINDS OF THINGS.
THERE WAS A VERY GRAND CAROUSEL...
AN ENORMOUS FERRIS WHEEL...
BUMPER CARS, OF COURSE...
A PETTING ZOO WITH FARM ANIMALS...
AND ALL KINDS OF VERY TASTY THINGS TO EAT...
Totally Fried Pickles
BACON WRAPPED
CHICKEN & WAFFLES
Cookies
GUACAMO
PB&J
CANDYAPPLES
POPCORN
COTTON CANDY
Funnel-Cakes
urros Lemonade FunnelCakes

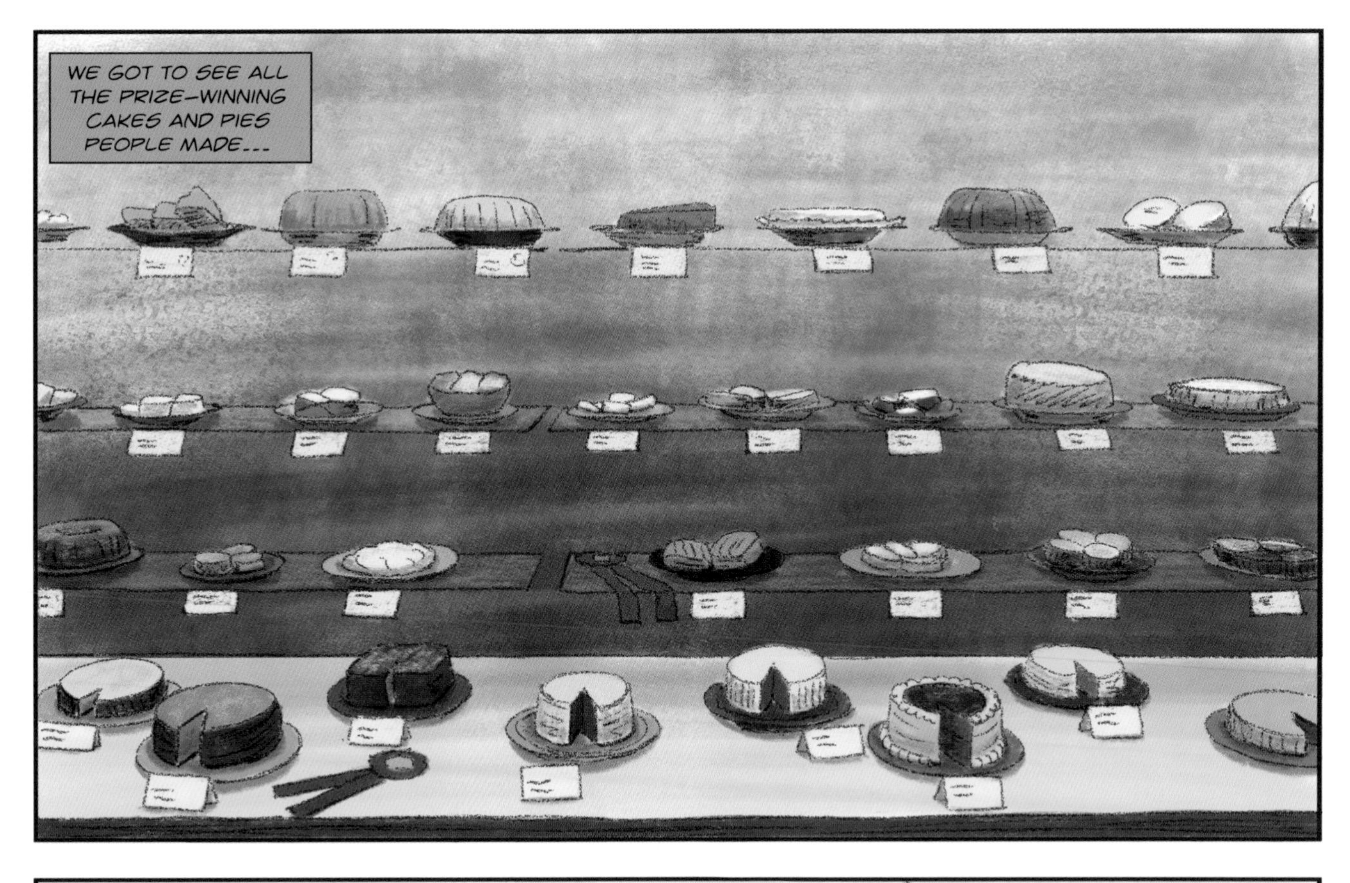
WE GOT TO SEE ALL THE PRIZE-WINNING CAKES AND PIES PEOPLE MADE...

AND AWARD-WINNING ARTS AND CRAFTS OF VARIOUS KINDS...

AND THERE WERE GAMES TO PLAY, TO WIN PRIZES...
WHAT KIND OF PRIZES?
MOSTLY STUFFED ANIMALS, AS I REMEMBER...
POOR PARENTS. THEY'RE THE ONES WHO END UP CARRYING THOSE GIANT BEARS, LIONS, AND BIRDS... THE WHOLE DAY.
IT WAS ONE OF THOSE GAMES THAT I STILL REMEMBER TO THIS DAY,
BECAUSE WHEN I SAW IT, EVEN WITH THE CROWD OF PEOPLE AROUND IT WANTING TO JOIN IN, I WAS SURE WE'D BE ABLE TO WIN.
WHAT WAS IT?!

GUESS THE NUMBER OF JELLY BEANS IN THE JAR!
THE ONE WHO GUESSES THE CLOSEST WINS THE POLAR BEAR!
I HAD AN IDEA ABOUT HOW WE COULD FIGURE IT OUT. WE WOULD NOT HAVE TO GUESS. WE'D BE TAKING THE POLAR BEAR HOME.
THE JAR WAS ESSENTIALLY A SIMPLE CYLINDER IN SHAPE.
OF ALL THE JARS HE COULD HAVE CHOSEN...
...THIS ONE WOULD BE THE EASIEST TO WORK WITH.
WHY?
WELL, I HAD JUST BEEN STUDYING DIFFERENT GEOMETRICAL SHAPES FOR MATH CLASS...
AND I'D LEARNED THE FORMULAS FOR THE VOLUME AND FOR THE SURFACE AREA OF VARIOUS OBJECTS...
OBJECTS LIKE THE SPHERE, THE CUBE, AND THE CYLINDER!
r
h
d
h
w
l
ALL WE NEEDED TO KNOW WAS THE HEIGHT OF THE CYLINDER AND THE WIDTH OF THE CIRCULAR CROSS-SECTION.
SO YOU'D NEED A RULER TO MEASURE IT?
BUT WOULDN'T HE BE SUSPICIOUS IF YOU ASKED TO TAKE MEASUREMENTS?
ACTUALLY, ALL WE NEEDED TO DO WAS A QUICK COUNT.
HUH?

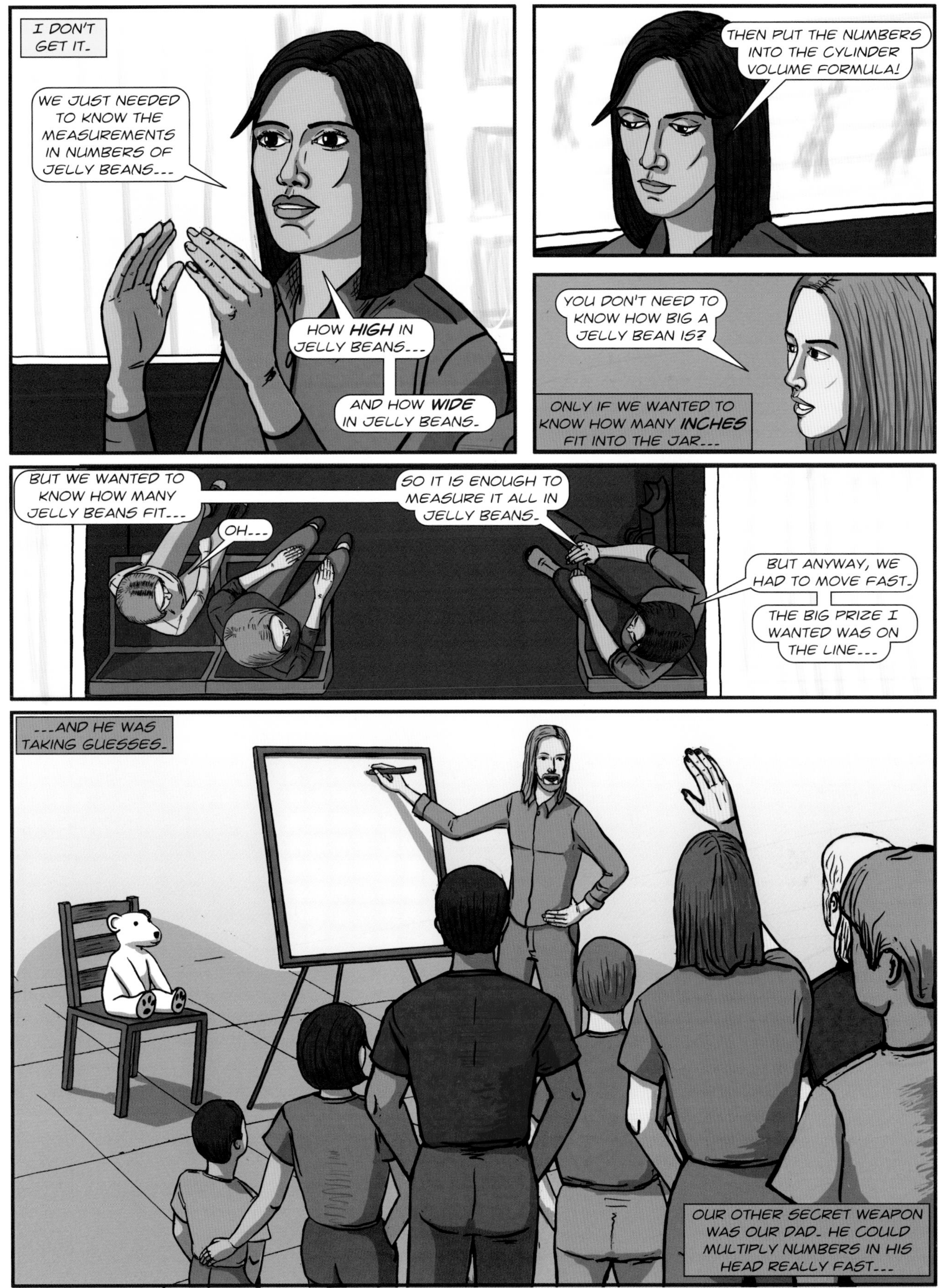
I DON'T GET IT.
WE JUST NEEDED TO KNOW THE MEASUREMENTS IN NUMBERS OF JELLY BEANS...
HOW HIGH IN JELLY BEANS...
AND HOW WIDE IN JELLY BEANS.
THEN PUT THE NUMBERS INTO THE CYLINDER VOLUME FORMULA!
YOU DON'T NEED TO KNOW HOW BIG A JELLY BEAN IS?
ONLY IF WE WANTED TO KNOW HOW MANY INCHES FIT INTO THE JAR...
BUT WE WANTED TO KNOW HOW MANY JELLY BEANS FIT...
SO IT IS ENOUGH TO MEASURE IT ALL IN JELLY BEANS.
OH...
BUT ANYWAY, WE HAD TO MOVE FAST.
THE BIG PRIZE I WANTED WAS ON THE LINE...
...AND HE WAS TAKING GUESSES.
OUR OTHER SECRET WEAPON WAS OUR DAD. HE COULD MULTIPLY NUMBERS IN HIS HEAD REALLY FAST...

SO MY BROTHER AND I GOT TO COUNTING THE HEIGHT AND WIDTH IN BEAN UNITS...
AND WE GAVE THE NUMBERS TO OUR DAD AND I QUICKLY TOLD HIM WHICH FORMULA TO USE!
HE CALCULATED...
2138!
HE WROTE DOWN OUR RESULT.
REALLY? YOU SURE? EXACTLY THAT?
NOT 2139?
EVERYONE LAUGHED AT ME, BUT I DID NOT MIND.
I WAS CONFIDENT THAT OUR ANSWER WOULD SURELY BE THE CLOSEST.
WE DID NOT GUESS. WE DID NOT NEED TO. IT WAS A QUESTION OF GEOMETRY. THERE WAS AN ANSWER TO IT AND WE'D WORKED IT OUT.
SO WHAT HAPPENED?!
YES, WHAT HAPPENED?

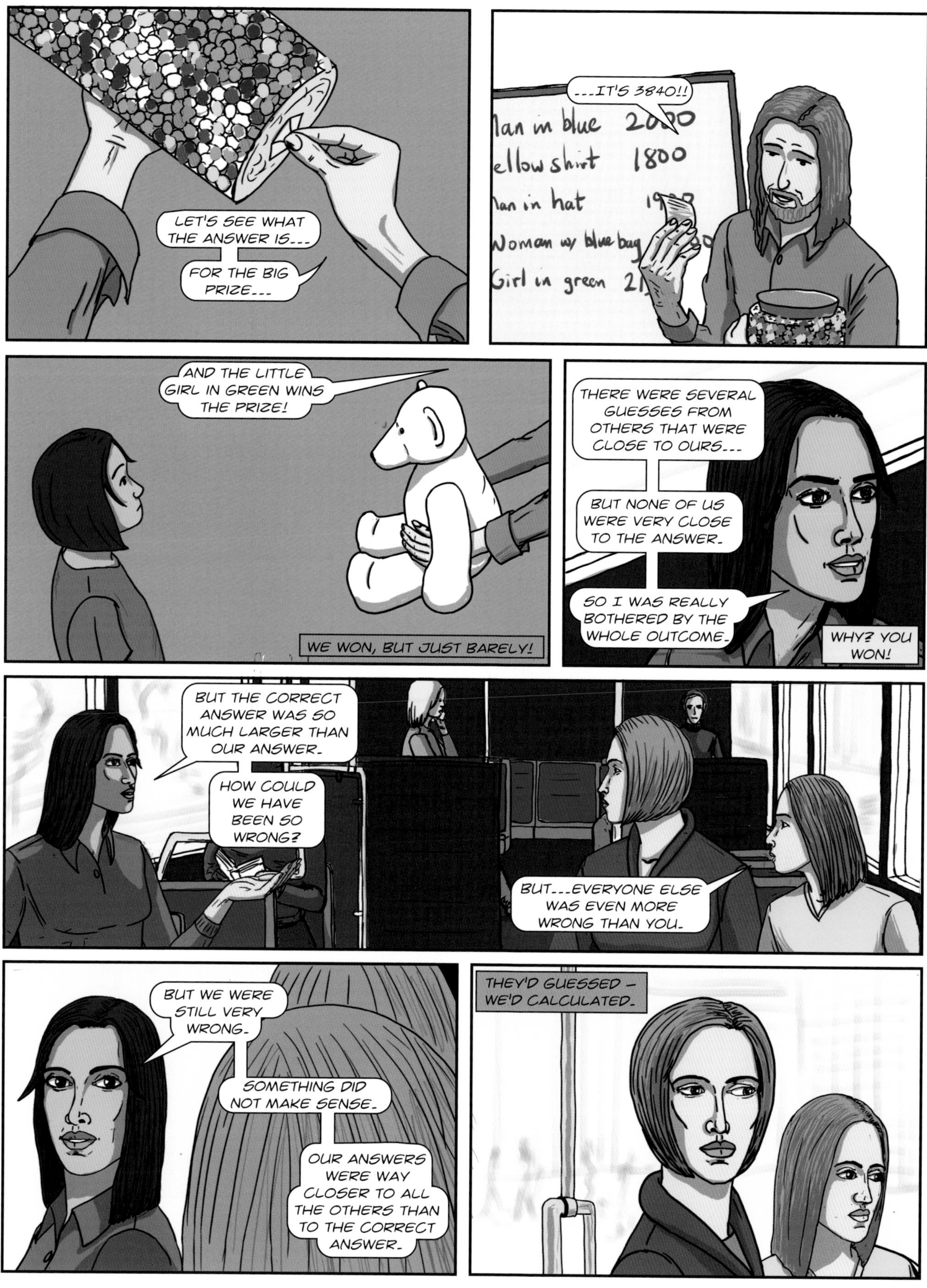
LET'S SEE WHAT THE ANSWER IS...
FOR THE BIG PRIZE...
...IT'S 3840!!
Woman w/ blue bag
Girl in green
AND THE LITTLE GIRL IN GREEN WINS THE PRIZE!
WE WON, BUT JUST BARELY!
THERE WERE SEVERAL GUESSES FROM OTHERS THAT WERE CLOSE TO OURS...
BUT NONE OF US WERE VERY CLOSE TO THE ANSWER.
SO I WAS REALLY BOTHERED BY THE WHOLE OUTCOME.
WHY? YOU WON!
BUT THE CORRECT ANSWER WAS SO MUCH LARGER THAN OUR ANSWER.
HOW COULD WE HAVE BEEN SO WRONG?
BUT...EVERYONE ELSE WAS EVEN MORE WRONG THAN YOU.
BUT WE WERE STILL VERY WRONG.
SOMETHING DID NOT MAKE SENSE.
OUR ANSWERS WERE WAY CLOSER TO ALL THE OTHERS THAN TO THE CORRECT ANSWER.
THEY'D GUESSED – WE'D CALCULATED.

I SPENT A LONG TIME THINKING ABOUT IT, WHILE WE VISITED OTHER PARTS OF THE FAIR.
IT KEPT BOTHERING ME...
THEN LATER, WHILE STILL AT THE FAIR...
I FIGURED IT OUT.
I KNEW THE ANSWER.

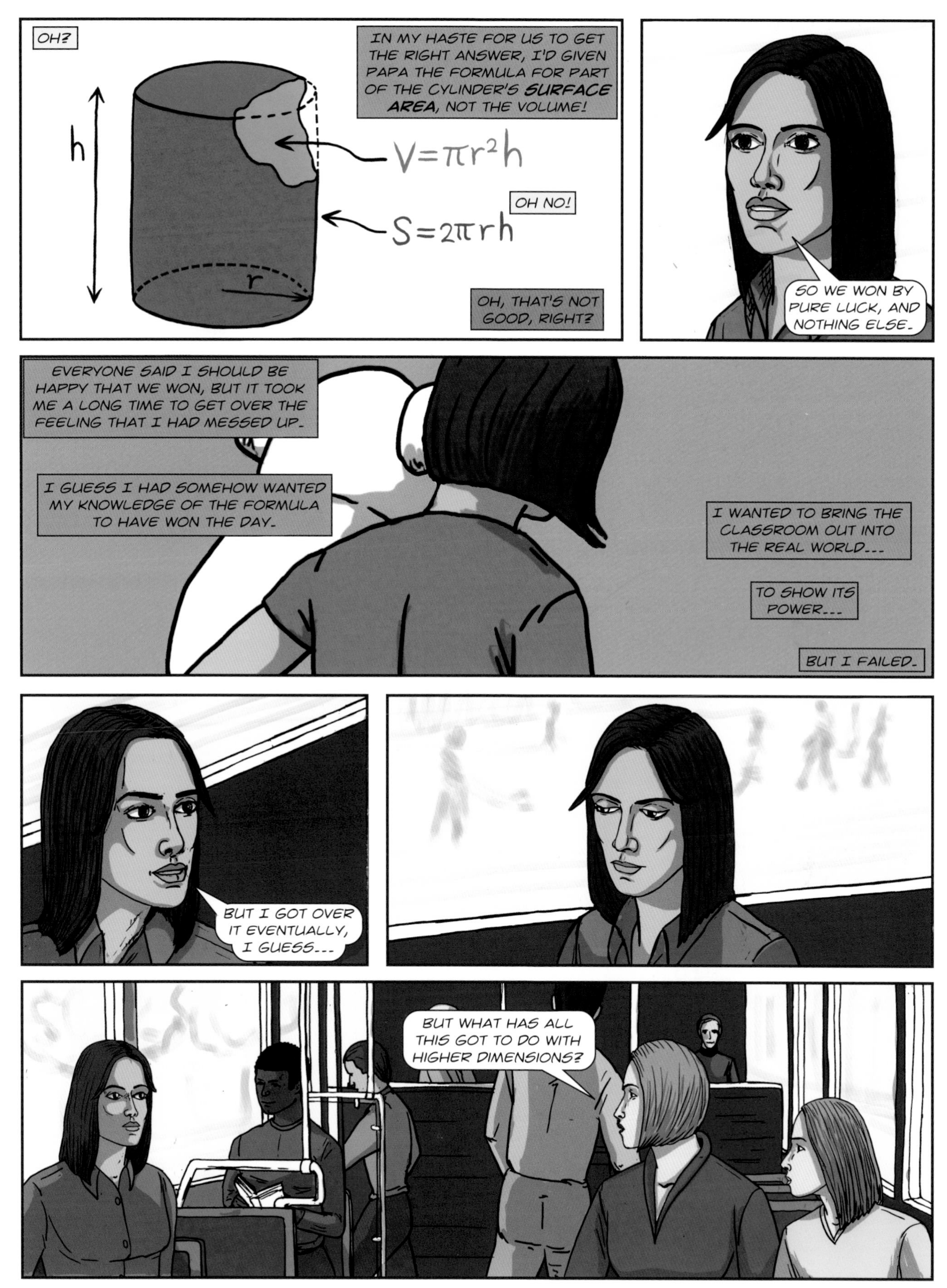
OH?
IN MY HASTE FOR US TO GET THE RIGHT ANSWER, I'D GIVEN PAPA THE FORMULA FOR PART OF THE CYLINDER'S SURFACE AREA, NOT THE VOLUME!
h
V=πr²h
OH NO!
S=2πrh
r
OH, THAT'S NOT GOOD, RIGHT?
SO WE WON BY PURE LUCK, AND NOTHING ELSE.
EVERYONE SAID I SHOULD BE HAPPY THAT WE WON, BUT IT TOOK ME A LONG TIME TO GET OVER THE FEELING THAT I HAD MESSED UP.
I GUESS I HAD SOMEHOW WANTED MY KNOWLEDGE OF THE FORMULA TO HAVE WON THE DAY.
I WANTED TO BRING THE CLASSROOM OUT INTO THE REAL WORLD...
TO SHOW ITS POWER...
BUT I FAILED.
BUT I GOT OVER IT EVENTUALLY, I GUESS...
BUT WHAT HAS ALL THIS GOT TO DO WITH HIGHER DIMENSIONS?

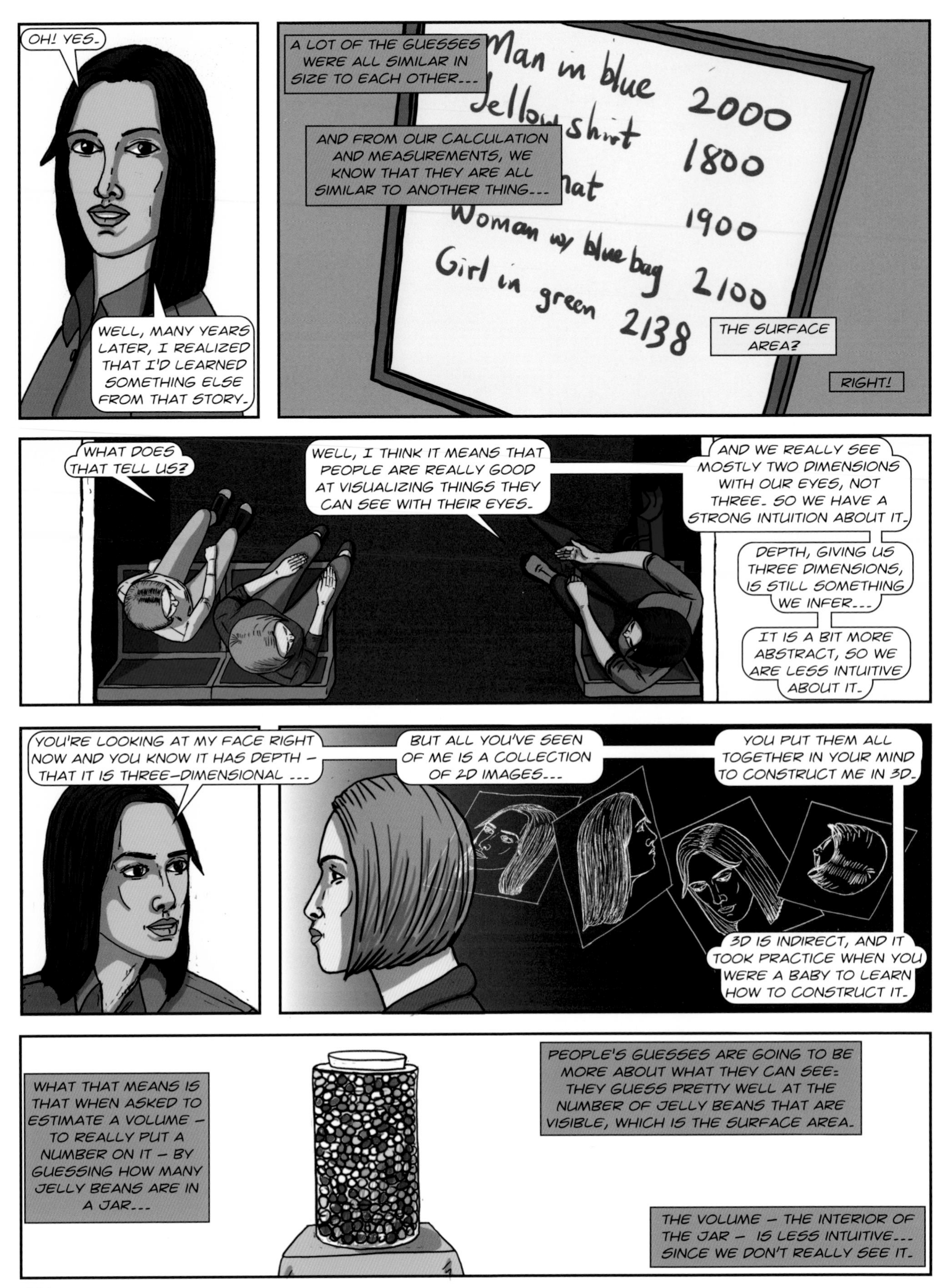
OH! YES.
WELL, MANY YEARS LATER, I REALIZED THAT I'D LEARNED SOMETHING ELSE FROM THAT STORY.
A LOT OF THE GUESSES WERE ALL SIMILAR IN SIZE TO EACH OTHER...
Man in blue 2000
Yellow shirt 1800
AND FROM OUR CALCULATION AND MEASUREMENTS, WE KNOW THAT THEY ARE ALL SIMILAR TO ANOTHER THING...
Woman w/ blue bag 2100
1900
Girl in green 2138
THE SURFACE AREA?
RIGHT!
WHAT DOES THAT TELL US?
WELL, I THINK IT MEANS THAT PEOPLE ARE REALLY GOOD AT VISUALIZING THINGS THEY CAN SEE WITH THEIR EYES.
AND WE REALLY SEE MOSTLY TWO DIMENSIONS WITH OUR EYES, NOT THREE. SO WE HAVE A STRONG INTUITION ABOUT IT.
DEPTH, GIVING US THREE DIMENSIONS, IS STILL SOMETHING WE INFER...
IT IS A BIT MORE ABSTRACT, SO WE ARE LESS INTUITIVE ABOUT IT.
YOU'RE LOOKING AT MY FACE RIGHT NOW AND YOU KNOW IT HAS DEPTH – THAT IT IS THREE-DIMENSIONAL ...
BUT ALL YOU'VE SEEN OF ME IS A COLLECTION OF 2D IMAGES...
YOU PUT THEM ALL TOGETHER IN YOUR MIND TO CONSTRUCT ME IN 3D.
3D IS INDIRECT, AND IT TOOK PRACTICE WHEN YOU WERE A BABY TO LEARN HOW TO CONSTRUCT IT.
WHAT THAT MEANS IS THAT WHEN ASKED TO ESTIMATE A VOLUME – TO REALLY PUT A NUMBER ON IT – BY GUESSING HOW MANY JELLY BEANS ARE IN A JAR...
PEOPLE'S GUESSES ARE GOING TO BE MORE ABOUT WHAT THEY CAN SEE: THEY GUESS PRETTY WELL AT THE NUMBER OF JELLY BEANS THAT ARE VISIBLE, WHICH IS THE SURFACE AREA.
THE VOLUME – THE INTERIOR OF THE JAR – IS LESS INTUITIVE... SINCE WE DON'T REALLY SEE IT.

I STILL DON'T SEE YOUR POINT.
MAYBE ALL YOU LEARNED IS THAT THE CROWD AT THE COUNTY FAIR DID NOT HAVE SOMEONE WITH THE RIGHT KIND OF BRAIN FOR SEEING HIGHER DIMENSIONS.
WELL, WHAT I'M SAYING IS THAT IT IS NOT ABOUT HAVING A SPECIAL KIND OF BRAIN.
IT'S MOSTLY ABOUT PRACTICE, AND ALSO...
WE HAVE TOOLS TO HELP US WITH THE DIMENSIONS THAT WE DON'T SEE DIRECTLY – LIKE THAT DEPTH THE CYLINDER HAS.
I REDID THE CALCULATION LATER AT HOME, USING THE VOLUME FORMULA, AND GOT A LOT CLOSER TO THE CORRECT ANSWER, SO IT WOULD HAVE WORKED.
I THINK THIS IS TRUE FOR THOSE EXTRA DIMENSIONS OF SPACE YOU'RE READING ABOUT IN YOUR BOOK.
DIRECTIONS BEYOND HEIGHT, WIDTH, AND DEPTH.
I'VE TALKED WITH SCIENTISTS WHO STUDY EXTRA DIMENSIONS. THEY'RE TRYING TO FIGURE OUT IF THEY EXIST OR NOT...
THEY DON'T HAVE SPECIAL BRAINS THAT ARE DIFFERENT FROM OURS.
WHAT THEY USE IS MATHEMATICS TO HELP THEM THINK ABOUT THOSE HIGHER DIIMENSIONS.
THAT'S THE TOOL?
EXACTLY! IT GIVES FORMULAS THAT HELP WORK OUT THOSE THINGS THAT ARE NOT INTUITIVE.
I GUESS MY POINT IS THAT THOSE SCIENTISTS ARE NOT ALIEN CREATURES...
THEY'RE REGULAR PEOPLE LIKE YOU AND ME...
ARE YOU A SCIENTIST?
NO, BUT I LOVE LEARNING ABOUT SCIENCE, LIKE YOU.
I'VE BEEN A FAN SINCE I WAS A GIRL.

SO ANYWAY, THAT'S MY STORY.
I'M SAYING THAT I THINK ABOUT SCIENTISTS NOT AS SOME KIND OF SPECIAL PEOPLE, BUT AS REGULAR PEOPLE DOING A SPECIAL THING – SCIENCE.
I GUESS I WANT TO MAKE SURE THAT YOU KNOW THAT IF IT TURNS OUT YOU WANT TO BE A SCIENTIST, YOU SHOULD GIVE IT A SHOT...
AND NOT WORRY ABOUT WHETHER YOU HAVE A SPECIAL KIND OF BRAIN OR SPECIAL TALENT OR NOT.
IT'S ALL ABOUT BEING INTERESTED, AND THEN PRACTICING AND WORKING HARD TO DEVELOP THE RIGHT SKILLS AND TOOLS...
THEN YOU GET TO EXPLORE THE WORLD, FOLLOW YOUR CURIOSITY...
AND ANSWER THE QUESTIONS YOU COME UP WITH...
HEY!
THAT'S MY STOP! I'VE GOT TO GO!
I HOPE YOU ENJOY THE REST OF THE BOOK!
THANKS FOR THE STORY!
'BYE!
GOODBYE!

# Notes

Building up intuition for thinking about higher (or lower) dimensions is mostly achieved like anything else: by practice. People get into the subject in a variety of ways, often because of recreational mathematics, or books that play with these ideas in fiction. It's all about finding a way in, and then maybe you get interested enough to stay, and perhaps to get hooked.

One of the all-time classic works of fiction exploring thinking in other dimensions is Edwin A. Abbott's *Flatland*. It is worth nothing that it is also intended to be a satire of the late nineteenth-century social attitudes (concerning gender, class, etc.) of the time. The focus of the book is two dimensions, and imagining how our three- (spatial) dimensional world looks to creatures who exist in two dimensions (such as insects living on the surface of a pond) provides lots of insights that help with thinking about worlds with dimensions different than our own.

There are lots of annotated versions that unpack the mathematical ideas explored in Abbott's *Flatland*. Here are two: Edwin A. Abbott, William F. Lindgren, and Thomas F. Banchoff, *Flatland; an Edition with Notes and Commentary* (Cambridge: Cambridge University Press, 2009); Edwin A. Abbott, *The Annotated Flatland: A Romance of Many Dimensions; Introduction and Notes by Ian Stewart* (New York: Basic Books, 2008). An extension of Abbott's *Flatland* is Ian Stewart, *Flatterland: Like Flatland Only More So* (New York: Basic Books, 2002).

Another notable book about lower-dimensional worlds was the one about the Planiverse in 1984. Here is a reprint edition: A. K. Dewdney, *The Planiverse: Computer Contact with a Two-Dimensional World* (New York: Copernicus, 2001). The project is discussed and compared to other works in A. K. Dewdney, "The Planiverse Project: Then and Now," *The Mathematical Intelligencer* 22 (2000): 46.

Beyond just thinking about other dimensions, there's lots of fun to be had with other recreational mathematics. The collections of articles written by Martin Gardner are a delight in this regard, and there are lots to be found. An excellent start is Martin Gardner, *The Colossal Book of Mathematics: Classic Puzzles, Paradoxes, and Problems* (New York: W. W. Norton & Co., 2001).

Here's a delightful biography of Donald Coxeter, whose work was full of delightful geometry, both for serious research and for pure recreation (and the large overlap between those two areas): Siobhan Roberts, *King of Infinite Space: Donald Coxeter, the Man Who Saved Geometry* (New York: Walker & Co., 2006).

A recent book with lots of new material, very much in the spirit of Gardner, and a delight to read as well, is Matt Parker, *Things to Make and Do in the Fourth Dimension: A Mathematician's Journey through Narcissistic Numbers, Optimal Dating Algorithms, at Least Two Kinds of Infinity, and More* (New York: Farrar, Straus and Giroux, 2015).